# LECTURES IN APPLIED MATHEMATICS
Proceedings of the Summer Seminar, Boulder, Colorado, 1957

### Probability and Related Topics in Physical Sciences
By Mark Kac with G. E. Uhlenbeck, A. R. Hibbs, and Balth. van der Pol

### Lectures on Fluid Mechanics
By Sydney Goldstein with J. M. Burgers

### Solid Mechanics
By R. S. Rivlin with W. Prager

### Partial Differential Equations
Edited by L. Bers and Fritz John with L. Gårding and A. N. Milgram

# LECTURES IN APPLIED MATHEMATICS
Proceedings of the Summer Seminar, Boulder, Colorado, 1957

AMERICAN MATHEMATICAL SOCIETY EDITORIAL COMMITTEE

ALSTON S. HOUSEHOLDER, *Chairman*
*Oak Ridge National Laboratory*

MARK KAC
*Department of Mathematics, Cornell University*

H. J. GREENBERG
*International Business Machines Corporation*

*"The subject has of late attracted increased attention in various countries."*

Horace Lamb, 1924

# LECTURES ON FLUID MECHANICS

**BY SYDNEY GOLDSTEIN**
*Division of Engineering and Applied Physics*
*Harvard University*

WITH SPECIAL LECTURES BY

**J. M. Burgers**
*Institute for Fluid Dynamics and Applied Mathematics*
*University of Maryland*

AMERICAN MATHEMATICAL SOCIETY
Providence, Rhode Island 02904

First Printing 1960 by Interscience Publishers, Inc.
Second Printing 1971 by the American Mathematical Society
Third Printing 1976 by the American Mathematical Society

INTERNATIONAL STANDARD BOOK NUMBER 0-8218-0048-5

LIBRARY OF CONGRESS CATALOG CARD NUMBER 60-12712

Copyright © 1960 by Interscience Publishers, Inc.

The Summer Seminar in Applied Mathematics of which these are the proceedings was arranged by the American Mathematical Society and the University of Colorado at Boulder, Colorado, June 23 to July 19, 1957. The Summer Seminar was conducted, and the proceedings prepared in part, under the following contracts and grants, each with (or to) the American Mathematical Society: (1) Contract AF 49(538)-59 with the United States Air Force, monitored by the Air Force Office of Scientific Research of the Air Research and Development Command. (2) Contract AT(30-1)-2012 with the United States Atomic Energy Commission. (3) Grant NSF-G3193 from the National Science Foundation. (4) Contract Nonr-2304(00) with the United States Navy, issued by the Office of Naval Research. (5) Contract DA-19-020-ORD-4373 with the Ordinance Corps, United States Army; Ordinance Project No. TB2-0001 (1823).

ALL RIGHTS RESERVED. Reproduction in whole or in part is permitted for any purpose of the United States Government, and the information contained herein is available to all persons without restrictions.

# Foreword

The present volume is the second of four which are to contain the proceedings of the Summer Seminar on Applied Mathematics, sponsored by the American Mathematical Society and held at the University of Colorado over the four weeks beginning June 23, 1957.

The purpose of the Seminar was tutorial, to present to mature mathematicians the current status of the theory in the several fields covered, and to pose for them some of the more pressing and interesting mathematical problems lying open in these fields. It could be described as an endeavor to promote cooperation of mathematicians with theoretical physicists. The publication of these volumes is intended to extend the same information to a much wider public than was privileged to actually attend the Seminar itself, while at the same time serving as a permanent reference for those who did attend.

The program of the Seminar was organized by a committee of the American Mathematical Society with the following membership:

> P. R. Garabedian
> A. S. Householder
> Mark Kac
> R. E. Langer
> C. C. Lin
> Wm. Prager
> J. J. Stoker
> M. H. Martin, *Chairman*.

Local arrangements, including the social and recreational program, were organized by a committee of the Department of

Applied Mathematics, University of Colorado, as follows:

> J. R. Britton
> R. Ben Kriegh
> L. W. Rutland
> L. C. Snively
> K. H. Stahl
> C. A. Hutchinson, *Chairman*.

The indefatigable energy and enthusiasm of the chairmen, and the cooperation of other members of the university staff, contributed immeasurably to the successful execution of the plans for the seminar and to the enjoyment of the participants.

The seminar opened Sunday evening, June 23, with an address by Professor Richard P. Feynman, California Institute of Technology, on the subject "The Relation of Mathematics to Physics." The formal, technical sessions were held in the mornings, leaving the afternoons free for study and for holding informal discussion groups on related special topics. Several such were organized and met regularly throughout the period of the Seminar.

A. S. HOUSEHOLDER

October 13, 1958

# Preface

These lectures were prepared for the Seminar in Applied Mathematics arranged by the American Mathematical Society and the University of Colorado at Boulder, Colorado, June 23—July 19, 1957.

The aim of the Seminar, as stated in the prospectus, was to give mature mathematicians an opportunity to become familiar with some of the major sectors in applied mathematics. There was a request that some aspect of Magnetohydrodynamics be included. Twelve lectures (plus two special lectures) were allotted to Fluid Mechanics. Some members of the audience were, in fact, already familiar with Fluid Mechanics; some were not. I did not believe myself capable of producing in twelve lectures a justified sense of familiarity with Fluid Mechanics in those who did not have it, so from the first I decided to compromise. Notes were reproduced and circulated, and most of them (though not all) were discussed afterwards in the lectures. Chapter 11, on the dynamics of inviscid gases, was not circulated before the lectures; it was added afterwards.

Nevertheless, many important topics had to be omitted entirely, or almost entirely. The choice in such cases is necessarily to some extent a personal one. Among the topics omitted I may mention turbulence, stability, surface and tidal waves, airfoil theory, and large parts of the dynamics of inviscid gases. It was with real regret that these were omitted; they are some of the most important and most fascinating topics in Fluid Dynamics. I doubt if it would have been worthwhile to start a discussion of turbulence without being able to continue at some considerable length. More space might have been given to some of the other omitted topics, but only by omitting instead some of the matter now included. I felt so strongly, however, about the omission of discussions of turbulence and stability that I drew special attention to it, and gave references, in Chapter 10. There are some notes on airfoil theory in Appendix III(c), and on the dynamics of inviscid gases in section 11.1; sections 11.2 and 11.3 are devoted to a particular treatment (slender-body theory) of a problem in inviscid-gas dynamics. On surface and tidal waves, nothing at all is said; but anyone who glances at any of the standard or "classical" texts on the subject, such as Lamb's *Hydrodynamics*, cannot fail to notice the importance of these topics.

Of course the discussions which are included have no pretensions to be complete accounts of the matters with which they deal. Always further reading and contemplation are required.

A notion of what is included can be seen from the list of contents. The first two chapters are concerned with the mathematical formulation of the fundamentals for any flowing continuum, the first chapter with those matters which do not involve forces, Newton's laws, or the use of a Newtonian frame of reference, and the second chapter with those that do. In Chapter 3, a new careful formulation is given of the basic equations of magnetohydrodynamics for a continuum, based on the electrodynamics of moving media. In Chapter 4, the basic results for an inviscid fluid are derived; with Appendices IV and V included, shock discontinuities and Prandtl-Meyer expansions, with some mention of "hypersonic" approximations, are also discussed, while Appendix III is devoted to some of those aspects of the irrotational motion of an incompressible fluid which I myself found most interesting. The Navier-Stokes equations for a Newtonian fluid, compressible or incompressible, and some known solutions, are discussed in Chapters 5 and 6, which also contain some discussion of the relation of the continuum equations to results from the kinetic theory of gases, and a discussion of dimensionless groups and dynamical similarity. Boundary-layer theory for incompressible fluids and for gases is discussed in Chapters 7 and 9. Chapter 8 is concerned with using the results of boundary-layer theory for an incompressible fluid as a starting point for an asymptotic expansion of a solution of the full Navier-Stokes equations, especially for a particular example, that of the flow past a semi-infinite flat plate along the stream. Chapters 10 and 11 have already been mentioned. Chapters 12 and 13, on mixtures, and high temperature effects in gases, and on longitudinal waves in an ionized gas (plasma) have been included to show that, difficult as it is to "solve" the Navier-Stokes equations, these equations certainly do not describe all the situations that are encountered. However, whereas Chapter 12 is purely descriptive, Chapter 13 contains a mathematical "solution", based on the use of the Boltzmann equation. Professor Burgers, in his special lectures, went deeply and in detail into two problems of magnetohydrodynamics, the propagation of a shock wave in a gas of high electrical conductivity in the presence of a magnetic field due to a dipole, and the meaning of the "tensor" electrical conductivity in an ionized gas. It was indeed fortunate, both for me and the audience, that Professor Burgers was able and willing to give us such an excellent account of his work on these important and interesting matters, and so to make the discussion on magnetohydrodynamics in this book so much more satisfactory.

To those who consider, as I do, that Applied Mathematics, when the term is used strictly, denotes all attempts to deepen and generalize our understanding of a science by the use of mathematical methods, I would add that, though I did not set out consciously to make it so, what is included in these lectures does, in fact, represent for me a typical example, except for one glaring and horrible omission — the omission of any comparison of theoretical with experimental results. Rightly or wrongly, I did not consider that such comparisons would be appropriate in these lectures.

I am indeed well aware of the deficiencies in these lecture notes, and beg the reader, particularly when he sees their incompleteness and their lack of cohesion, to recall that they are not presented as anything more than they were and are: a set of notes for twelve lectures, with a little additional reading. If any readers later begin to work through the fuller published accounts of the various branches of the subject, and especially if any reader who has already done so still finds anything of interest in these notes, the aim of this publication will have been fully achieved.

If the American Mathematical Society and the University of Colorado had not arranged the Seminar at Boulder, and if I had not been invited to give the lectures on Fluid Mechanics, the material here presented would not have been prepared. Many of my colleagues read the manuscript or the proof. To all of them, to the co-operating members of the staff of Interscience Publishers, and to the faithful and responsive audience at Boulder, I record my thanks. I would mention especially Dr. J. D. Murray, Dr. G. S. S. Ludford, and Dr. D. H. Michael, who pointed out to me many errors and misprints in the script and the proof, and to whom I give special thanks.

SYDNEY GOLDSTEIN

December, 1959

# Contents

**Introduction. Notation, etc.** .................. 1

## 1. Kinematics ..................... 3

1.1 Lagrangian and Eulerian specifications. ......... 3
1.2 Local, convective, and total rates of change. Acceleration 4
1.3 Equation of continuity. ................ 5
1.4 Boundary conditions. ................. 10
1.5 Axes in relative motion ............... 11
1.6 The analysis of the motion in a neighborhood. The rate-of-strain components and the vorticity. (*See also Appendix I.*) .......................... 13
1.7 The momentum and angular momentum of a small portion of fluid. ...................... 16
1.8 Vortex lines and tubes. Circulation. Vortex sheets. Line vortices ........................ 18
1.9 Rate of change of circulation ............. 23
1.10 Irrotational or potential motions. The velocity potential. Incompressible fluids. ................ 25
1.11 Vector potentials and stream functions. (*See also Appendix II.*) .......................... 27

## 2. Dynamics of the General Fluid. ............. 35

2.1 The stress tensor ................... 35
2.2 The equations of momentum and angular momentum. . 38
2.3 The equation of energy ................ 39
2.4 Hydrostatic pressure. Removal or neglect of gravity body forces ......................... 44
2.5 Impulsive generation of motion in a fluid of constant density 46

## 3. Electric and Magnetic Forces. ............. 47

3.1 Introduction. Maxwell's equations. Physical preliminaries 47
3.2 Electrodynamics of a medium at rest ......... 50
3.3 Moving media. .................... 51
3.4 The equations of momentum and energy in fluid dynamics, with electric and magnetic forces ........... 58
3.5 Equations for **B** and $\rho_e$ ................ 60

## 4. Inviscid Fluids . . . . . . . . . . . . . . . . . . 63

4.1 Integration of the equation of motion. Conditions for steady motion. Bernoulli's equation. The equation for the velocity potential in the irrotational motion of a gas with constant entropy . . . . . . . . . . . . . . . . . . . . . . . 63
4.2 The theorems of Kelvin, Helmholtz, and Lagrange. Cauchy's vorticity equations. (*See also Appendix III*) . . . . . . 70
4.3 The $\omega$, **B** analogy . . . . . . . . . . . . . . . . . 76
4.4 Shock waves. Remarks on plane sound waves of finite amplitude.
(a) Shock waves. (*See also Appendix IV* and *Appendix V*) 78
(b) Plane waves of finite amplitude, and shock-wave formation . . . . . . . . . . . . . . . . . . . . . 82

## 5. Newtonian Fluids and the Navier-Stokes Equations. . . 87

5.1 Relations between $\pi_{ij}$ and $e_{ij}$. Equations of momentum 87
5.2 Dissipation of energy . . . . . . . . . . . . . . . . . 89
5.3 Rate of change of circulation, and equations for the vorticity 91
5.4 Results from the kinetic theory of gases . . . . . . . 92
5.5 Dimensionless groups. Dynamical similarity . . . . . . 101

## 6. Exact Solutions of the Navier-Stokes Equations . . . . 107

6.1 Incompressible fluids of constant properties . . . . . . 107
6.2 Compressible fluids. . . . . . . . . . . . . . . . . . 110

## 7. Boundary-Layer Theory for Incompressible Fluids . . . 115

7.1 Two-dimensional boundary-layer equations . . . . . . 116
7.2 Motion symmetrical about an axis . . . . . . . . . . 121
7.3 Flow along a flat plate . . . . . . . . . . . . . . . 123
7.4 Solutions with similar velocity distributions. Flow near a stagnation point. . . . . . . . . . . . . . . . . . . 127

## 8. Extension of Boundary-Layer Theory for Incompressible Fluids . . . . . . . . . . . . . . . . . . . . . . . . . 131

8.1 Introduction. . . . . . . . . . . . . . . . . . . . . 131
8.2 Asymptotic solution of the Navier-Stokes equations for flow past a flat plate . . . . . . . . . . . . . . . . 137

## 9. Boundary Layers in Gases . . . . . . . . . . . . . . . 147

9.1 Boundary-layer equations for two-dimensional flow . . . 149
9.2 Boundary-layer equations for axisymmetrical flow. Mangler's transformation . . . . . . . . . . . . . . 152

| | | |
|---|---|---|
| 9.3 | von Mises's transformation for steady two-dimensional motion | 154 |
| 9.4 | An integral of the energy equation for $Pr = 1$, and for a heat-insulated surface or zero main-stream acceleration | 156 |
| 9.5 | The Illingworth-Stewartson theorem | 158 |
| 9.6 | Remarks on flow past a flat plate $(dU/dx = 0)$ | 162 |
| 9.7 | Conclusion | 165 |

## 10. Turbulence. Stability . . . . . . . . . . . . . . . . . . 167

| | | |
|---|---|---|
| 10.1 | Turbulence | 167 |
| 10.2 | Stability | 169 |

## 11. Dynamics of Inviscid Gases . . . . . . . . . . . . . . 171

| | | |
|---|---|---|
| 11.1 | Introduction | 171 |
| 11.2 | Slender-body theory. Subsonic flow | 183 |
| | (a) Flow past a body of revolution at zero yaw by the method of sources | 183 |
| | (b) Boundary condition at the surface | 190 |
| | (c) Fourier transforms | 193 |
| | (d) Example | 195 |
| 11.3 | Slender-body theory. Supersonic flow | 197 |
| | (a) Flow past a body of revolution at zero yaw | 198 |
| | (b) Laplace transforms | 200 |
| | (c) Plane airfoils of small aspect ratio (subsonic or supersonic flow) | 201 |

## 12. Mixtures, and High Temperature Effects in Gases . . . 207

## 13. Longitudinal (Electrostatic) Waves in an Ionized Gas . 215

| | | |
|---|---|---|
| Appendix I. | The Rate-of-Strain Components and the Vorticity | 223 |
| Appendix II. | Sources, Doublets, and Line Vortices in an Incompressible Fluid: Potential Flow | 229 |
| Appendix III. | Some Remarks on the Irrotational Motion of an Incompressible Fluid | 233 |
| | (a) Flows without discontinuities. Sources and doublets for a body of revolution | 233 |
| | (b) Kinetic energy, impulse, forces | 236 |
| | (c) Flows with vortex sheets. Airfoil theory | 244 |
| Appendix IV. | Formulae for Shock Waves in Perfect Gases with Constant Specific Heats | 253 |
| | General equations for perfect gases with constant specific heats | 253 |

|  | |
|---|---|
| Stationary shock waves | 254 |
| Equations in terms of the pressure-ratio $p_2/p_1$ | 255 |
| Equations in terms of $M_1^2 \cos^2\alpha_1$ | 256 |
| Equations for the stagnation pressure (or total head) | 257 |
| Equations for the gain in entropy | 258 |
| Formulae for $\gamma = 1.4$ | 259 |
| Deflection and downstream Mach number for an oblique shock wave | 260 |
| Weak oblique shock waves. Hypersonic approximations for small deflections | 262 |

Appendix V. **Formulae for a Prandtl-Meyer Expansion in a Perfect Gas with Constant Specific Heats** . . 265

Book List . . . 271

Special Lectures: Some Problems of Magneto-Gasdynamics, by J. M. Burgers

Introduction . . . 273

1. Basic equations . . . 273
2. Equations for the vector potential $A$ when the conductivity of the gas is infinite . . . 276
3. Explicit expressions for the magnetic field . . . 278
4. Case where the gas ahead of the shock front is nonconducting . . . 280
5. Influence of a small value of the resistivity of the gas . . . 281
6. More accurate results for the region just ahead of the shock front, when account is taken of the reaction of the magnetic forces on the motion of the gas . . . 283
7. Hugoniot equations for a shock in a magnetic field . . . 285
8. Discussion of Equation (32) . . . 287
9. Final remarks on the shock wave problem . . . 291
10. The calculation of the electric current in an ionized gas subjected simultaneously to an electric and a magnetic field . . . 293
11. Basic formulas of the theory of diffusion . . . 294
12. Application to an ionized gas . . . 297
13. Discussion and simplification of the formulas obtained . . . 300

Index of Authors . . . 303

Index of Subjects . . . 305

## Introduction. Notation, etc.

A system of mutually orthogonal axes in three dimensions will be used. The coordinates will be denoted by $x_1$, $x_2$, $x_3$, or $x_i$ ($i = 1, 2, 3$). The summation convention for repeated subscripts will be used; if any subscript such as $i$, $j$, $k$, or $l$, occurs twice of an expression, the sum is to be taken for all the possible values in the subscript; these values will usually be 1, 2, 3; thus $A_i B_i$ means $A_1 B_1 + A_2 B_2 + A_3 B_3$. When the subscripts may take values other than 1, 2, 3 (relativistic electrodynamics) this will be specially mentioned. An expression in which the same subscript occurs more than twice is meaningless; such expressions are to be avoided.

Vectors will be shown in bold type. $\nu$ will be used in a general way for the outwardly-directed normal to a surface; $A_\nu$ will be the component of a vector **A** in the direction of such a normal, and $\partial \phi / \partial \nu$ will be the component of grad $\phi$ in such a direction; $\boldsymbol{\nu}$ will be the unit vector along the normal, with components $\nu_i$, so $\nu_i$ will denote the direction cosines of the normal.

The scalar product of two vectors will be denoted by a dot, as in $\mathbf{A} \cdot \mathbf{B}$, and the vector product by a cross, as in $\mathbf{A} \times \mathbf{B}$. Also $\boldsymbol{\nabla}$ may occasionally be used for the vector operation $\partial / \partial x_i$, with $\boldsymbol{\nabla} \phi$ for the gradient of a scalar $\phi$ and $\boldsymbol{\nabla} \cdot \mathbf{v}$ for the divergence of a vector **v**. It will not be found convenient or necessary to use $\boldsymbol{\nabla} \times \mathbf{v}$ for curl **v**.

For kinematics, we may choose any frame of reference, which may be moving in any manner relative to any other frame. But when we come to the dynamics of the subject, unless specific mention is made to the contrary, we suppose our frame to be a Newtonian, or "fixed," frame, in which Newton's law of motion for a particle of constant mass is $m\ddot{\mathbf{x}} = \mathbf{X}$, where **X** is the force acting on the particle.

Mathematical conditions on the functions we shall encounter which are sufficient for the correctness of the statements made and theorems used, and which are, physically, not unduly onerous, will usually be well known or easily found. They will not usually be mentioned unless some physical significance is attached to them.

Our business is with the dynamics of fluids treated as continua, but, especially in gases, we shall need to know the thermodynamic properties, and to lean heavily both on the kinetic theory of gases and on such experimental evidence as is available for information on such matters as the variation of viscosity with the thermodynamic state of the gas. In certain extreme, but interesting, circumstances, studies are still necessary on the physics and mechanics of the gas particles, and on their statistics, in order to obtain sufficiently correct "continuum" or macroscopic equations, when such equations exist. The "classical" equations of fluid dynamics may need revision in such circumstances, and the necessary studies are by no means completed. But we leave such matters aside to begin with, and study the formulation in mathematical terms of the "classical" dynamics of a flowing continuum.

# CHAPTER 1

# Kinematics

Results in this section are valid for any fluid that may be considered as a continuum and that is moving in any manner, except where (in Section 1.8, 1.10, and 1.11) we expressly stipulate that we are assuming an inviscid fluid, or an incompressible fluid, or a fluid moving without vorticity.

## 1.1. Lagrangian and Eulerian specifications

In the Lagrangian method of specification in a field of fluid flow, we attach a "label" to each fluid element; this label is its position vector **a** relative to the origin at some given time $t_0$, which we may, if we wish, choose the same for all fluid elements and take as the initial time, $t = 0$. Then any quantity $F$ associated with the fluid is considered to be a function, not of a point in space and of time, but of a fluid element and time, i.e., of a material point moving with the fluid, and of time. Then $F$ is a function of **a** and of $t$. The coordinates $x_i$ of a fluid element are themselves functions of the $a_i$ and $t$; its velocity components $v_i$ and acceleration components $f_i$ are then

$$v_i = \partial x_i/\partial t, \quad f_i = \partial v_i/\partial t = \partial^2 x_i/\partial t^2.$$

In the Eulerian method, quantities associated with the fluid are taken as functions of a point in space and of time, i.e., as functions of **x** and $t$, and we dispense with the need to consider the history of each element, i.e., to find the $x_i$ as functions of the $a_i$ and $t$. In particular, the velocity components $v_i$ are taken as functions of the $x_i$ and $t$. When they have been found, to pass to the results of the Lagrangian method, the equations $dx_i/dt = v_i$, with the initial conditions $x_i = a_i$, would have to be solved; this solution, which is usually difficult to complete, would give us the particle paths.

In particular problems, a Lagrangian solution would usually

give more than is needed, and an Eulerian solution is all that is wanted. Occasionally (e.g. in certain parts of acoustics) a Lagrangian solution is convenient, and the method is sometimes useful for the understanding of fundamental theory.[1]

Note also that the streamlines (defined as lines whose tangent at any point is in the direction of the velocity at that point) have equations

$$\frac{dx_1}{v_1} = \frac{dx_2}{v_2} = \frac{dx_3}{v_3},$$

so that at any instant a particle path touches the streamline through the point at which the particle is. The two sets of lines are identical only if the motion is *steady* — i.e., if **v** is independent of $t$ in the Eulerian notation—when the streamlines are fixed lines in space. (Strictly speaking, the magnitude of the velocity may be multiplied by a function of the time—the same function everywhere—so long as its direction stays constant everywhere.) Note also that whether a motion is steady or not may depend on the frame of reference; it may not be steady relative to any frame, but if it is steady relative to one frame, it is not steady relative to any other frame in relative translational motion (except when the velocity components are independent of one coordinate, and the relative velocity of the frames is along that coordinate axis). If fluid is flowing smoothly past a solid body, the motion may be steady relative to axes fixed in the body, but is not steady relative to axes fixed with respect to the undisturbed fluid.[2]

### *1.2. Local, convective, and total rates of change. Acceleration*

In the Eulerian system, $\partial \mathbf{v}/\partial t$ gives the local rate of change of **v** at a particular point of space, not the rate of change of the velocity of a particle—i.e., not its acceleration. In general, if $F$ is a function of the $x_i$ and $t$, and the $x_i$ themselves change as $t$ changes so as always to refer to the same fluid element, then the rate of change of $F$, denoted in the literature by $DF/Dt$, is found by temporarily

---

[1] Both methods are, in fact, due to Euler. See the references in Lamb's *Hydrodynamics*, Section 3.

[2] On particle paths, streamlines, and filament lines (German: Streichlinie), see Prandtl and Tietjens, Vol. 1, Sections 37 and 38.

## KINEMATICS

expressing the $x_i$ as functions of the $a_i$ and $t$, so
$$DF/Dt = \partial F/\partial t + v_j\, \partial F/\partial x_j$$
The first term is the local rate of change, the sum of the remaining three the convective rate of change, and the whole the total rate of change.

The component $f_i$ of the acceleration of the element is found by substituting $v_i$ for $F$:

(1) $$f_i = Dv_i/Dt = \partial v_i/\partial t + v_j\, \partial v_i/\partial x_j.$$

If $F$ is a scalar,

(2) $$DF/Dt = \partial F/\partial t + \mathbf{v}\cdot\boldsymbol{\nabla} F$$

is in invariant form, but for the total rate of change of a vector it is best to transform the convective terms to express the result in vector form. For the acceleration a small calculation shows that

(3) $$\mathbf{f} = \partial\mathbf{v}/\partial t + \boldsymbol{\nabla}(\tfrac{1}{2}\mathbf{v}^2) + \boldsymbol{\omega}\times\mathbf{v},$$

where $\boldsymbol{\omega} = \operatorname{curl}\mathbf{v}$.

From this form, the components of $\mathbf{f}$ are easily found for any system of curvilinear orthogonal coordinates.

{The generalization to the rate of change of any vector $\mathbf{A}$ of this well-known formula for the acceleration is
$$D\mathbf{A}/Dt = \partial\mathbf{A}/\partial t + \tfrac{1}{2}[\boldsymbol{\nabla}(\mathbf{v}\cdot\mathbf{A}) + \boldsymbol{\omega}\times\mathbf{A} + \operatorname{curl}\mathbf{A}\times\mathbf{v}$$
$$- \operatorname{curl}(\mathbf{v}\times\mathbf{A}) + \mathbf{v}(\boldsymbol{\nabla}\cdot\mathbf{A}) - \mathbf{A}(\boldsymbol{\nabla}\cdot\mathbf{v})].\}$$

### 1.3. Equation of continuity

The equation of continuity is usually obtained as the equation of the conservation of mass by equating the rate of change of the mass of a small rectangular volume element, with edges parallel to the axes, to the rate of mass flux across the boundary. The result is

(4) $$\partial\rho/\partial t + \boldsymbol{\nabla}\cdot(\rho\mathbf{v}) = 0,$$

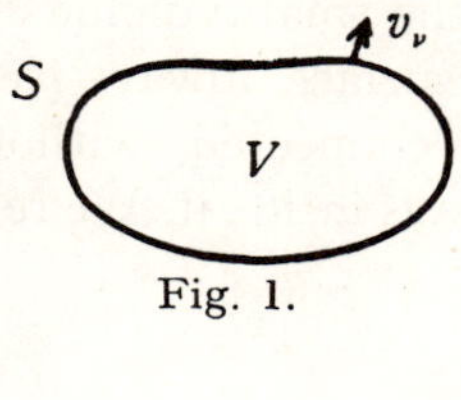

Fig. 1.

where $\rho$ is the fluid density. In fact, if $S$ is a surface fixed in space, the rate of increase of the mass inside it is

$$\frac{\partial}{\partial t}\int_V \rho d\tau = \int_V \frac{\partial \rho}{\partial t} d\tau,$$

where $d\tau$ is a volume element, and the integral is throughout the interior of $S$. The rate of mass flux inwards through $S$ is

$$-\int \rho v_\nu dS$$

over the surface $S$, which by the divergence theorem is

$$-\int_V \mathbf{\nabla} \cdot (\rho \mathbf{v}) d\tau.$$

This must equal the rate of increase of mass, i.e.,

$$\int_V \left[\frac{\partial \rho}{\partial t} + \mathbf{\nabla} \cdot (\rho \mathbf{v})\right] dS = 0;$$

this is true for any volume, so the integrand vanishes at every point at which it is continuous.

Equation (4) is (since $\rho \neq 0$)

(5) 
$$\frac{1}{\rho}\frac{\partial \rho}{\partial t} + \frac{1}{\rho}\frac{\partial}{\partial x_i}(\rho v_i) = \frac{1}{\rho}\frac{\partial \rho}{\partial t} + \frac{v_i}{\rho}\frac{\partial \rho}{\partial x_i} + \frac{\partial v_i}{\partial x_i}$$
$$= \frac{1}{\rho}\frac{D\rho}{Dt} + \mathbf{\nabla} \cdot \mathbf{v} = 0,$$

and it is more instructive to obtain the equation directly in this form, in the manner of Euler, by letting our small rectangular volume element move with the fluid. If $\delta\tau$ is the volume of the element, it is shown (Lamb, Section 7) that as $\delta\tau$ tends to a point, then in the limit

(6) $$(\delta\tau)^{-1} D(\delta\tau)/Dt = \mathbf{\nabla} \cdot \mathbf{v},$$

which could serve as an illuminating definition of div $\mathbf{v}$. Note that when we say that the small volume element moves with the fluid, we imply that its surface, always passing through the same fluid particles, remains connected, without holes or folds, and that the fluid inside it stays inside it. We return to this point later.

## KINEMATICS

The classic name for the equation is the equation of continuity, not the equation of conservation of mass. The "continuity" is here physical, not mathematical, continuity: the fluid stays a continuum. This is probably the more important, certainly the more subtle, point: to express conservation of mass is trivial. The point is, mathematically, made clearer by considering the Lagrangian representation. If the fluid which occupies a volume $V$ at time $t$ occupied a volume $\Omega$ at time $t = 0$, if $\rho_0$ was its density when $t = 0$, and we use $d(\mathbf{a})$ and $d(\mathbf{x})$ for volume elements in the $a_i$-space and the $x_i$-space, respectively, the equation of conservation of mass is

$$\iiint \rho_0 \, d(\mathbf{a}) = \iiint \rho \, d(\mathbf{x}).$$

Since the fluid was originally, and remains, a continuum, there are no holes and no two portions of fluid occupy the same space. There is therefore a complete uniform (1, 1) correspondence of the points of the volumes $V$ and $\Omega$, and the Jacobian

$$J = \frac{\partial(x_1, x_2, x_3)}{\partial(a_1, a_2, a_3)}$$

never vanishes or becomes infinite. At $t = 0$, $x_i = a_i$, $J = 1$, so $J$ is always positive. Hence the formula for the transformation of triple integrals gives

$$0 = \iiint \rho_0 \, d(\mathbf{a}) - \iiint \rho \, d(\mathbf{x}) = \iiint (\rho_0 - \rho J) d(\mathbf{a}),$$

and since this holds for any volume,

(7) $\qquad \rho J = \rho_0 = $ constant  for each element for all time.

This is an integral of (5); we recover (5) by operating with $\partial/\partial t$ in the Lagrangian (or $D/Dt$ in the Eulerian) representation. In fact, by a small calculation we verify that

$$J^{-1}(DJ/Dt) = \nabla \cdot \mathbf{v},$$

which is the mathematical form of (6).

A fluid is said to be *incompressible* if the density of any element remains constant as the time changes during the motion considered, i.e., if

(8) $$D\rho/Dt = 0.$$

Hence the equation of continuity for an incompressible fluid is

(9) $$\nabla \cdot \mathbf{v} = 0.$$

(This is so whether $\rho$ is constant throughout the fluid or not.)

Note also that, according to physical continuity, there may be surfaces in the fluid at which the tangential velocity is discontinuous, but that $\rho v_\nu$ must be continuous. Discontinuities of tangential velocity may, as we shall see, persist in an ideal fluid, devoid of viscosity, but not in a real viscous fluid.

Note also that if a surface $S$ moves with the fluid, so as always to pass through the same fluid particles, it remains a connected surface, without holes or folds, and every fluid element inside it at any time remains inside it.

Now consider the rate of change of the total amount of any quantity associated with the fluid which is inside a surface $S$ at any instant. The quantity may be a component of momentum, or of angular momentum, or it may be energy, or dissolved matter for which diffusion may be neglected. Let $F(\mathbf{x}, t)$ be the amount per unit mass, so the amount in a volume element $d\tau$ is $\rho F \, d\tau$. If the quantity is the $x_1$-component of momentum, $F = v_1$; if it is angular momentum about the $x_1$-axis, $F = v_3 x_2 - v_2 x_3$; if it is kinetic energy, $F = \tfrac{1}{2}\mathbf{v}^2$; and so on. We require

$$\frac{D}{Dt}\int_V \rho F \, d\tau,$$

as $S$ moves with the fluid. If we change this into an integral over the mass, by writing $\rho \, d\tau = dm$, then $dm$ is invariable, so we expect to find

$$\frac{D}{Dt}\int \rho F d\tau = \int \frac{DF}{Dt} dm = \int \frac{DF}{Dt} \rho d\tau.$$

We may also consider the matter by letting the surface $S$ move with the fluid. After a time $\delta t$, the fluid is inside a surface

$S'$ (Fig. 2), whose volume is obtained from that of $S$ by adding a surface layer of thickness $v_\nu \delta t$ (correct to the first order in $\delta t$). To that order, the amount of the quantity has become

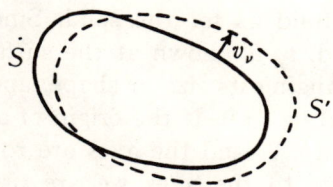

Fig. 2.

$$\int_V [\rho F + \frac{\partial}{\partial t}(\rho F)\delta t]\, d\tau + \delta t \int \rho F v_\nu\, dS,$$

and the rate of increase is

$$\frac{D}{Dt}\int \rho F\, d\tau = \int \frac{\partial}{\partial t}(\rho F)\, d\tau + \int \rho F v_\nu\, dS.$$

The equality of the two expressions is easily established, since by the divergence theorem the second is

$$\int_V \left[\frac{\partial}{\partial t}(\rho F) + \nabla \cdot (\rho F \mathbf{v})\right] d\tau$$

$$= \int_V \left[F\left\{\frac{\partial \rho}{\partial t} + \nabla \cdot (\rho \mathbf{v})\right\} + \rho\left\{\frac{\partial F}{\partial t} + \mathbf{v} \cdot \nabla F\right\}\right] d\tau.$$

The first term in the integrand is zero by the equation of continuity. The second is $\rho DF/Dt$. So

(10) $$\frac{D}{Dt}\int \rho F\, d\tau = \int \rho \frac{DF}{Dt}\, d\tau = \int \frac{\partial}{\partial t}(\rho F)\, d\tau + \int \rho F v_\nu\, dS.$$

For a quantity given as $G(x_i, t)$ per unit volume, we have similarly

(11) $$\frac{D}{Dt}\int G\, d\tau = \int \left\{\frac{DG}{Dt} + G\nabla \cdot \mathbf{v}\right\} d\tau = \int \frac{\partial G}{\partial t}\, d\tau + \int G v_\nu\, dS.$$

## 1.4. Boundary conditions

At any point on the boundary of a fluid formed by an impermeable solid surface, physical continuity requires that the component velocity along the normal to the surface should be the same for the fluid as for the solid. Since this is supposed known for the solid, $v_\nu$ is known at the surface for the fluid. If the solid is not changing its size or shape, and we use axes fixed relative to the solid, $v_\nu = 0$. If the origin of such "body" axes is moving with velocity **V**, and the axes are rotating with angular velocity $\boldsymbol{\Omega}$, relative to the axes we are using, then $v_\nu = \boldsymbol{\nu} \cdot (\mathbf{V} + \boldsymbol{\Omega} \times \mathbf{x})$.

We shall have occasion to require the boundary condition at the surface of a deforming body. Let $f(\mathbf{x}, t) = 0$ be the equation of the surface of such a body, $R$ denote the positive square root of $(\boldsymbol{\nabla} f)^2$, and $f_t$ denote $\partial f/\partial t$. Then the direction cosines of the normal to the surface are the components of

$$\boldsymbol{\nu} = \boldsymbol{\nabla} f / R.$$

If **u** is the velocity of the point **x** of the solid surface at time $t$, then to the first order in $\delta t$

$$f(\mathbf{x} + \mathbf{u}\, \delta t,\ t + \delta t) = 0,$$

so

$$\mathbf{u} \cdot \boldsymbol{\nabla} f + f_t = 0.$$

Let $u_\nu$ be the component of **u** along the normal. Since $\boldsymbol{\nabla} f$ is along the normal

$$u_\nu = -f_t/R,$$

But $v_\nu = u_\nu$ at the boundary. The required boundary condition is

(12) $$v_\nu = -f_t/R.$$

Since $v_\nu = \mathbf{v} \cdot \boldsymbol{\nu} = \mathbf{v} \cdot \boldsymbol{\nabla} f / R$, the condition may also be put into the form

(13) $$f_t + \mathbf{v} \cdot \boldsymbol{\nabla} f = \frac{Df}{Dt} = 0.$$

This result also follows from the argument that the velocity of a fluid particle on the surface must be wholly tangential or zero, so a particle on the surface remains on the surface for a time $\delta t$ (to the first order in $\delta t$), and $f$ stays zero to the first order in $\delta t$ if $\mathbf{x}$ remains the position vector of a particular fluid particle.

If a solid surface is especially made not impermeable, to allow for suction or blowing, then it is usual to assume that the suction velocity is known, so $v_\nu$ is again known on the boundary.

All these boundary conditions on the normal velocity component $v_\nu$ are purely kinematical, resulting from physical continuity. In a real viscous fluid (no matter how small the viscosity) there is also a physical condition on the tangential velocity. For the most part, we may postulate, with great accuracy, that there is no relative tangential velocity, or velocity of slip, at a solid boundary. In some cases, e.g. for a moderately rarefied gas, it may be better to allow for a slip velocity. We return to the matter later; until further notice, the boundary condition of no relative tangential velocity will be postulated for a real fluid.

## 1.5. *Axes in relative motion*

Consider two sets of axes, to which we shall refer as $s$ and $b$, in relative motion, the origin of $b$ moving with velocity $\mathbf{V}$, and the axes of $b$ rotating with angular velocity $\mathbf{\Omega}$, relative to $s$. (In applications, we consider $s$ as axes "fixed" in space, and $b$ as axes fixed in a moving body) At any instant let the directions of the axes of $s$ coincide with those of $b$, so that space derivatives are the same; but the directions will not continue to coincide, and time rates-of-change will require consideration.

Up to a point, we shall follow the ideas used in the kinematics of a point mass. Relative motion of axes does not affect any scalar property of a point mass, but does affect its velocity and acceleration relative to the axes chosen. Let $\mathbf{v}_s$, $\mathbf{v}_b$ denote the velocity, and $\mathbf{f}_s$, $\mathbf{f}_b$ the acceleration, of a fluid particle, relative to the axes $s$ and $b$, respectively. From the ordinary kinematics of moving axes,

(14) $$\mathbf{v}_s = \mathbf{v}_b + \mathbf{V} + \boldsymbol{\Omega} \times \mathbf{x}.$$

Also
$$\left(\frac{D}{Dt}\right)_s = \left(\frac{\partial}{\partial t}\right)_s + \mathbf{v}_s \cdot \boldsymbol{\nabla}, \quad \left(\frac{D}{Dt}\right)_b = \left(\frac{\partial}{\partial t}\right)_b + \mathbf{v}_b \cdot \boldsymbol{\nabla},$$

The equation of continuity must be independent of the axes chosen; we shall have both

$$\rho^{-1}(D\rho/Dt)_s + \boldsymbol{\nabla} \cdot \mathbf{v}_s = 0, \quad \rho^{-1}(D\rho/Dt)_b + \boldsymbol{\nabla} \cdot \mathbf{v}_b = 0.$$

In fact both $\boldsymbol{\nabla} \cdot \mathbf{v}$ and $D\rho/Dt$ are each separately independent of the axes. (The total rate of change of any scalar must be independent of the axes). All this may easily be verified in detail if one wishes.

We turn to the acceleration. From the usual kinematical formula

$$\mathbf{f}_s = (D\mathbf{v}_s/Dt)_b + \boldsymbol{\Omega} \times \mathbf{v}_s$$

Here we depart from the ideas of the kinematics of a point mass, in which one substitutes for $\mathbf{v}_s$ to find $\mathbf{f}_s$ in terms of $\mathbf{x}$ and $\mathbf{v}_b$ and its rate of change—i.e., of $\mathbf{x}$, $(d\mathbf{x}/dt)_b$, $(d^2\mathbf{x}/dt^2)_b$. We note that curl $\mathbf{v}_b$ = curl $\mathbf{v}_s - 2\boldsymbol{\Omega}$, and our primary applications will be to cases in which we take curl $\mathbf{v}_s$ (not curl $\mathbf{v}_b$) to be zero, so we transform $\mathbf{f}_s$ into a form which retains $\mathbf{v}_s$. For the $i$th component

$$(Dv_{si}/Dt)_b = (\partial v_{si}/\partial t)_b + \mathbf{v}_b \cdot \boldsymbol{\nabla} v_{si}$$
$$= (\partial v_{si}/\partial t)_b + (\mathbf{v}_s - \mathbf{V} - \boldsymbol{\Omega} \times \mathbf{x}) \cdot \boldsymbol{\nabla} v_{si}.$$

Since $\mathbf{V}$ is independent of the coordinates, $\boldsymbol{\nabla} V_i = 0$, curl $\mathbf{V} = 0$, and we have, exactly as in the transformation for the acceleration in (3), that $(\mathbf{v}_s - \mathbf{V}) \cdot \boldsymbol{\nabla} v_{si}$ are the components of

$$\boldsymbol{\nabla} \tfrac{1}{2}(\mathbf{v}_s - \mathbf{V})^2 + \boldsymbol{\omega}_s \times (\mathbf{v}_s - \mathbf{V}),$$

where $\boldsymbol{\omega}_s$ = curl $\mathbf{v}_s$. Also, it is a straightforward matter to show that $(\boldsymbol{\Omega} \times \mathbf{x}) \cdot \boldsymbol{\nabla} v_{si}$ are the components of

$$\boldsymbol{\Omega} \times \mathbf{v}_s - \boldsymbol{\nabla}(\boldsymbol{\Omega} \cdot (\mathbf{v}_s \times \mathbf{x})) + \boldsymbol{\omega}_s \times (\boldsymbol{\Omega} \times \mathbf{x}).$$

Hence finally

(15) $\quad \mathbf{f}_s = (\partial \mathbf{v}_s/\partial t)_b + \nabla\{\tfrac{1}{2}(\mathbf{v}_s - \mathbf{V})^2 + \boldsymbol{\Omega}\cdot(\mathbf{v}_s \times \mathbf{x})\} + \boldsymbol{\omega}_s \times \mathbf{v}_b,$

and the relation between $\mathbf{v}_b$ and $\mathbf{v}_s$ is given in (14).
We shall apply the formula later.

## 1.6. The analysis of the motion in a neighborhood. The rate-of-strain components and the vorticity

Let $P$ be at $x_i$ and $Q$ at $x_i + \delta x_i$. We work to the first order only in the $\delta x_i$. The velocity of $Q$ relative to $P$ has components

$$\delta v_i = (\partial v_i/\partial x_j)\, \delta x_j.$$

$\partial v_i/\partial x_j$ is, of course, a second-order (cartesian) tensor. If we transform to a different set of axes $x_i'$, and denote by $v_i'$ the velocity components along the new axes, and if $\alpha_{ij}$ is the cosine of the angle between the axes of $x_i'$ and $x_j$, then

$$x_k' = \alpha_{ki} x_i, \quad x_j = \alpha_{lj} x_l', \quad v_k' = \alpha_{ki} v_i, \quad v_j = \alpha_{lj} v_l',$$

and

$$\partial v_k'/\partial x_l' = \alpha_{ki}\, \partial v_i/\partial x_l' = \alpha_{ki}\alpha_{lj}\, \partial v_i/\partial x_j.$$

Split $\partial v_i/\partial x_j$ into its odd and even parts by writing

(16) $\quad e_{ij} = \partial v_j/\partial x_i + \partial v_i/\partial x_j, \quad \xi_{ij} = \partial v_j/\partial x_i - \partial v_i/\partial x_j,$

so

(17) $\quad e_{kl}' = \alpha_{ki}\alpha_{lj} e_{ij}, \quad \xi_{kl}' = \alpha_{ki}\alpha_{lj}\, \xi_{ij},$

and

(18) $\quad \delta v_i = \tfrac{1}{2} e_{ij}\, \delta x_j - \tfrac{1}{2} \xi_{ij}\, \delta x_j.$

$e_{ij}$ is a symmetric second-order tensor; $\xi_{ij}$ is an antisymmetric second-order tensor, or polar vector, for which $\xi_{ij} = -\xi_{ji}$, and $\xi_{ij} = 0$ if $i = j$. Hence we may write $\xi_{23} = -\xi_{32} = \omega_1$, $\xi_{31} = -\xi_{13} = \omega_2$, $\xi_{12} = -\xi_{21} = \omega_3$. The transformation formula for $\xi_{ij}$ reduces to

$$\omega_k' = \beta_{ki}\omega_i,$$

where $\beta_{ki}$ is the cofactor of $\alpha_{ki}$ in the determinant of the matrix $\|\alpha_{ij}\|$. From known results for the direction cosines of an orthogonal triad (or from the application of the vector-product rule to pairs of unit vectors along the $x'$ axes) this formula is equivalent to

$$\omega'_k = \pm \alpha_{ki}\omega_i,$$

where the positive sign is to be taken if the axes of $x'$ may be obtained from those of $x$ by a pure rotation, and the negative sign is to be taken if, in addition to any possible rotation, a reflexion of axes is involved in the transformation. This is the usual transformation formula for a polar vector.

From (16), we now have

(19) $$\boldsymbol{\omega} = \text{curl } \mathbf{v}.$$

$\boldsymbol{\omega}$ is called the vorticity. The second term in (18) is seen to be the $i$th component of $\frac{1}{2}\boldsymbol{\omega} \times \delta\mathbf{x}$. This is the velocity at $Q$ resulting from a purely rigid-body rotation with angular velocity $\frac{1}{2}\boldsymbol{\omega}$.

The first term in (18) may now be seen to represent a rate of distortion of the fluid element in the neighborhood of $P$. To begin with, we may find the significance of each $e_{ij}$ separately by taking the $\xi_{ij}$ to be zero, and all of the $e_{ij}$ to be zero except the one under consideration. In this way, it is a simple matter to see that $e_{11}$ is twice the rate of extensional strain of a line element in the direction of the axis of $x_1$, and similarly for $e_{22}$ and $e_{33}$; $e_{23}$ is a rate of shearing strain, being the rate of change of the angle between two lines of particles which at the instant under consideration lie along the axes of $x_2$ and $x_3$. Similarly for $e_{31}$ and $e_{12}$. (We return to this matter in Appendix I).

Now write

$$\Phi = e_{ij}\delta x_i \delta x_j,$$

and regard $P$ at $\mathbf{x}$ as an origin of coordinates, the current coordinates being denoted by $\delta x_i$. Then $\Phi = $ constant is the equation of a quadric surface with its center at $P$, called the rate-of-strain quadric. If the quadric be referred to another set of axes, parallel to the axes of $x'_i$, $\Phi$ becomes

$$\alpha_{ki}\alpha_{lj}\, e_{ij}\delta x'_k\, \delta x'_l = e'_{kl}\delta x'_k \delta x'_l.$$

The principal axes of this quadric, which are the principal axes for the tensor $e_{ij}$, are called the principal axes of rate-of-strain.

## KINEMATICS

With them as axes, the rates of shear are zero; twice the rates of extension along them, called the principal rates of extension, will be denoted by $e_1$, $e_2$, $e_3$.

We may now give a first physical definition of vorticity. At any time there is, through any point of a fluid, a triad of three mutually orthogonal straight lines (the principal axes of rate-of-strain) such that, if they move with the fluid and continue to pass through the same fluid particles, then after a short time $\delta t$ the angles between them remain right angles to the first order in $\delta t$ (since the rates of shear are zero). The vorticity $\omega$ is then twice the angular velocity of this triad of lines relative to its instantaneous position.

If $\frac{1}{2}e'$ is the rate of extension of a line element in any direction, and $r$ the length of the radius vector from the origin to the rate-of-strain quadric in this direction, it is easy to prove that $e'r^2$ is constant for all directions. Hence the rate of extension of a line element drawn in any direction through $P$ is inversely proportional to the square of the radius vector to the rate-of-strain quadric in that direction, and is therefore stationary, for changes in direction, along the principal axes.

Also $\partial \Phi / \partial (\delta x_i) = 2e_{ij}\delta x_j$, and the first term in (18) is $\frac{1}{4}\nabla\Phi$, the gradient being taken with respect to $\delta x_i$. Hence, if $\mathbf{v}_P$ is the velocity at $P$, the velocity at $Q$ is

(20) $\qquad \mathbf{v}_Q = \mathbf{v}_P + \frac{1}{2}\boldsymbol{\omega} \times \delta\mathbf{x} + \frac{1}{4}\nabla\Phi.$

curl $\mathbf{v}_Q$, with respect to the coordinates $\delta x_i$, is $\boldsymbol{\omega}$, so

$$\mathbf{v}_Q = \mathbf{v}_P + \tfrac{1}{2}\operatorname{curl} \mathbf{v}_Q \times \delta\mathbf{x} + \tfrac{1}{4}\nabla\Phi.$$

For the volume element in the neighborhood of $P$, with $\delta x_i$ as current coordinates, the first two terms represent the most general rigid-body motion, and the last term represents a motion of distortion which, since it is represented by a gradient and has zero curl, is called a potential or irrotational motion of distortion.

In Appendix I, the actual displacement of two line elements originally at right angles is computed. Such a calculation drives home the physical meaning of the rate-of-strain components and the vorticity, finds the transformation formulae in a physical, rather than a mathematical way, and introduces a method which enables us easily to find expressions for the rate-of-strain components in curvilinear orthogonal coordinates, in a physical manner, and in a suitable form for applications. These expressions are also found in Appendix I. The results are easily generalized to the case when the two line elements are initially not at right angles; the more general results are stated in Appendix I.

We have given two physical interpretations of the vorticity, one in this section and another in Appendix I. A more vividly descriptive property may be found from the result of the next section.

### 1.7. The momentum and angular momentum of a small portion of fluid

We now calculate the momentum and angular momentum about the center of mass of a small portion of fluid. We take the point $P$ of the previous section at the mass center. The velocity at any point $Q$ is given by (20). For convenience we write $\mathbf{V} = \mathbf{v}_P$ and $\eta_i = \delta x_i$, so $\eta_i$ are coordinates with the mass center as origin, and $\mathbf{V}$, the velocity of the mass center, is independent of the $\eta_i$. The first two terms in (20) are the same as for a rigid-body motion, and will give the known results for the momentum and angular momentum. To include the effects of the third term, it is convenient to use as axes the principal axes of rate-of-strain. Then for the velocity components of $Q$ we have

$$v_1 = V_1 + \tfrac{1}{2}(e_1\eta_1 + \omega_2\eta_3 - \omega_3\eta_2)$$

and two similar expressions for $v_2$, $v_3$. Since $P$ is at the mass center, if $dm$ is an element of mass,

$$\int \eta_i \, dm = 0,$$

and the linear momentum is still simply $M\mathbf{V}$, where $M$ is the total mass. The deformation does not alter the linear momentum.

Now let $I_{ij}$ be the inertia tensor for the distribution of mass,

$$I_{ij} = \int (\eta_k \eta_k \delta_{ij} - \eta_i \eta_j) \, dm,$$

where

(21) $$\delta_{ij} = \begin{cases} 1 \text{ if } i = j \\ 0 \text{ if } i \neq j \end{cases}$$

The component of angular momentum about the $\eta_1$-axis is

$$\int \{v_3 \eta_2 - v_2 \eta_3\} dm.$$

A straightforward calculation shows that this is

$$\tfrac{1}{2}\{I_{11}\omega_1 + I_{12}\omega_2 + I_{13}\omega_3 + I_{23}(e_2 - e_3)\}$$
$$= \tfrac{1}{2}\{I_{1j}\omega_j + I_{23}(e_2 - e_3)\}.$$

Similarly the components about the other two axes are

$$\tfrac{1}{2}\{I_{2j}\omega_j + I_{31}(e_3 - e_1)\} \text{ and } \tfrac{1}{2}\{I_{3j}\omega_j + I_{12}(e_1 - e_2)\}.$$

Now suppose that the small portion of fluid is instantaneously solidified (made rigid) without any change in momentum, angular momentum, or mass distribution. The expression for the linear momentum shows that the mass center of the solid begins to move with the same velocity **V** as the mass center of the fluid. If its angular velocity is $\Omega$, the formulae for the components of its angular momentum are obtained from those of the fluid by writing $\Omega_i$ for $\tfrac{1}{2}\omega_i$, and putting $e_1 = e_2 = e_3 = 0$. Equating angular momentum components gives three simple linear equations for the $\Omega_i$. Now, in general, at any point of a fluid, $e_1$, $e_2$, $e_3$ each differ in value from either of the other two; we see at once that if this is so, then $\Omega = \tfrac{1}{2}\omega$, if, and only if, the products of inertia, $-I_{23}$, $-I_{31}$, $-I_{12}$ are zero, i.e., if the principal axes of inertia of the solid coincide with the principal axes of rate-of-strain of the fluid at the mass center. In particular, for a uniform sphere any three mutually perpendicular axes through the center are principal axes of inertia; so we may define the vorticity at $P$ as twice the initial angular velocity of the solid sphere when an infinitesimally small sphere of fluid, with its center at $P$, is suddenly solidified without change of angular momentum.

## 1.8. Vortex lines and tubes. Circulation. Vortex sheets. Line vortices

A vortex line is defined as a line drawn in the fluid such that the tangent to it at any point has the direction of the axis of the vorticity at the point. The vortex lines through every point of a small closed curve form a tube called a vortex tube. For such a vortex tube, of small cross-section, the product of the magnitude $\omega$ of the vorticity and the area $\delta S_\nu$ of a normal cross-section has the same value all along the tube; the constant product $\omega \delta S_\nu$ is called the strength of the tube. The constancy of $\omega \delta S_\nu$ follows at once from the fact that, since $\omega = \operatorname{curl} \mathbf{v}$,

(22) $$\nabla \cdot \omega = 0.$$

Hence, from the divergence theorem, $\int \omega_\nu dS = 0$ for any closed surface. Apply this to the surface formed by two cross-sections, and the "wall" of the tube between them. The result follows. (In some approximate calculations, there are "kinks" in the vortex lines, and $\omega$ is discontinuous; for a proof by Stokes's theorem applicable to such cases see Lamb, Section 145; the proof is due to Kelvin.)

The product of the area $\delta S$ of any cross-section of a vortex tube and the magnitude $\omega_\nu$ of the component of $\omega$ normal to the cross-section is equal to $\omega \delta S_\nu$, the strength of the tube.

As a consequence of the constancy of the strength, vortex tubes—and therefore also vortex lines—cannot begin or end in the fluid. The vortex lines form closed curves, or begin or end on the boundaries of the fluid, or—if we suppose the fluid unbounded — go to infinity.

The circulation, $C$, of any vector field $\mathbf{A}$ round a closed curve or "circuit" $\Gamma$ drawn in the fluid is

(23) $$C = \oint_\Gamma \mathbf{A} \cdot d\mathbf{x}.$$

If $\mathbf{A}$ is $\mathbf{v}$, this is called simply the circulation round the circuit:

(24a) $$K = \oint_\Gamma \mathbf{v} \cdot d\mathbf{x}.$$

If $S$ is any open two-sided surface bounded by $\Gamma$, then by

Stokes's theorem

(24b) $$K = \int_S \omega_\nu \, dS.$$

The circulation is equal to the total strength of all the vortex tubes that thread through the circuit; and the difference of the circulations in any two circuits is equal to the total strength of all the vortex tubes that pass between them. If there is no vorticity in any region of a fluid, the circulation must be the same in any two circuits that can be made to coincide with one another without passing out of that region; more generally, if there is vorticity, the circulation is the same in any two circuits that can be made to coincide without cutting across any vortex line.

If there is no vorticity anywhere in the fluid, the circulation must be zero in every circuit if the space occupied by the fluid is simply-connected, as for the space outside a sphere or an airplane. But for a doubly-connected space, as for the space outside an infinite cylinder or an anchor ring, we can assert that the circulation is zero only for reducible circuits—circuits that may be reduced to a point without passing out of the fluid. For irreducible circuits all we can assert is (i) that the circulation must have the same value for all circuits that can be reconciled with one another, such as circuits that go once round an infinite cylinder in the same (positive) sense; this we call the circulation round the cylinder; and (ii) that the circulation in any irreducible circuit is a (positive a negative) integral multiple of some fixed circulation; circuits that go $n$ times round an infinite cylinder will have a circulation $\pm nK$, according as they are taken in the positive or negative sense, where $K$ is the circulation in a circuit that goes round once in the positive sense. All this is for zero vorticity.

In an inviscid fluid, vorticity may be imagined concentrated on sheets or along lines. To consider the first case, begin by considering a shearing motion, with the velocity along the axis of $x_1$, and with its magnitude a function of $x_2$ only. The vorticity is about the axis of $x_3$, in the direction shown by the arrow (Fig. 3a), and its magnitude is $|dv_1/dx_2|$. If we suppose that the vorticity is zero except between the planes whose traces are AB and CD, then if AB and CD are taken of unit length, the total strength of all the vortex tubes in the layer, per unit length, is the circulation

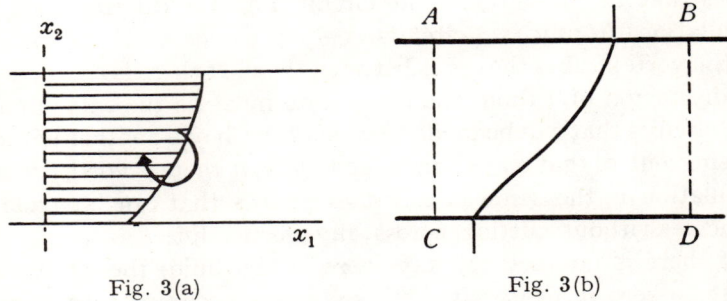

Fig. 3(a)　　　　　Fig. 3(b)

round ABDC, which is the difference in the velocity between AB and CD; it is called the strength of the layer. If now we suppose the velocity difference to remain constant, and the thickness to become zero, we obtain a surface at which the tangential velocity is discontinuous, called a vortex sheet. The strength of the sheet is numerically equal to the discontinuity in the velocity, and the vorticity is at right angles to the difference of the velocity on the two sides.

Any surface at which the normal velocity is continuous and the tangential velocity discontinuous is a vortex sheet; the vorticity at any point of the sheet is tangential to the sheet and at right angles to the velocity difference on the two sides, and the magnitude of the discontinuity in velocity is the magnitude of the strength of the vortex sheet. To sum up the matter in one equation, at a vortex sheet

$$(\mathbf{v}_1 - \mathbf{v}_2) \times \boldsymbol{\nu} = \boldsymbol{\omega}'$$

where $\omega'$ is the strength of the sheet; in Fig. 4, if $\mathbf{v}_1$ and $\mathbf{v}_2$ at $P$ are in the plane of the paper, with $v_1 > v_2$, $\omega'$ is upwards.

Vortex sheets may be regarded as idealizations for inviscid fluids of thin layers of vorticity in real fluids of small (but not zero) viscosity. Also, in the motion of such fluids of small viscosity, cases may often be observed in which the vorticity is sensibly concentrated into narrow tubes, called simply vortices. Such concentrated vortices may be observed if a blade is dipped into

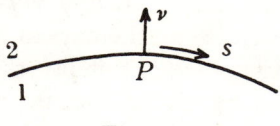

Fig. 4

running water, or the point of a spoon moved across the surface of a liquid, observation in these cases being facilitated by the depression of the free surface where the vortex meets it. But without a free surface, such vortices often exist behind solid bodies in motion relative to a fluid. In the theory of inviscid fluids such concentrated vortices are often idealized by supposing the vorticity concentrated along a line, the total strength remaining finite. In this way we arrive at the concept of a line vortex with a finite strength; the circulation in any circuit threaded by the line vortex (and no other vortex lines) is equal to its strength.

In a real viscous fluid, the vorticity in a vortex sheet or line vortex would diffuse, even if one could be created, so it would not persist as a discontinuity or singularity. In an ideal inviscid fluid, any vortex lines or vortex sheets that may exist in the fluid will, as we shall see, move with the fluid and persist indefinitely. However, the strength of a vortex sheet at any point changes; a vortex sheet is, in fact, unstable, and will roll up in some manner, so that in some sense the appearance of the flow approximates more and more to that due to a set of concentrated vortices. A thin vortex layer in a real fluid of small viscosity would also be unstable.

We may put the matter in a different way. In an inviscid fluid, there is no physical argument at all for assuming that the velocity components are continuous (apart from conditions arising from physical continuity, such as the mathematical continuity of $\rho v_\nu$ at a surface). It is sometimes pointed out that the existence of a surface discontinuity of tangential velocity (a vortex sheet) implies that the molecules of the fluid are being torn apart, so such a discontinuity will not arise. This argument is probably valid in some sense for a viscous fluid; at any rate, no discontinuity can persist in a viscous fluid, and any vortex sheets which may arise instantaneously in a mathematical theory for a viscous fluid must be idealizations of very thin layers if molecular properties are allowed for. But no argument from the molecular constitution of a fluid is in any way valid for an inviscid fluid; if the molecular constitution is to be brought in, then all the physical properties arising from molecular transfer, such as viscosity and heat conduction, must also be brought in. We may try to regard an inviscid fluid, not as one with zero viscosity, but as the limit of a real fluid of small viscosity as the viscosity tends to zero. Clearly singularities may appear in the limit. It is, as we shall see later in more detail, of the very essence of fluid dynamics that such limits are non-uniform.

Now clearly if we knew nothing at all about the discontinuities of $\mathbf{v}$, the theory of inviscid fluid motion would be in a hopeless position. We therefore proceed as follows. We first assume that $\mathbf{v}$ is continuous in the interior of the fluids; we also completely neglect the physical condition of zero slip (relative tangential velocity) at a bounding solid surface. When the problems so presented have been solved (if a solution exists; *cf*. Section 4.4), we may allow for zero slip by assuming that there is a vortex sheet, through which the relative tangential velocity falls to zero, coincident with each bounding surface. The results of such calculations may be compared with observation in a fluid of small viscosity. Usually, the results are found to be correct, or nearly correct, only in part of the velocity field—often in only a comparatively small part—and to be quite wrong elsewhere. The mathematical model must be refined. We may include the effects of viscosity, but the mathematical theory becomes very difficult. We may include discontinuities in the velocity field, and the most fruitful investigations along these lines are (apart from the occurrence of shock waves, Section 4.4) those which include vortex

**KINEMATICS**

sheets and line vortices—especially vortex sheets. At this stage, we cannot construct the mathematical model by strict logic; plausible physical arguments, often backed by some observational knowledge, must be used. In any case, the mathematical investigation would still remain difficult, and further (mathematical) approximations are often made. But enough has been said to explain why the concepts of vortex sheets and line vortices in invicid fluids are of importance. Moreover, calculations along these lines in inviscid fluids are often rather surprisingly effective in predicting the state of affairs for many motions of practical importance.

### 1.9. Rate of change of circulation

We now write down preliminary formulae for the rate of change of the circulation $K$ round a circuit $\Gamma$ (equation 24), as the circuit "moves with the fluid," always passing through the same fluid particles. Probably the most convincing way to start is to return to the Lagrangian symbolism. We first suppose $\Gamma$ to be a closed unicursal curve; the result may be extended immediately to a closed curve made up of any finite number of unicursal curves, not closed.

At time $t = 0$, we suppose the coordinates $a_i$ of points on $\Gamma$ to be functions of a parameter $\theta$, which increases from $\theta_0$ to $\theta_1$. Then a given value of $\theta$ always corresponds to a particular particle. At time $t$, the coordinates $x_i$ of the points of $\Gamma$ will be functions of $\theta$ and $t$, and

$$K = \int_{\theta_0}^{\theta_1} v_i \frac{\partial x_i}{\partial \theta} d\theta,$$

the integral being taken at a fixed $t$. Then

$$\frac{DK}{Dt} = \frac{\partial}{\partial t} \int_{\theta_0}^{\theta_1} v_i \frac{\partial x_i}{\partial \theta} d\theta = \int_{\theta_0}^{\theta_1} \frac{\partial v_i}{\partial t} \frac{\partial x_i}{\partial \theta} d\theta + \int_{\theta_0}^{\theta_1} v_i \frac{\partial^2 x_i}{\partial t \partial \theta} d\theta.$$

The second integral is

$$\int_{\theta_0}^{\theta_1} v_i \frac{\partial v_i}{\partial \theta} d\theta = \int_{\theta_0}^{\theta_1} \frac{\partial}{\partial \theta} (\tfrac{1}{2} \mathbf{v}^2) d\theta = 0,$$

since $\theta_0$ and $\theta_1$ refer to the same point and $\mathbf{v}^2$ must be single-valued. Hence

(25) $$\frac{DK}{Dt} = \oint_\Gamma f_i \, dx_i = \oint_\Gamma \mathbf{f} \cdot d\mathbf{x}.$$

If $C$ is the circulation of a vector field $\mathbf{A}$ (equation 23), and

$$\mathbf{B} = \operatorname{curl} \mathbf{A},$$

then

$$C = \int_S B_\nu \, dS,$$

where $S$ is any open two-sided surface bounded by $\Gamma$. We now cite the following result for the rate of change of the flux of any vector field $\mathbf{B}$ through a surface $S$ moving with the fluid:[3]

(26)
$$\frac{DC}{Dt} = \frac{D}{Dt} \int_S B_\nu \, dS = \int_S B_\nu^* \, dS,$$

where
$$\mathbf{B}^* = \frac{\partial \mathbf{B}}{\partial t} + \mathbf{v}(\nabla \cdot \mathbf{B}) + \operatorname{curl}(\mathbf{B} \times \mathbf{v}).$$

For $K$, substitute $\mathbf{v}$ for $\mathbf{A}$ and $\boldsymbol{\omega}$ for $\mathbf{B}$. Since $\nabla \cdot \boldsymbol{\omega} = 0$,

$$\frac{DK}{Dt} = \int_S \omega_\nu^* \, dS,$$

where
$$\boldsymbol{\omega}^* = \partial \boldsymbol{\omega}/\partial t + \operatorname{curl}(\boldsymbol{\omega} \times \mathbf{v}).$$

From (3),
$$\boldsymbol{\omega}^* = \operatorname{curl} \mathbf{f},$$

so

(27) $$\frac{DK}{Dt} = \int_S (\operatorname{curl}_\nu \mathbf{f}) \, dS,$$

as it must, from (25).

[3] Sommerfeld's *Mechanics of Deformable Bodies*, Section 18.1. See also M. Abraham and R. Becker, *The Classical Theory of Electricity and Magnetism*, Blackie and Son, London and Glasgow, 1937, Chap. 2, Section 12.

## 1.10. Irrotational or potential motions. The velocity potential. Incompressible fluids

If we suppose that throughout any region $\omega = \operatorname{curl} \mathbf{v} = 0$, then $v_i dx_i$ is a perfect differential, and a velocity potential $\phi$ exists such that

$$\mathbf{v} = \nabla \phi.\ ^{4}$$

If, in addition, the fluid is incompressible,

(28) $$\nabla \cdot \mathbf{v} = \nabla^2 \phi = 0,$$

where $\nabla^2$ is the Laplace operator $\partial^2/\partial x_i \partial x_i$. If $\omega = 0$ throughout a region extending to any solid boundaries, $\partial \phi/\partial \nu$ is given at the boundaries by the kinematical boundary condition (Section 1.4). If the region of irrotational motion is not completely bounded, it remains to consider boundary conditions at an infinite distance. Here we consider only cases in which the whole field of flow is irrotational, and bounded internally by one or more solid surfaces, and the fluid is incompressible.

The circulation in any circuit is equal to the change in $\phi$ as we go round the circuit. In a singly-connected space $\phi$ must be single-valued, since the circulation must be zero. In a doubly-connected space, such as the space outside an infinite cylinder, there may be a circulation $K$ in every simple irreducible circuit; then $\phi$ increases by $K$ as we go once round such a circuit, and $\phi$ is not single-valued. An example is a circulation $K$ round a circular cylinder, where $\phi = K\theta/2\pi$, $r$ and $\theta$ being polar coordinates in a plane perpendicular to the axis of the cylinder, with the origin on the axis.

Then it may be proved that, if there are no singularities, for a three-dimensional motion in which the fluid is at rest at infinity, and the space is singly-connected, $\phi$ is of the order $R^{-2}$, and the velocity components of order $R^{-3}$, at infinity, where $R$ is distance from any finite point. If the fluid has a given velocity field at infinity, these statements refer to the disturbance. (The elementary

---

[4] In some textbooks, such as Lamb and Milne-Thomson, $\phi$ has the opposite sign; $\mathbf{v} = -\nabla \Phi$.

singularities in potential flow are discussed in Appendix 2.) It is assumed that the solids in the fluid, whose surfaces form the internal boundary, are not being deformed in such a way that their volume is altering; if it is, the effect is that of a source.

Three-dimensional motions in doubly-connected spaces are of some mathematical, but no great physical interest. For references see Lamb, Section 116.

For two-dimensional flow, if we are considering the motion outside an infinite cylinder, the space is doubly-connected; outside two cylinders, it is triply-connected, with two cyclic constants in $\phi$, and so on. If there is no circulation, then for the disturbance at infinity $\phi$ is of order $r^{-1}$ and the velocity of order $r^{-2}$, where $r$ is distance in the plane from any finite point. Similar remarks apply about the volume of contained solids. If there is more than one cylinder, the circulation in a circuit embracing all of them is the sum of the circulations round each separately, and the result above holds if the total circulation is zero. If it is not, but is equal to $K$, there is a term $K\theta/2\pi$ in the asymptotic expansion of $\phi$ at infinity, where $r$, $\theta$ are polar coordinates (with any finite origin), and a term $K/(2\pi r)$ in the transverse velocity $v_\theta$; apart from these, which are determined if $K$ is known, the above remarks apply to the remainders.

With known circulations for multiply-connected spaces, the solution is uniquely determined.

An irrotational flow, so calculated, satisfies the kinematic boundary condition on the normal component of velocity at a solid surface, but cannot satisfy the physical condition on the tangential velocity component. The solution may be supposed completed by assuming vortex sheets, coincident with the boundary surfaces, through which the tangential velocity falls to zero. Such a solution, is, in fact, correct, as we shall see, for the initial motion of an incompressible fluid after an impulsive start. However, in a fluid whose viscosity is finite, no matter how small, such a vortex sheet will immediately start to diffuse. In most cases the final steady velocity distribution will, over much of the velocity field, be quite different from that predicted by a theory of ir-

KINEMATICS

rotational flow, as above; the explanation why this is so in these cases is one of the accomplishments of boundary-layer theory, even though that theory is at present incapable of forecasting the actual velocity field in a real fluid in such cases.

### 1.11. Vector potentials and stream functions

If the flow is steady, so that $\partial \rho/\partial t = 0$, from the equation of continuity (4), $\nabla \cdot (\rho \mathbf{v}) = 0$. There is therefore a vector field $\mathbf{A}$ such that

$$\rho \mathbf{v} = \operatorname{curl} \mathbf{A}.$$

For an incompressible fluid, whether the flow is steady or not, $\nabla \cdot \mathbf{v} = 0$, and there is a vector field $\mathbf{A}$ such that

$$\mathbf{v} = \operatorname{curl} \mathbf{A}.$$

In either case, since the gradient of an arbitrary scalar function may be added to $\mathbf{A}$ without altering its curl, we may stipulate also that

$$\nabla \cdot \mathbf{A} = 0.$$

It then follows for the second case that

(29) $\qquad \nabla^2 A_i = \nabla_i \nabla \cdot \mathbf{A} - \operatorname{curl}_i \operatorname{curl} \mathbf{A} = -\omega_i.$

If the motion is two-dimensional, parallel to the $(x_1, x_2)$ plane, with $v_3 = 0$, and all quantities independent of $x_3$, $\mathbf{A}$ may be taken to have only one component, $A_3$, which is independent of $x_3$. For the steady flow of a compressible fluid we may denote this component by $\rho_0 \psi$, where $\rho_0$ is some standard density, and then

(30) $\qquad \rho v_1 = \rho_0 \partial \psi/\partial x_2, \quad \rho v_2 = -\rho_0 \partial \psi/\partial x_1.$

Similarly, for the flow of an incompressible fluid in two dimensions, we may take $A_1 = A_2 = 0$, and write $\psi$ for $A_3$, where $\psi$ is independent of $x_3$; then

(31) $\qquad v_1 = \partial \psi/\partial x_2, \quad v_2 = -\partial \psi/\partial x_1,$

and, from (29),

(32) $\qquad \nabla_1^2 \psi = -\omega_3,$

where $\nabla_1^2$ is the two-dimensional Laplace operator, $\partial^2/\partial x_1^2 + \partial^2/\partial x_2^2$.

The velocity magnitude is the magnitude of $\nabla\psi$, and its direction is obtained from that of $\nabla\psi$ by rotating in the $(x_1, x_2)$ plane through a right angle in the negative sense. The volume of fluid crossing any curve $AP$, per unit breadth per unit time, in the sense of $v_\nu$ (Fig. 5), is therefore

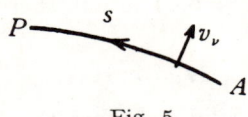

Fig. 5

$$\int_A^P v_\nu \, ds = \int_A^P \frac{\partial \psi}{\partial s} ds = \psi_P - \psi_A,$$

and the total flux out of a closed curve is the increase $[\psi]$ in $\psi$, as we go round the curve in the positive sense. Also the streamlines, $dx_1/v_1 = dx_2/v_2$, are simply $\psi =$ constant.

Since $v_\nu$ is given at a solid boundary, $\partial\psi/\partial s$ is given on the boundary, so $\psi$ is known apart from an additive constant. If the surface is impermeable, then over a stationary surface, $v_\nu = 0$, so $\psi$ is constant. At the surface of a solid cylinder for which a point fixed in the cylinder section has component velocities $(U_1, U_2)$, and which is rotating about a parallel to the $x_3$-axis with angular velocity $\Omega$,

$$v_\nu = \partial\psi/\partial s = \nu_1(U_1 - \Omega x_2) + \nu_2(U_2 + \Omega x_1)$$
$$= (U_1 - \Omega x_2) \, dx_2/ds - (U_2 + \Omega x_1) dx_1/ds,$$

$ds$ being a length element of a section of the boundary. Hence

(33) $\qquad \psi = U_1 x_2 - U_2 x_1 - \tfrac{1}{2}\Omega(x_1^2 + x_2^2) +$ constant

on the boundary.

If the two-dimensional motion of an incompressible fluid is also irrotational, $\omega_3 = 0$, and

(34) $\qquad\qquad\qquad \nabla_1^2 \psi = 0.$

In fact, there is then both a velocity potential and a stream function, and

$$v_1 = \partial\phi/\partial x_1 = \partial\psi/\partial x_2, \ v_2 = \partial\phi/\partial x_2 = -\partial\psi/\partial x_1,$$

so $\phi + i\psi = w$ is a function of the complex variable $z = x_1 + ix_2$. Also

(35) $$dw/dz = v_1 - iv_2 = q \exp(-i\chi),$$

where $q$ is the velocity magnitude, and $\chi$ the angle that the velocity vector makes with the $x_1$-axis.

For the steady two-dimensional flow of a compressible fluid, the magnitude of the velocity is $(\rho_0/\rho)|\nabla\psi|$, and its direction is obtained from that of $\nabla\psi$ by a rotation through a right angle in the negative sense. The *mass* of fluid crossing a curve $AP$, per unit breadth per unit time, is

$$\int_A^P \rho v_\nu \, ds = \rho_0 \int_A^P \frac{\partial\psi}{\partial s} ds = \rho_0(\psi_P - \psi_A).$$

The streamlines are again $\psi = $ constant.

If $v_3 \neq 0$, but all quantities are still independent of $x_3$, $A_1$ and $A_2$ cannot both be zero, but we may still write $A_3 = \psi$ (or $\rho_0\psi$), and the velocity components $v_1$, $v_2$ may still be derived from a stream function, as before. The velocity $v_3$ has to be added, and, if $v_3$ varies with $x_1$ and $x_2$, there are vorticity components $\omega_1$ and $\omega_2$, but our statements are otherwise still correct ($\psi = $ constant being now the projection of the streamlines on the ($x_1$, $x_2$) plane).

If a three-dimensional motion is symmetrical about an axis, we take the axis of $x_3$ along the axis of symmetry, and let $r$, $\vartheta$, and $x_3 = x$ be cylindrical polar coordinates. All quantities are then independent of $\vartheta$. In that case, we write, for the steady flow of a compressible fluid, $rA_\vartheta = \rho_0\psi$, and then

(36) $$r\rho v_r = -\rho_0 \partial\psi/\partial x, \ r\rho v_x = \rho_0 \partial\psi/\partial r.$$

$\psi$ is called Stokes's stream function. For the flow of an incompressible fluid, write $rA_\vartheta = \psi$; then

(37) $$rv_r = -\partial\psi/\partial x, \ rv_x = \partial\psi/\partial r.$$

The velocity component $v_\vartheta$ need not be zero, but is taken to be independent of $\vartheta$. Then

$$\omega_\vartheta = \partial v_r/\partial x - \partial v_x/\partial r = -r^{-1} D^2 \psi,$$

where

(38) $$D^2 = \partial^2/\partial r^2 - r^{-1}\, \partial/\partial r + \partial^2/\partial x^2.$$

If $\omega_\vartheta = 0$, the equation for $\psi$ is

(39) $$D^2 \psi = 0.$$

In any meridian plane through the axis of symmetry, the product of $r$ and the velocity in that plane is equal to the gradient of $\psi$ in magnitude, and its direction is obtained from that of $\nabla \psi$ by rotating through a right angle in the negative sense, the sense from the direction of $r$ increasing to the direction of $x$ increasing by the shorter way. Hence, if $AP$ is any arc in a meridian plane,

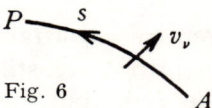

Fig. 6

the flux in the direction of $v_\nu$, (Fig. 6) across the surface formed by rotating the arc about the axis of symmetry through $2\pi$ is

$$2\pi \int_A^P r v_\nu\, ds = 2\pi (\psi_P - \psi_A).$$

If $v_\vartheta = 0$, the streamlines lie in a meridian plane, and are given by

$$dx/v_x = dr/v_r,$$

i.e., by

$$\psi = \text{constant}.$$

For the steady flow of a compressible fluid, the corresponding *mass* flux is $2\pi \rho_0 (\psi_P - \psi_A)$, and if $v_\vartheta = 0$ the streamlines are $\psi = $ constant.

In spherical polar coordinates $R$, $\theta$, $\vartheta$, equation (37) becomes

(40a) $$v_R = \frac{1}{R^2 \sin\theta} \frac{\partial \psi}{\partial \theta}, \quad v_\theta = -\frac{1}{R \sin\theta} \frac{\partial \psi}{\partial R}$$

(there is a similar transformation of (36)), and

(40b) $$D^2 = \frac{\partial^2}{\partial R^2} + \frac{\sin\theta}{R^2}\frac{\partial}{\partial\theta}\left(\frac{1}{\sin\theta}\frac{\partial}{\partial\theta}\right),$$

with $\omega_\vartheta = -(R\sin\theta)^{-1} D^2\psi$.

We now return to the equation (29) for the vector potential in the flow of an incompressible fluid. If the velocity and vorticity fields are supposed to occupy all space, and sufficient conditions are satisfied at infinity, then from (29), the value of **A** at a point $P$ is given by

$$A_i = \frac{1}{4\pi}\int \frac{\omega_i}{r}\,d\tau,$$

i.e.,

(41) $$\mathbf{A} = \frac{1}{4\pi}\int \frac{\omega}{r}\,d\tau,$$

where $r$ is now the distance of $P$ from the volume element $d\tau$ at (say) $Q$, which means that the coordinates of $Q$ are the integration variables when the integral is expressed as a triple integral.

In fact, we may here cite the following fundamental theorem of vector analysis.[5]

"A continuous vector field **V**, defined everywhere in space and vanishing at infinity together with its first derivatives, can be represented as the sum of an irrotational field $\mathbf{V}_1$ and a solenoidal field $\mathbf{V}_2$, i.e., $\mathbf{V} = \mathbf{V}_1 + \mathbf{V}_2$, where curl $\mathbf{V}_1 = 0$, div $\mathbf{V}_2 = 0$. $\mathbf{V}_1$ can be derived from a scalar potential in the form $\mathbf{V}_1 = \nabla\phi +$ constant, and $\mathbf{V}_2$ from a vector potential in the form $\mathbf{V}_2 =$ curl $\mathbf{A} +$ constant, where $\nabla \cdot \mathbf{A} = 0$, and

$$4\pi\phi = -\int \frac{\nabla \cdot \mathbf{V}}{r}\,d\tau, \quad 4\pi\mathbf{A} = \int \frac{\operatorname{curl} \mathbf{V}}{r}\,d\tau,$$

Apart from the additive constants, the solution is unique."

We do not wish to enter here into the representation of any motion by distributions of sources, doublets, and vortices.[6] We

---

[5] Sommerfeld's *Mechanics of Deformable Bodies*, Section 20. The sign of $\phi$ has been altered, and the wording changed only slightly.

[6] In this connection compare also Lamb, Sections 56–58, 148, 151.

are content to use (41) to show how to calculate the velocity field due to a vortex tube in an unbounded fluid. To apply the results in the presence of solid boundaries, further contributions to the velocity field must be added to satisfy the boundary conditions. (These may sometimes be obtained by the method of images.)

If $\sigma$ is the (small) cross-section of a vortex tube, and $ds$ an element of length along it, we replace $d\tau$ in (41) by $\sigma ds$. Then $\omega$ and $d\mathbf{s}$ are in the same direction, so

$$\omega d\tau = \sigma \omega ds = \sigma \omega d\mathbf{s} = K\, d\mathbf{s},$$

where $K$ is the strength of the vortex tube. Hence the value of $\mathbf{A}$ at $P$ is

$$\mathbf{A} = \frac{K}{4\pi} \int \frac{d\mathbf{s}}{r}$$

taken along the vortex tube, where $r$ is the distance of $P$ from $d\mathbf{s}$, and

$$\mathbf{v} = \operatorname{curl} \mathbf{A} = \int d\mathbf{v},$$

where

(42b) $$d\mathbf{v} = \frac{K}{4\pi} \nabla \frac{1}{r} \times d\mathbf{s} = -\frac{K}{4\pi r^3} \mathbf{r} \times d\mathbf{s} = \frac{K}{4\pi r^3} d\mathbf{s} \times \mathbf{r},$$

This contribution to $\mathbf{v}$ is of magnitude

$$\frac{K}{4\pi r^2} ds \sin \chi,$$

where $\chi$ is the angle between $d\mathbf{s}$ and the vector $\mathbf{r}$ drawn from $d\mathbf{s}$ to $P$, and its direction is at right angles to $d\mathbf{s}$ and $\mathbf{r}$ and corresponds with a right-handed screw (for right-handed axes) about the direction of $d\mathbf{s}$.

This is mathematically analogous to the law of Biot and Savart for finding the magnetic induction due to a line current in a medium of constant magnetic permeability. The analogy extends further. There is a mathematical analogy between $\mathbf{v}$ and $\boldsymbol{\omega}$ on the one hand, and the magnetic field $\mathbf{H}$ and current vector **J** (in *MKSQ* units, or $4\pi$**j** in electromagnetic units) on the other. The equations connecting them are the same, even to the analogy between the condition in Section 1.8 satisfied at a vortex sheet, and the condition on the magnetic field at a surface carrying a surface current, and also to the results that a line current has an "equivalent" magnetic shell and a vortex an "equivalent" layer of doublets of strength $K/4\pi$.

CHAPTER 2

# Dynamics of the General Fluid

## 2.1. The stress tensor

The analysis of stress is considered in many works, including, in particular, Prof. Rivlin's lectures on Solid Mechanics in this series, and will therefore be dealt with very briefly.

If we imagine a surface $S$ drawn in a fluid, then forces (due to molecular action) are exerted between the portions of the fluid close to any part of $S$ on its two sides; these sets of forces are equal and opposite, in the nature of an action and reaction. The set of forces on the fluid on one side of any portion of $S$ is equivalent to a force and a couple. We assume that the resultant force exerted across a vanishingly small area is ultimately proportional to the area; that is, that the ratio force/area has a finite non-zero limit when the area shrinks to a point. It follows that the ratio of the resultant couple to the area tends to zero as the area shrinks to a point. The intensity of the action at any point of the surface is therefore specified by the limit of the ratio of force to area; this may be called the force per unit area at the point, and is, by definition, the stress at the point. In general, whatever the surface, the stress may be in any direction. For a fluid at rest it is normal to the surface, and is in the nature of a pressure. For fluids in motion, however, tangential stresses occur. The existence of these tangential stresses constitutes the phenomenon known as viscosity or internal friction in fluids.

For the present, we consider only cases in which the forces, other than stresses, acting on any portion of fluid, are body forces, such as the force of gravity, specified as forces **F** per unit mass, i.e., forces $\rho\mathbf{F}$ per unit volume. The stresses are specified as forces per unit area on the boundary of the portion of fluid. Each part of the fluid is in motion, but we consider the dynamics of its motion on the basis of statics by the use of D'Alembert's principle, associating with each mass element a force equal to the product of the mass and its acceleration, with the sign changed, i.e., with its direction reversed; the reversed mass-accelerations are called inertia forces. If the inertia forces are included, all the forces acting on any portion of fluid are in statical equilibrium; they have zero resultant force and zero resultant moment about any point.

Consider first the forces exerted across a surface of contact, $S$, of a fluid and a solid body. Forces are exerted between the parts of the solid body and the fluid close to $S$, and there is a resultant force and torque on the body and a stress at any point of its surface. By including the inertia forces and considering the "equilibrium" of a vanishingly thin layer of fluid next to the solid body, and using the equality of action and reaction, it follows, since the effects of the body forces and inertia forces are vanishingly small, that the force across any portion of $S$ is equal to the force exerted across the same portion of an infinitely near surface in the fluid; hence the stress at any point on the surface of a solid body in a fluid is the same as the stress at a neighboring internal point of the fluid. If the fluid and the solid body are at rest, the solid body is subject to normal pressures only; but if they are in relative motion, there are also tangential stresses on the solid body, giving rise to the phenomenon called skin-friction.

Since in the specification of a stress one magnitude and two directions enter, namely the direction of the normal to the surface over which we are taking the stress, and the direction of the stress itself, we expect the stress to be a tensor. When the direction of the unit vector $\boldsymbol{\nu}$ normal to the surface is specified, the stress may be in any direction, and is a vector $\mathbf{p}_\nu$, with components $p_{\nu i}$ along the coordinate axes. Thus $p_{ij}$ is the component along the $x_j$-axis of the stress over a surface normal to the $x_i$-axis, and is taken to be the stress exerted on the fluid on the side on which $x_i$ is (algebraically) the smaller; when $i = j$, $p_{ij}$ is a normal tension or pressure, with a tension reckoned as positive and a pressure as negative.

It is now easy to show that the resultant force on any small plane area is approximately equal to the product of the area and the stress at the centroid of the area, acting through the centroid, the error in the force and the moment about the centroid being each of the order of the fourth power of the linear dimensions of the area.

If we now consider the equilibrium, with inertia forces included, of a small rectangular volume element with its sides parallel to the axes, equate to zero the resultant moment about the centroid of the volume, and take the limit as the lengths of the sides tend to zero, we easily prove that $p_{ij} = p_{ji}$.

We next consider the equilibrium, with inertia forces included, of the fluid within a small tetrahedron, three of whose edges are in the directions of the coordinate axes, and have lengths with fixed ratios. The slant face through the nonconcurrent ends of these edges has then a normal in a fixed direction. Let $\boldsymbol{\nu}$ be unit vector in this direction. Equate to zero the resultant force on this small portion of fluid, and take the limit as the lengths of the edges tend to zero. This leads to the equations

$$(43) \qquad p_{\nu j} = \nu_i p_{ij}$$

We now transform from the axes of $x_i$ to axes of $x'_i$, with $\alpha_{ij}$ as the cosine of the angle between the axes of $x'_i$ and $x_j$. Since $\mathbf{p}_\nu$ is a vector,

$$(44) \qquad p'_{\nu l} = \alpha_{lj} p_{\nu j}.$$

Hence, if we take $\boldsymbol{\nu}$ in the direction of the axis of $x'_k$, so that $\nu_i = \alpha_{ki}$, we obtain

$$(45) \qquad p'_{kl} = \alpha_{ki} \alpha_{lj} p_{ij}.$$

Thus $p_{ij}$ is a symmetrical second-order tensor. We may therefore (in the same way as for the rate-of-strain tensor) define a stress quadric, principal axes of stress, and principal normal stresses, which we denote by $p_1$, $p_2$, $p_3$. If $i = j$, the $p_{ij}$ are pressures if negative (tensions if positive); if $i \neq j$, the $p_{ij}$ are shearing stresses. With axes along the principal axes, the shearing stresses are zero. The planes perpendicular to the principal axes of stress are called the principal planes of stress; the stress across each is purely normal.

## 2.2. The equations of momentum and angular momentum

The equations of momentum and angular momentum for any portion of fluid may be found by considering the fluid inside a surface $S$ fixed in space, using D'Alembert's principle, and equating to zero the resultant force and moment for all the forces acting on the fluid, the inertia forces included. We denote by **F** the body force per unit mass. The surface integrals arising from the total force and moment due to the stresses on the surface are transformed into volume integrals by using (43) and the divergence theorem. Thus for the force component along the $x_i$-axis

$$\int p_{\nu i}\, dS = \int \nu_j p_{ji}\, dS = \int \frac{\partial p_{ji}}{\partial x_j}\, d\tau,$$

and for the moment about the axis of $x_k$ (with $i, j, k$ in cyclic order),

$$\int (x_i p_{\nu j} - x_j p_{\nu i}) dS = \int (\nu_l x_i p_{lj} - \nu_l x_j p_{li}) dS$$

$$= \int \frac{\partial}{\partial x_l}(x_i p_{lj} - x_j p_{li}) d\tau = \int \left( x_i \frac{\partial p_{lj}}{\partial x_l} - x_j \frac{\partial p_{li}}{\partial x_l} + p_{ij} - p_{ji} \right) d\tau.$$

Then the equation of equilibrium for the force component along the $x_i$-axis is

$$0 = \int \rho(f_i - F_i) d\tau - \int p_{\nu i}\, dS = \int \left( \rho f_i - \rho F_i - \frac{\partial p_{ji}}{\partial x_j} \right) d\tau.$$

This holds for any volume of integration, so the integrand vanishes at every point at which it is continuous. The equation of linear momentum is therefore

(46) $$\rho f_i = \rho F_i + \partial p_{ji}/\partial x_j.$$

The equations of equilibrium for the component moments may be similarly written down. When equation (46) is used, the moment equations reduce to $p_{ij} = p_{ji}$, for all pairs $i$ and $j$. Thus we have verified that, as was to be expected, satisfaction of the moment equation for an infinitesimally small rectangular volume element, leading to the symmetry of the stress tensor, ensures the satisfaction of the moment equation for any portion of the fluid if the force (linear momentum) equations are satisfied.

We obtain exactly the same result by letting the surface $S$ move with the fluid, and equating the rate of change of momentum to the total force. For this gives

$$\frac{D}{Dt}\int \rho v_i \, d\tau = \int \rho F_i \, d\tau + \int p_{vi} \, dS,$$

and, from (10),

$$\frac{D}{Dt}\int \rho v_i \, d\tau = \int \rho \frac{Dv_i}{Dt} \, d\tau = \int \rho f_i \, d\tau.$$

This is similarly correct for the angular momentum. This method is correct because the linear momentum is "tied" to a portion of fluid, and not to a part of space. For the equation of momentum, it is recommended that a fixed surface be used, as above.

For the stress tensor we write

(47) $$p_{ij} = -p\delta_{ij} + \pi_{ij},$$

where $p$ is the thermodynamic pressure, and $\delta_{ij}$ is defined in (21). Expressions for $\pi_{ij}$ in particular fluids will be considered later. The equations (46) become

(48) $$\rho f_i = \rho F_i - \partial p/\partial x_i + \partial \pi_{ji}/\partial x_j.$$

### 2.3. The equation of energy

In discussing the energy balance for a portion of the fluid, we assume the existence of the usual thermodynamic variables, such that the relations among them provided by classical reversible thermodynamics are satisfied. (See L. Howarth, ed., *Modern Developments; High Speed Flow*, Chap. 2, Section 3). We have anticipated this to the extent that the thermodynamic pressure, so defined, was introduced in (47). To begin with, however, we assume only the existence of an internal energy function, $\mathfrak{E}$ for the fluid; we use $\mathfrak{E}$ for the internal energy per unit mass. We shall also include for the present a term $D$ to represent the rate of heat energy addition by chemical action, radiation, electromagnetic action, and also (since we cannot foresee all the processes that may possibly take place in a moving fluid in the most general circumstances) by such other processes as have not been specifically mentioned. Electromagnetic action will be considered in a future section; whenever any process is specifically allowed for, it is to be dropped from $D$. We shall, for the present, suppose that the rate of heat addition $D$ is given by $R$ per unit mass.

In writing the energy balance, we must include the rate at which work is done by the body forces and the stresses over the surface. To obtain the equation of energy, it is therefore best to write the energy balance for a definite portion of the fluid, bounded by a surface $S$ which moves with the fluid, i.e., whereas for momentum a surface fixed in space should be used, for energy a surface moving with the fluid should be used. Then the equation of energy balance is

$$A + B + C + D = G + H,$$

where $A$ is the rate at which the stresses on the surface do work, $B$ the rate at which the body forces do work, $C$ the rate at which heat energy is conducted inwards through the boundary, and $G$ and $H$ are the rates at which the kinetic and internal energies are increasing. In the usual theory of heat conduction, $C$ is the integral over the boundary of the product of the thermal conductivity, $k$, and the normal gradient of temperature $\partial T/\partial \nu$; for the present, however, we shall insert a general heat-flux vector, $\mathbf{q}$, such that $C$ is the surface integral of $q_\nu$, since the relation $\mathbf{q} = k\mathbf{\nabla} T$ is not the most general relation, and according to the kinetic theory of gases, is, for example, on the same level of approximation in a gas as the expressions we shall later introduce for $\pi_{ij}$; in fact, it is, quite generally, best to introduce specific expressions for $\mathbf{q}$ and $\pi_{ij}$ at the same time. (Note that although $\nu$ is along the outward normal, $q_\nu dS$ is the rate of heat energy conducted inwards across a surface element $dS$, so later (p. 90) we take $q_\nu = +k\, \partial T/\partial \nu$.)

## DYNAMICS OF THE GENERAL FLUID

Then the equation of energy balance is

$$\int p_{\nu i} v_i \, dS + \int \rho F_i v_i \, d\tau + \int q_\nu \, dS + \int \rho R \, d\tau$$

$$= \frac{D}{Dt} \int \rho (\tfrac{1}{2} v_i v_i + \mathfrak{E}) d\tau = \int \rho \frac{D}{Dt} (\tfrac{1}{2} v_i v_i + \mathfrak{E}) d\tau$$

$$= \int \rho \left( v_i f_i + \frac{D\mathfrak{E}}{Dt} \right) d\tau.$$

But

$$\int p_{\nu i} v_i \, dS = \int \nu_j p_{ji} v_i \, dS = \int v_i \frac{\partial p_{ji}}{\partial x_j} d\tau + \int p_{ji} \frac{\partial v_i}{\partial x_j} d\tau$$

$$= \int v_i (\rho f_i - \rho F_i) d\tau + \int \pi_{ji} \frac{\partial v_i}{\partial x_j} d\tau - \int p \frac{\partial v_i}{\partial x_i} d\tau.$$

$\pi_{ij}$ is a symmetrical tensor, so $\pi_{ij} \partial v_i / \partial x_j = \pi_{ij} \partial v_j / \partial x_i$, and we may write

(49) $\quad \Phi = \pi_{ij}(\partial v_i/\partial x_j) = \pi_{ij}(\partial v_j/\partial x_i) = \tfrac{1}{2} \pi_{ij} e_{ij}.$

From the equation of continuity,

$$\partial v_i / \partial x_i = \mathbf{\nabla} \cdot \mathbf{v} = -\rho^{-1}(D\rho/Dt) = \rho D\rho^{-1}/Dt.$$

Also

$$\int q_\nu \, dS = \int \mathbf{\nabla} \cdot \mathbf{q} \, d\tau.$$

Our equation becomes

$$\int \left[ \rho \left\{ \frac{D\mathfrak{E}}{Dt} + p \frac{D}{Dt} \left( \frac{1}{\rho} \right) \right\} - \Phi - \mathbf{\nabla} \cdot \mathbf{q} - \rho R \right] d\tau = 0,$$

and the required equation of energy is

$$\rho \left[ \frac{D\mathfrak{E}}{Dt} + p \frac{D}{Dt} \left( \frac{1}{\rho} \right) \right] = \Phi + \mathbf{\nabla} \cdot \mathbf{q} + \rho R.$$

The left side of this equation may be transformed into various forms by use of the equations of thermodynamics. The most important involve the enthalpy, $I$, per unit mass, and the entropy, $S$, per unit mass. We have

$$I = \mathfrak{E} + p/\rho,$$

so
$$\frac{DI}{Dt} = \frac{D\mathfrak{E}}{Dt} + p\frac{D}{Dt}\left(\frac{1}{\rho}\right) + \frac{1}{\rho}\frac{Dp}{Dt}$$

Also, if $T$ is the temperature, and $c_p$, $c_v$ the specific heats at constant pressure and constant volume,
$$\frac{DI}{Dt} = T\frac{DS}{Dt} + \frac{1}{\rho}\frac{Dp}{Dt},$$
and
$$\frac{DS}{Dt} = \frac{c_p}{T}\frac{DT}{Dt} - \left[\frac{\partial}{\partial T}\left(\frac{1}{\rho}\right)\right]_p \frac{Dp}{Dt}$$
$$= \frac{c_v}{T}\frac{DT}{Dt} + \left(\frac{\partial p}{\partial T}\right)_\rho \frac{D}{Dt}\left(\frac{1}{\rho}\right).$$

Hence
$$\rho\left[\frac{D\mathfrak{E}}{Dt} + p\frac{D}{Dt}\left(\frac{1}{\rho}\right)\right] = \rho\frac{DI}{Dt} - \frac{Dp}{Dt} = \rho T\frac{DS}{Dt}$$

(50a)
$$= \rho c_p \frac{DT}{Dt} + \frac{T}{\rho}\left(\frac{\partial \rho}{\partial T}\right)_p \frac{Dp}{Dt} = \rho c_v \frac{DT}{Dt} + T\boldsymbol{\nabla}\cdot\mathbf{v}\left(\frac{\partial p}{\partial T}\right)_\rho$$
$$= \Phi + \boldsymbol{\nabla}\cdot\mathbf{q} + \rho R.$$

For a perfect gas, we may write
$$p = \mathfrak{R}\rho T,$$
where $\mathfrak{R}$ is an absolute constant divided by the molecular weight of the gas. For such a gas, the last two expressions on the left in (50a) reduce to
$$\rho c_p\, DT/Dt - Dp/Dt = \rho c_v\, DT/Dt + p\boldsymbol{\nabla}\cdot\mathbf{v}.$$

For a truly incompressible fluid, $\boldsymbol{\nabla}\cdot\mathbf{v} = 0$. But care must be taken for the motion of a gas even at low speeds, when the fractional changes in density are not large, since the thermodynamic pressure $p$ is usually quite large. In fact, as we shall see later, in such cases there are important problems in which it is $Dp/Dt$ (and also $\Phi$) which are to be neglected, not $p\boldsymbol{\nabla}\cdot\mathbf{v}$, and the operative specific heat is $c_p$, not $c_v$.

The quantity $\Phi$ is the rate of dissipation of mechanical energy into heat, per unit time per unit volume. For we retain the first law of thermodynamics

$$\delta Q = d\mathfrak{E} + p d(\rho^{-1})$$

(with $\delta Q$ in mechanical work units) by asserting that, whether the fluid be at rest or not, whenever small changes of state occur the energy which has to the supplied as heat to provide for these changes is $d\mathfrak{E} + p d(\rho^{-1})$. The rate of heat supply needed for a fluid element is, therefore,

$$D\mathfrak{E}/Dt + p D\rho^{-1}/Dt$$

per unit mass, i.e.,

$$\rho[D\mathfrak{E}/Dt + p D\rho^{-1}/Dt] = \Phi + \nabla \cdot \mathbf{q} + \rho R$$

per unit volume. The second and third terms on the right give the heat supplied by conduction, and by chemical action, radiation, etc. Thus $\Phi$ represents the heat gained by the dissipation of mechanical energy by internal friction. This agrees well with the result that $\Phi$ is a linear function of the $\pi_{ij}$, which represent the deviation of the stress tensor, due to the combined results of internal friction and distortion, from the isotropic thermodynamic pressure.

If we multiply the equations (48) (with $f_i = Dv_i/Dt$) by $v_i$ and sum, and add to (50a), we obtain an interesting variation of the equation of energy. Let the forces $F_i$ be conservative, with $F_i = -\partial \Omega/\partial x_i$, where $\Omega$ is the potential energy per unit mass. Then we obtain

(50b)
$$\rho \frac{D}{Dt}[I + \tfrac{1}{2} \mathbf{v}^2 + \Omega] = \rho \frac{\partial \Omega}{\partial t} + \frac{\partial p}{\partial t} + \rho R + \frac{\partial}{\partial x_j}(v_i \pi_{ij}) + \frac{\partial q_j}{\partial x_j}.$$

The sum of the enthalpy and kinetic and potential energies per unit mass, $I + \tfrac{1}{2}\mathbf{v}^2 + \Omega$, will be denoted by $\mathfrak{U}$. For a steady flow for which $R = 0$, the right side of equation (50b) is a divergence, and the rate of change of the sum of the enthalpy, kinetic, and potential energies in any volume is

$$\frac{D}{Dt}\int \rho \mathfrak{U}\, d\tau = \int \rho \frac{D\mathfrak{U}}{Dt}\, d\tau = \int \rho \mathfrak{U} v_\nu\, dS,$$

and is equal to

$$\int q_\nu\, dS + \int v_j v_i \pi_{ij}\, dS,$$

where the surface integrals are over the boundary of the volume. The result vanishes if the boundary is a solid boundary at rest, on which $\mathbf{v} = 0$, which is also diathermal, or heat insulated, with no flux of heat, $q_\nu = 0$, or if the surface lies entirely in a part of the fluid in which the heat conduction and viscous stresses may be ignored, or is made up partly by the one or partly by the other.

## 2.4. Hydrostatic pressure. Removal or neglect of gravity body forces

If a fluid is permanently at rest, the stress tensor is an isotropic pressure, $p_{ij} = -p_0 \delta_{ij}$, and the acceleration is zero. Hence, with body forces $\mathbf{F}$ per unit mass and a hydrostatic density $\rho_0$,

(52a) $$\rho_0 \mathbf{F} = \nabla p_0.$$

For an incompressible fluid of constant density, $\mathbf{F}$ must be conservative if there is to be equilibrium. In any motion of such a fluid, if we put

(52b) $$p = p_0 + p',$$

the equation of motion becomes

(53) $$\rho_0 f_i = -\partial p'/\partial x_i + \partial \pi_{ji}/\partial x_j.$$

The body force has been eliminated by substituting for the pressure the difference of the pressure from the hydrostatic. For flow past an immersed solid body, for example, the only effect of substituting $p'$ for $p$, and dropping $\mathbf{F}$, in the equation of motion, is that in the end we must remember to add the hydrostatic Archimedes force to the force on the solid body due to the pressures $p'$ on its surface; for in such a problem the pressure $p$ does not enter into the boundary conditions. On the other hand (although in these lectures we shall not study surface gravity waves), we may note that for waves at the interface of two fluids which do not mix, the total stress must be continuous if surface tension is neglected; in any case the boundary conditions at the interface involve the total stress, so the total pressure $p$ enters. For motion under gravity, $g$ is a significant parameter in such problems.

Matters are much more complicated for compressible fluids (gases). To determine $p_0$ and $\rho_0$ it is usual to assume a relation between $p_0$ and $\rho_0$, such as that for constant entropy[7]; we may note that, strictly, for a completely still atmosphere, we have the equation (52a), an equation of state connecting pressure, density, and temperature, and, from the equation of energy, the equation expressing that the rate of conduction of heat into any element is zero (since all other terms in the equation of energy are zero). However, we cannot in any case eliminate the body force in the same manner as before, since the density now varies when the fluid moves.

In most problems, the only body force is the force of gravity, and it is usual in many problems to neglect the effect of the gravity force. There appears to be no adequately complete and rigorous discussion of this matter, but the following remarks may help. Clearly the effect of gravity must be retained if we are studying such diffusive processes as the mixing of two gases of very different densities, or free convection due to large temperature differences. Otherwise, if the processes with which we are concerned take place over a height of order $h$, the effect of retaining the gravity term is to introduce terms of order $\exp(gh/a^2)$, where $a$ is the velocity of sound; these are just equal to 1 when $g$ is neglected. So one criterion is that $gh$ must be $\ll a^2$. If $h$ is large, as in the study of large-scale motions in the atmosphere, the condition may be violated. With some exceptions that are usually unimportant,[8] the effect of gravity may be neglected in the problems of aerodynamics.

[7] See, for example, H. Lamb, *Statics*, Cambridge University Press, 1928, Section 120.

[8] See C. R. Illingworth, *Proc. Roy. Soc. (London)* **A199** (1949), pp. 533–558; A. D. Young, in L. Howarth, ed., *Modern Developments in Fluid Dynamics; High Speed Flow, Chap. X*, Section 6.

## 2.5. Impulsive generation of motion in a fluid of constant density

At any moment let impulsive body forces act on an incompressible fluid of constant density, and let the boundary conditions be suddenly altered, as when an immersed solid body is suddenly set in motion with a finite velocity, and there is an impulsive pressure $\tilde{\omega}$ in the fluid. The velocity $\mathbf{v}$ stays finite, and so do $\nabla(\tfrac{1}{2}\mathbf{v}^2)$, $\omega$, and the viscous stress $\pi_{ij}$ and its divergence $\partial \pi_{ji}/\partial x_j$, but $\mathbf{v}$ changes abruptly, say from $\mathbf{v}_1$ to $\mathbf{v}_2$, so that $\partial \mathbf{v}/\partial t$ and $\mathbf{f}$ are infinite. If the impulse takes place at time $t$, we may define the impulsive body force $\tilde{\mathbf{F}}$ and the impulsive pressure by

$$\tilde{\mathbf{F}} = \lim_{\tau \to 0} \int_t^{t+\tau} \mathbf{F}\, dt, \qquad \tilde{\omega} = \lim_{\tau \to 0} \int_t^{t+\tau} p\, dt.$$

Integrate the equation of motion from $t$ to $t + \tau$, and take the limit as $\tau \to 0$. Since $\rho$ does not change,

(54) $$\mathbf{v}_2 - \mathbf{v}_1 = \tilde{\mathbf{F}} - \rho^{-1}\nabla\tilde{\omega}.$$

If there are no impulsive body forces,

(55) $$\mathbf{v}_2 - \mathbf{v}_1 = -\rho^{-1}\nabla\tilde{\omega}.$$

Since $\nabla \cdot \mathbf{v}_1 = \nabla \cdot \mathbf{v}_2 = 0$, and $\rho$ is constant, $\nabla^2 \tilde{\omega} = 0$.

For a motion so started impulsively from rest, the initial velocity is $\mathbf{v} = -\nabla(\tilde{\omega}/\rho)$. Hence the initial motion is irrotational, with a single-valued velocity potential $\phi = -\tilde{\omega}/\rho + \text{constant}$. This is true for a viscous as well as for an inviscid fluid. For a viscous fluid, with the motion caused by the impulsive start of an immersed solid body, for example, we must suppose there is initially a vortex sheet coincident with the boundary of the solid (and similarly at any other solid boundaries that may be present) through which the relative tangential velocity falls to zero.

It is borne out by observation that the initial motion in such cases is irrotational, without circulation.

In a gas, an impulsive pressure at any point would produce wave motion. We do not here consider the matter further. In an incompressible fluid, the wave velocity would be infinite, and an impulsive pressure at the boundary of a solid suddenly set in motion would be communicated instantaneously to every point of the liquid.

# CHAPTER 3

# Electric and Magnetic Forces

## 3.1. Introduction. Maxwell's equations. Physical preliminaries

We proceed to consider electrodynamic effects in an electrically conducting fluid. For the presentation of the electrodynamic theory, I shall take Sommerfeld's *Electrodynamics* as a reference text, and shall (later) base the discussion on Minkowski's theory of the electrodynamics of moving media.

In the presentation of mechanics, there was no need to specify explicitly the units used; the equations have the same form with any consistent set of units. However, in presenting the equations of electrodynamics, we must specify the units we use, since questions of dimensions are also involved. Practically, at the present time, the only systems that come into question are the Gaussian system, the system of electromagnetic units, and the MKSQ (meter-kilogram-second-coulomb) system. We here follow Sommerfeld in using the MKSQ system, and refer to his book for a discussion of the question of units and dimensions.

In a medium at rest in any chosen frame of reference, Maxwell's equations are

(56) $$\text{curl } \mathbf{E} = -\partial \mathbf{B}/\partial t, \quad \text{curl } \mathbf{H} = \partial \mathbf{D}/\partial t + \mathbf{J},$$
$$\nabla \cdot \mathbf{D} = \rho_e, \quad \nabla \cdot \mathbf{B} = 0,$$

where $\mathbf{E}$ is the electric field strength, $\mathbf{D}$ the "displacement" (or "excitation"), $\mathbf{B}$ the magnetic induction, $\mathbf{H}$ the magnetic "excitation," $\mathbf{J}$ the electric current density, and $\rho_e$ the electric charge density; also

$$\mathbf{D} = \varepsilon \mathbf{E}, \quad \mathbf{B} = \mu \mathbf{H}, \quad \mathbf{J} = \sigma \mathbf{E},$$

where $\varepsilon$ is the dielectric constant, $\mu$ the permeability, and $\sigma$ the electric conductivity. (For anisotropic bodies this direct proportionality fails; for crystals, for example, the expressions are replaced by linear vector functions.)

We assume $\varepsilon$ and $\mu$ constant throughout the medium. (This should be emphasized and borne in mind throughout the whole of this section. Immediately after this lecture was given, Mr. Boa-Teh Chu began work on the thermodynamics of an electrically conducting fluid medium when $\varepsilon$ and $\mu$ may depend on the state of the medium. As in the classical theories of electrostriction and magnetostriction, for example, the enquiry begins with a consideration of the free energy of the system. See Brown Univ. Div. Eng. Rept., W.A.D.C. TN 57–350, Dec. 1957.)

In a vacuum $\mathbf{D} = \varepsilon_0 \mathbf{E}$, $\mathbf{B} = \mu_0 \mathbf{H}$, where $\varepsilon_0 \mu_0 = 1/c^2$, $c$ being the velocity of light, about $3.10^8$ meters per second. If $\Omega$ is an ohm,

$$\varepsilon_0 = (10^{-9}/36\pi) \, S/\Omega M, \quad \mu_0 = 4\pi.10^{-7} \Omega S/M.$$

The approximate hydrodynamic effects to be expected may be anticipated from fairly simple physical considerations, and we first summarize these results. Later we proceed to a rather full mathematical discussion. We shall require results for the "fixed" Newtonian frame of reference which we use for mechanics; this we shall call for convenience the laboratory frame.

From the theory of relativity, Maxwell's equations (56) will apply unaltered. Since we shall not use relativistic mechanics, we shall disregard all terms of order $v^2/c^2$, where $v$ is the velocity magnitude at any point and time. We then hope that we may still put

(57) $$\mathbf{D} = \varepsilon \mathbf{E}, \quad \mathbf{B} = \mu \mathbf{H}$$

(but we shall see that this involves an approximation other than neglecting $v^2/c^2$). On the other hand, in the laboratory frame of reference, the total current $\mathbf{J}$ is composed of two parts, the convection current $\rho_e \mathbf{v}$, and the conduction current $\mathbf{J} - \rho_e \mathbf{v}$. Moreover, a moving charge, just like a current, experiences a force from the magnetic field, and the effective "driving" force per unit charge for a velocity $\mathbf{v}$ becomes $\mathbf{E} + \mathbf{v} \times \mathbf{B}$. Hence we anticipate that the third constitutive relation will be replaced by

(58) $$\mathbf{J} - \rho_e \mathbf{v} = \sigma(\mathbf{E} + \mathbf{v} \times \mathbf{B}).$$

The ponderomotive force per unit volume due to the electric and magnetic fields is $\rho_e \mathbf{E} + (\mathbf{J} \times \mathbf{B})$, so we anticipate that a term $\rho_e E_i + (\mathbf{J} \times \mathbf{B})_i$ must be added to the right of Eq. (48). Also we expect the rate of dissipation of electromagnetic energy into heat, the Joule heat, to be $(\mathbf{J} - \rho_e \mathbf{v})^2/\sigma$, so if electromagnetic effects are not included in $R$ in (50), we expect that $(\mathbf{J} - \rho_e \mathbf{v})^2/\sigma$ must be added to $\Phi$.

We proceed to put these results on a more satisfactory mathematical basis. The chief difficulty occurs in the equation of energy, where certain questions of principle are involved. These questions will be enunciated when we come to them, after we have set out for reference the mathematical basis of the electrodynamic theory of moving media. Even though we shall drop $v^2/c^2$, in order to establish the theory on a mathematically satisfactory basis, to answer with some confidence the questions involved in deriving the equation of energy, and to see clearly the approximations involved, it seems most satisfactory to start, before approximating, from a relativistically invariant theory.

Before we begin, we may state explicitly that the present continuum theory takes no account of the net result of certain microscopic effects in an ionized gas, for example. (In general, such effects may give rise to quite complicated phenomena.) We neglect, for example, the relative diffusion of the electrons and ions in a gas expansion, due to the greater thermal velocities of the electrons, an effect which is independent of the presence of a magnetic field. We neglect the results of the "drift" of the electrons (and ions) in the direction at right angles to the electric and magnetic fields (the "Hall current"). Most important of all, we neglect the effects of the spiralling of the electrons and ions about the magnetic lines of force, and assume the collision frequency large compared with the Larmor frequency (the frequency of rotation about the magnetic lines of force); this is a serious restriction which will not be applicable at all if the density is too low (or the magnetic field too intense). At reduced density in an ionized gas, at least the scalar conductivity $\sigma$ must be replaced by a tensor conductivity.

### 3.2. Electrodynamics of a medium at rest

We commence the mathematical discussion with a summary of the electrodynamics of a medium at rest in some frame of reference. In addition to Maxwell's equations, and the constitutive equations $\mathbf{D} = \varepsilon \mathbf{E}$, $\mathbf{B} = \mu \mathbf{H}$, $\mathbf{J} = \sigma \mathbf{E}$, we have the following relations.

The electromagnetic energy per unit volume is

(59) $$W = \tfrac{1}{2}(\mathbf{E} \cdot \mathbf{D} + \mathbf{H} \cdot \mathbf{B}),$$

the Poynting vector is

(60) $$\mathbf{N} = \mathbf{E} \times \mathbf{H},$$

the electromagnetic momentum is

(61) $$\mathbf{G} = \mathbf{D} \times \mathbf{B},$$

and the Maxwell stress tensor is

(62) $$T_{ij} = D_i E_j + B_i H_j - W \delta_{ij}.$$

Because of the constitutive relations (57) these may be taken in other forms ($E_i D_j = D_i E_j = \varepsilon E_i E_j$, etc.). We have chosen the forms we shall finally adopt for a moving medium. The order of the subscript in $T_{ij}$ has, however, been inverted from the order adopted in Sommerfeld's book, and other works on electrodynamics, to accord with the usage adopted for the stress tensor $p_{ij}$ in hydrodynamics.

The magnetic part of the stress tensor is easily seen to represent a tension equal in magnitude to the magnetic energy $\tfrac{1}{2}\mathbf{H} \cdot \mathbf{B}$ in the direction of the lines of magnetic force, and a pressure of the same magnitude in the perpendicular directions. A similar remark applies to the electrical portion.

We compute

$$k_i = \partial T_{ji}/\partial x_j.$$

Since $W = \tfrac{1}{2}(\varepsilon \mathbf{E}^2 + \mu \mathbf{H}^2)$,

$$\frac{\partial W}{\partial x_j} = D_k \frac{\partial E_k}{\partial x_j} + B_k \frac{\partial H_k}{\partial x_j}$$

and

$$k_i = \frac{\partial E_i}{\partial x_j} D_j + E_i \frac{\partial D_j}{\partial x_j} + \frac{\partial H_i}{\partial x_j} B_j + H_i \frac{\partial B_j}{\partial x_j} - D_k \frac{\partial E_k}{\partial x_i} - B_k \frac{\partial H_k}{\partial x_i}$$

$$= \rho_e E_i + (\operatorname{curl} \mathbf{E} \times \mathbf{D})_i + (\operatorname{curl} \mathbf{H} \times \mathbf{B})_i$$

$$= \rho_e E_i - \left(\frac{\partial \mathbf{B}}{\partial t} \times \mathbf{D}\right)_i + \left(\frac{\partial \mathbf{D}}{\partial t} \times \mathbf{B}\right)_i + (\mathbf{J} \times \mathbf{B})_i$$

Hence

(63) $\qquad \mathbf{k} = \rho_e \mathbf{E} + (\mathbf{J} \times \mathbf{B}) + \partial \mathbf{G}/\partial t,$

and when we allow for the rate of change of the electromagnetic momentum, the stresses give rise to the force $\rho_e \mathbf{E} + (\mathbf{J} \times \mathbf{B})$ per unit volume, as we know they should.

Next compute the rate of change of electromagnetic energy in a given volume. Since the medium is at rest, the bounding surface $S$ is taken as fixed in space; the rate of decrease of $W$ is

$$-\frac{\partial}{\partial t} \int W \, d\tau = -\int \frac{\partial W}{\partial t} d\tau = -\int (\mathbf{E} \cdot \dot{\mathbf{D}} + \mathbf{H} \cdot \dot{\mathbf{B}}) d\tau$$

(64) $\quad = -\int [\mathbf{E} \cdot (\operatorname{curl} \mathbf{H} - \mathbf{J}) - \mathbf{H} \cdot \operatorname{curl} \mathbf{E}] d\tau$

$$= \int \mathbf{E} \cdot \mathbf{J} \, d\tau + \int \mathbf{\nabla} \cdot \mathbf{N} \, d\tau = \int \mathbf{E} \cdot \mathbf{J} \, d\tau + \int N_\nu \, dS.$$

The first term on the right gives the rate of dissipation of energy, the Joule heat, $\mathbf{E} \cdot \mathbf{J}$ per unit volume. The second term gives the flux of energy out through the boundary.

### 3.3. Moving media

We turn now to a moving medium, in which the velocity at a point $P$ at time $t$ is $\mathbf{v}\,(P, t)$, relative to the laboratory frame. $\mathbf{v}$ may be variable both in space and time; all we require is that, before we average over the particles of the medium to consider it a continuum, the motion of any electrically charged particle should have been electrodynamically quasi-steady, in the sense that the fractional change in the velocity shall be very small in the time taken by an electromagnetic, or light, wave to sweep over the particle. This is no restriction at all for our present purposes. (The restrictions previously mentioned for ionized gases are, in practice, much more serious.)

We require the formulae of transformation from the laboratory frame to a local frame at any moving point $P$ at any time $t$ by a general Lorentz transformation. This may be accomplished by rotating the axes so that the axis of $x_1$ is in the direction of $\mathbf{v}(P, t)$, applying a special Lorentz transformation, and rotating the axes back again. The special Lorentz transformation involved may be thought of this way. Write $x_4 = ict$; and consider an orthogonal transformation in space-time, for which $x'_i = \alpha_{ij} x_j$, the subscripts now taking values from 1 to 4. Then for the special Lorentz transformation involved, $x_2$ and $x_3$ remain the same;

(65) $$\alpha_{11} = \alpha_{44} = (1 - \beta^2)^{-\frac{1}{2}}, \; \alpha_{14} = -\alpha_{41} = i\beta(1 - \beta^2)^{-\frac{1}{2}},$$
$$\alpha_{22} = \alpha_{33} = 1$$

where $\beta = v/c$, and all the other $\alpha$'s are zero. It is easily verified that the transformation is then orthogonal. The transformation is applied only locally at $P$, so $v$ is not varied; a different transformation to a local frame applies at each point of space-time.

The transformation formula for a vector is the same as that for $\mathbf{x}$. The transformation formula for a tensor under this or any other orthogonal transformation is exactly the same as in three dimensions,

(66) $$T'_{kl} = \alpha_{ki} \alpha_{lj} T_{ij},$$

but now the subscripts take the values 1 to 4, and summation over all these values is intended when a subscript is repeated.

For a rotation of the space axes only, $\alpha_{44} = 1$, and any other $\alpha$ with a subscript 4 in either place is zero.

Maxwell's equations are now written for a local frame of reference at $P$. In the local frame[9] they are

(67) $$\operatorname{curl}' \mathbf{E}' = -\partial \mathbf{B}'/\partial t', \; \operatorname{curl}' \mathbf{H}' = \partial \mathbf{D}'/\partial t' + \mathbf{J}',$$
$$\nabla' \cdot \mathbf{D}' = \rho'_e, \; \nabla' \cdot \mathbf{B}' = 0,$$

[9] Sommerfeld's *Electrodynamics*, Section 34. For Maxwell's equations in four-dimensional form, see Section 26.

where the primes on curl and $\nabla$ indicate that the differentiations are with respect to the space coordinates of the local system at $(P, t)$, and $t'$ is the time in that system. If we define

(68) $\quad \tilde{\mathbf{E}} = \mathbf{E} + \mathbf{v} \times \mathbf{B}, \quad \tilde{\mathbf{B}} = \mathbf{B} - (\mathbf{v} \times \mathbf{E})/c^2$
$\quad \tilde{\mathbf{D}} = \mathbf{D} + (\mathbf{v} \times \mathbf{H})/c^2, \quad \tilde{\mathbf{H}} = \mathbf{H} - \mathbf{v} \times \mathbf{D}, \quad \tilde{\mathbf{J}} = \mathbf{J} - \rho_e \mathbf{v}$

then it is known (Sommerfeld, *loc. cit.*) that (with $\parallel$ and $\perp$ signifying parallel and perpendicular to the velocity $\mathbf{v}$)

$$E'_{\parallel} = \tilde{E}_{\parallel}, \quad E'_{\perp} = \frac{\tilde{E}_{\perp}}{(1-\beta^2)^{\frac{1}{2}}}, \quad B'_{\parallel} = \tilde{B}_{\parallel}, \quad B'_{\perp} = \frac{\tilde{B}_{\perp}}{(1-\beta^2)^{\frac{1}{2}}}$$

(69) $\quad D'_{\parallel} = \tilde{D}_{\parallel}, \quad D'_{\perp} = \frac{\tilde{D}_{\perp}}{(1-\beta^2)^{\frac{1}{2}}}, \quad H'_{\parallel} = \tilde{H}_{\parallel}, \quad H'_{\perp} = \frac{\tilde{H}_{\perp}}{(1-\beta^2)^{\frac{1}{2}}}$

$$J'_{\parallel} = \frac{\tilde{J}_{\parallel}}{(1-\beta^2)^{\frac{1}{2}}}, \quad J'_{\perp} = \tilde{J}_{\perp}, \quad \rho'_e = \frac{\rho_e - (\mathbf{v} \cdot \mathbf{J})/c^2}{(1-\beta^2)^{\frac{1}{2}}}.$$

In the local frame the constitutive relations are

(70) $\quad\quad\quad \mathbf{D}' = \varepsilon \mathbf{E}', \quad \mathbf{B}' = \mu \mathbf{H}', \quad \mathbf{J}' = \sigma \mathbf{E}';$

the constants $\varepsilon$, $\mu$, $\sigma$ have the same values as for a medium at rest relative to the laboratory frame, since the medium at $(P, t)$ knows nothing of its motion.

When the inverse Lorentz transformation is carried out, (67) give us (56), but (57) are now altered; from (68) and (69) we have, for both parallel and perpendicular components,

(71) $\quad\quad\quad \tilde{\mathbf{D}} = \varepsilon \tilde{\mathbf{E}}, \quad \tilde{\mathbf{B}} = \mu \tilde{\mathbf{H}}.$

Also

(72) $\quad\quad\quad \tilde{J}_{\parallel}/(1-\beta^2)^{\frac{1}{2}} = \sigma \tilde{E}_{\parallel}, \quad \tilde{J}_{\perp} = \sigma \tilde{E}_{\perp}/(1-\beta^2)^{\frac{1}{2}}$

When we drop $v^2/c^2$,

(73) $\quad\quad\quad \mathbf{E}' = \tilde{\mathbf{E}}, \quad \mathbf{D}' = \tilde{\mathbf{D}}, \quad \mathbf{B}' = \tilde{\mathbf{B}}, \quad \mathbf{H}' = \tilde{\mathbf{H}}, \quad \mathbf{J}' = \tilde{\mathbf{J}}.$

Also, if we eliminate $\mathbf{B}$ in (71) to obtain the expression for $\mathbf{D}$ in terms of $\mathbf{E}$ and $\mathbf{H}$, and then eliminate $\mathbf{D}$ to obtain the expression for $\mathbf{B}$ in terms of $\mathbf{E}$ and $\mathbf{H}$ (still dropping $v^2/c^2$) the resulting equations are

(74) $$\mathbf{D} = \varepsilon\mathbf{E} + \left(\frac{\varepsilon\mu}{\varepsilon_0\mu_0} - 1\right)\frac{\mathbf{v} \times \mathbf{H}}{c^2}, \quad \mathbf{B} = \mu\mathbf{H} - \left(\frac{\varepsilon\mu}{\varepsilon_0\mu_0} - 1\right)\frac{\mathbf{v} \times \mathbf{E}}{c^2}.$$

Also

$$\mathbf{J} - \rho_e\mathbf{v} = \sigma(\mathbf{E} + \mathbf{v} \times \mathbf{B}).$$

(58 bis)

The second terms on the right in (74) are small in most of the problems that are usually considered. We shall therefore drop these terms and not attempt to work out their consequences *in these lectures*. Thus, in a laboratory frame, we return to equations (56), (57), and (58). So far, the only result of our consideration of the equations in relativistic form is to show that, in addition to dropping $v^2/c^2$, we have dropped the second terms in (74), and to check (58). However, (and this is more important), we shall use the relativistic equations in explaining Minkowski's theory of the Maxwell stress tensor, energy, Poynting vector, and electromagnetic momentum. As explained, the forms of these for a medium at rest in the chosen frame are not unique;(59)–(62) provide a form valid at each point in space-time in a local frame, but we shall need to work entirely in a laboratory frame. Which form shall we choose in each local frame, before transforming to a laboratory frame?

We choose a form which is relativistically invariant; but the stress tensor by itself is not a legitimate physical quantity in the sense of the theory of relativity, where we are concerned with transformations, not in three-dimensional space, but in four-dimensional space-time. We must consider a fourth-order tensor, and we start from a known result, that in a vacuum the correct extension is to a fourth-order "stress energy" tensor, where $T_{jk}$ has the values previously given (in 62), when $j$ and $k$ take the values 1, 2, 3, $T_{44} = W$, and $T_{4j} = T_{j4} = -iN_j/c$ for $j = 1, 2, 3$. ($i$, when it does not occur as a subscript, denotes the usual square root of $-1$). In a vacuum $T_{4j}$ may also be written $-icG_j$; and this form is preferable, because with it the fourth term in $\partial T_{jm}/\partial x_j$ just cuts out the rate of change of the electromagnetic momentum, since $x_4 = ict$. We may now easily prove that if we take the stress-energy tensor in the form (62) for $i$ and $j = 1, 2, 3$, with $W$, $\mathbf{N}$, and $\mathbf{G}$ as in (59), (60), and (61), and with $T_{44} = W$, $T_{4j} = -ic\,G_j$, $T_{j4} = -iN_j/c$, and if we use the primed symbols, as given by (69), in a local frame, then the complete fourth-order tensor is invariant in form for any rotation of the space axes and for the simple Lorentz transformation summarized in (65). If we start with the given form in the laboratory frame, then the form is unaltered by any rotation of the space axes; the effect of the Lorentz transformation is exactly to replace the unprimed by the primed symbols, as given by (69); and this form is again unaltered by any rotation of the space axes. We may also effect the reverse transformation, from a local to the laboratory frame, by using in place of (65) the inverse Lorentz transformation, obtained by interchanging the primed and unprimed coordinates, and replacing $v$ by $-v$, $\beta$ by $-\beta$. The result is now, of course, that if we start with the primed symbols $\mathbf{E}'$, $\mathbf{D}'$, etc., we end up with the unprimed ones.

In a vacuum, since $\varepsilon_0\mu_0 = 1/c^2$, $-icG_j = -iN_j/c$, but this is not true in a medium where $\varepsilon\mu \neq \varepsilon_0\mu_0$. As a result, the Minkowski stress-energy tensor is not symmetrical. It is true that, because of the first two constitutive relations in (70), the third-order stress tensor is itself symmetrical in a local frame, but the asymmetry will appear in the stress tensor itself after transformation to any other frame. The effects are small and scarcely observable, and, since we are neglecting the second terms on the right in (74), we are in any case neglecting the asymmetry in these lectures. Nevertheless, before leaving the subject here, the following remarks may be in order. There seems to be no experimental evidence against the asymmetry, nor any *a priori* reason against it. Nevertheless, it was felt for a long time that the asymmetry represented a real difficulty in Minkowski's theory, and other (symmetrical) forms of the stress-energy-momentum tensor were suggested, especially by M. Abraham.[10]

---

[10] See W. Pauli, *Encyklopädie der mathematischen Wissenschaften*, B. G. Teubner, Leipzig and Berlin, Band V, 19, Relativitätstheorie, pp. 539–775, especially pp. 665–667.

The practical differences are all very small, and there are (quantitatively small) somewhat questionable features about Abraham's theory. The objection to Minkowski's theory seems to have been based on the argument that the macroscopic theory must be derivable by an averaging process from the theory of electrons, and that since the microscopic stress-energy tensor (the tensor in a vacuum) is symmetrical, the macroscopic tensor would be. But it now appears that the argument is probably incorrect,[11] and, although the matter is not completely settled, Minkowski's theory is gaining acceptance.

Before we return to our mechanics, there are two questions to be answered about the equation of energy balance. First, we must decide whether the electric and magnetic forces are to be thought of as body forces, or, in the manner of Faraday and Maxwell, as surface forces — tensions along the lines of force and pressures across them — because extra terms $T_{ji}\,(\partial v_i/\partial x_j)$, analogous to $p_{ji}(\partial v_i/\partial x_j)$, appear in the energy equation in the second case and not in the first. For a continuum theory, as we are concerned with here, we anticipate that the second concept will be correct, and we verify below that this leads to consistent results. The second question is this: if we are using a laboratory frame, we view the stress tensor and energy density in such a frame, but if we write down the energy equation for the volume inside a surface in motion relative to such a frame, should not the Poynting vector, which is the rate of flux of energy through such a surface, be taken in the local frame at each point of the surface at the right time? We expect that it should. To show that in this way we obtain a simple and consistent scheme for moving surfaces, we consider the electromagnetic energy balance (only) for the volume within a moving surface, for a medium devoid of electric charge and currents.

Let **N** and **N'** be the values of the Poynting vector in the laboratory and a local frame. **N** is given by (60). With $v^2/c^2$

---

[11] See C. Møller, *The Theory of Relativity*, Oxford University Press, 1952, Section 75.

neglected, **E**' and **H**' are given by (73) and (68). Hence, again with $v^2/c^2$ neglected,

$$\begin{aligned}
\mathbf{N}' = \mathbf{E}' \times \mathbf{H}' &= (\mathbf{E} + \mathbf{v} \times \mathbf{B}) \times (\mathbf{H} - \mathbf{v} \times \mathbf{D}) \\
&= \mathbf{N} - \mathbf{E} \times (\mathbf{v} \times \mathbf{D}) - \mathbf{H} \times (\mathbf{v} \times \mathbf{B}) \\
&= \mathbf{N} - \mathbf{v}\,(\mathbf{E} \cdot \mathbf{D} + \mathbf{H} \cdot \mathbf{B}) + (\mathbf{E} \cdot \mathbf{v})\mathbf{D} + (\mathbf{H} \cdot \mathbf{v})\mathbf{B} \\
&= \mathbf{N} + (\mathbf{E} \cdot \mathbf{v})\mathbf{D} + (\mathbf{H} \cdot \mathbf{v})\mathbf{B} - 2W\mathbf{v}.
\end{aligned} \quad (75)$$

Now let $S$ be a closed surface moving in any manner. The rate at which the stresses on the surface are doing work is

$$\int T_{\nu i}\, v_i\, dS,$$

and

$$T_{\nu i} v_i = \nu_j T_{ji} v_i = \nu_j (E_i D_j v_i + H_i B_j v_i - W v_j) \\
= (\mathbf{E} \cdot \mathbf{v}) D_\nu + (\mathbf{H} \cdot \mathbf{v}) B_\nu - W v_\nu = N'_\nu - N_\nu + W v_\nu. \quad (76)$$

The rate of increase of electromagnetic energy within the volume is

$$\int \frac{\partial W}{\partial t}\, d\tau + \int W v_\nu\, dS = \int W v_\nu\, dS - \int N_\nu\, dS,$$

from (64) (with $J = 0$). Hence

$$\int T_{\nu i} v_i\, dS = \int (N'_\nu - N_\nu + W v_\nu)\, dS = \int N'_\nu\, dS \\
+ \int \frac{\partial W}{\partial t}\, d\tau + \int W v_\nu\, dS.$$

If we take the Poynting vector as **N**', we have the correct energy balance — the rate at which work is done is equal to the sum of the rate of increase of energy within the volume and the rate of flux of energy outwards through its boundary. Note that $S$ is any moving surface, and that **v** here simply refers to the velocity of any point of the surface at any time; $S$ is not necessarily a surface "moving with the medium"; the medium may, in fact, be at rest; $S$ may be moving through the medium in any way. Nor have we assumed that the energy is "tied" to each particular portion of the medium; we have made no assumptions at all about the relative motions of the medium and the lines of force, etc.

## 3.4. The equations of momentum and energy in fluid dynamics, with electric and magnetic forces

We now return to our mechanics, and use a laboratory frame. (Note that if the electric and magnetic fields are given in one frame, and we change to a frame moving relatively to it, even with uniform velocity, the electric and magnetic quantities in this second frame, must, with $v^2/c^2$ omitted, according to (73) be replaced by the quantities in (68).) The energy density, Poynting vector, electromagnetic momentum, and stress tensor are given by (59)–(62), and because we are neglecting the second terms in (74) the constitutive equations are simply (57) and (58).

With these we use Maxwell's equations (56). With our assumptions the tensor $T_{ij}$ is now symmetrical.

With $v^2/c^2$ neglected the Lorentz transformation (65) to a frame (not now a local frame) moving with a constant velocity $v$ relative to the laboratory frame becomes $x_1' = x_1 - vt$, $t' = t - vx/c^2$. With these formulae, together with (68), Maxwell's equations are invariant. In the mechanical equations, it may be shown that if the term $vx/c^2$ in $t'$ is dropped, and a Galilean transformation is used, the difference is of order $v^2/c^2$ only, and is to be neglected. However, as long as the electric and magnetic fields are correctly expressed in the laboratory frame, it is *not* necessary for all our equations to be invariant under a Galilean transformation: in particular, the equations apply whether or not the displacement current and charge accumulation are neglected (*cf*. Section 3.5).

We take the electric and magnetic forces to be surface forces, not body forces, with a stress tensor $T_{ij}$. Our previous formulae for the tensor $p_{ij}$ in Section 2.1 now apply, by the arguments there given, in the first place only to the sum, $p_{ij} + T_{ij}$, but since $T_{ij}$ is itself a symmetrical second-order tensor, they still apply to $p_{ij}$.

The equation of momentum is now found by considering the momentum of the fluid inside a *fixed* surface $S$. If we include the inertia forces, and the action of the Maxwell stress over the surface, and equate the resultant to the rate of change of the electromagnetic momentum inside the surface, we obtain (using (63))

(77) $$\begin{aligned}\rho f_i &= \rho F_i + \partial p_{ji}/\partial x_j + \partial T_{ji}/\partial x_j - \partial G_i/\partial t \\ &= \rho F_i - \partial p/\partial x_i + \rho_e E_i + (\mathbf{J} \times \mathbf{B})_i + \partial \pi_{ji}/\partial x_j.\end{aligned}$$

We now repeat our previous calculation of the energy balance, allowing for the work done by the electromagnetic stress, the rate of increase of electromagnetic energy, and the rate of flux of electromagnetic energy, for a conducting fluid medium which may have both electric charge and currents. We now consider a volume inside a surface $S$ moving with the fluid. We equate the sum of the work done, the heat conducted in, and the energy added by chemical action, radiation, etc. to the sum of the rate of increase of kinetic, internal, and electromagnetic energy and the flux of electromagnetic energy outwards. (Compare the equations preceding (49).) The equation expressing the energy balance is

$$\int T_{\nu i} v_i \, dS + \int p_{\nu i} v_i \, dS + \int \rho F_i v_i \, d\tau + \int \mathbf{\nabla} \cdot \mathbf{q} \, d\tau + \int \rho R \, d\tau$$

$$= \frac{D}{Dt} \int \rho(\tfrac{1}{2} v_i v_i + \mathfrak{E}) \, d\tau + \int \frac{\partial W}{\partial t} \, d\tau + \int W v_\nu \, dS + \int N'_\nu \, dS$$

$$= \int \rho \left( v_i f_i + \frac{D\mathfrak{E}}{Dt} \right) d\tau - \int \mathbf{E} \cdot \mathbf{J} \, d\tau + \int (N'_\nu - N_\nu + W v_\nu) \, dS,$$

(by (64)). Now

$$\int p_{\nu i} v_i \, dS = \int v_j p_{ji} v_i \, dS = \int v_i \frac{\partial p_{ji}}{\partial x_j} \, d\tau + \int p_{ji} \frac{\partial v_i}{\partial x_j} \, d\tau$$

$$= \int v_i [\rho f_i - \rho F_i - \rho_e E_i - (\mathbf{J} \times \mathbf{B})_i] d\tau - \int p \frac{\partial v_i}{\partial x_i} \, d\tau$$

$$+ \int \Phi \, d\tau.$$

(Equations (77) and (49) have been used.) Now use also (76) and the equations following (49). The result is

$$\int \left[ \rho \left\{ \frac{D\mathfrak{E}}{Dt} + p \frac{D}{Dt}\left(\frac{1}{\rho}\right) \right\} - \mathbf{\nabla} \cdot \mathbf{q} - \rho R - \Phi - \mathbf{E} \cdot \mathbf{J} \right.$$
$$\left. + \rho_e \mathbf{v} \cdot \mathbf{E} + \mathbf{v} \cdot (\mathbf{J} \times \mathbf{B}) \right] d\tau = 0.$$

But $\mathbf{v} \cdot (\mathbf{J} \times \mathbf{B}) = -\mathbf{J} \cdot (\mathbf{v} \times \mathbf{B})$, and, since $\mathbf{v} \cdot (\mathbf{v} \times \mathbf{B}) = 0$,
$$\mathbf{E} \cdot \mathbf{J} - \rho_e \mathbf{v} \cdot \mathbf{E} - \mathbf{v} \cdot (\mathbf{J} \times \mathbf{B}) = (\mathbf{E} + \mathbf{v} \times \mathbf{B}) \cdot (\mathbf{J} - \rho_e \mathbf{v})$$
$$= \tilde{\mathbf{E}} \cdot \mathbf{J} = \sigma^{-1} \tilde{\mathbf{J}}^2.$$

$\tilde{\mathbf{J}}$ is the conduction current, $\mathbf{J} - \rho_e \mathbf{v}$. This is exactly what we expected for the Joule heat, and our equation of energy is

(78) $$\rho[D\mathscr{E}/Dt + pD\rho^{-1}/Dt] = \rho DI/Dt - Dp/Dt = \rho T DS/Dt$$
$$= \Phi + (\mathbf{J} - \rho_e \mathbf{v})^2/\sigma + \nabla \cdot \mathbf{q} + \rho R.$$

In terms of the fields
$$(\mathbf{J} - \rho_e \mathbf{v})^2/\sigma = \sigma\{\mathbf{E} + \mathbf{v} \times \mathbf{B}\}^2.$$

Note that if we multiply (77) by $v_i$ and sum, and add to (78), we obtain equation (50b) with $\mathbf{E} \cdot \mathbf{j}$ added to the right side.

### 3.5. Equations for B and $\rho_e$

Before leaving this subject for the present, we may note some immediate electromagnetic consequences of the equations (56), (57), and (58) alone. It is a straightforward matter to obtain the following equations for $\mathbf{B}$ and $\rho_e$ alone:

(79) $$\partial \mathbf{B}/\partial t + \text{curl } (\mathbf{B} \times \mathbf{v}) = -\eta \left[ \text{curl curl } \mathbf{B} + \varepsilon\mu \, \partial^2 \mathbf{B}/\partial t^2 \right.$$
$$\left. - \mu \text{ curl } (\rho_e \mathbf{v}) \right]$$

(80) $$\partial \rho_e/\partial t + \rho_e/(\eta \varepsilon \mu) + \nabla \cdot (\rho_e \mathbf{v}) = (\eta\mu)^{-1} \nabla \cdot (\mathbf{B} \times \mathbf{v})$$

where
$$\eta = (\mu\sigma)^{-1},$$

and is called the magnetic diffusivity. Unless we have oscillations of a high frequency, comparable to that of electromagnetic oscillations, the second term on the right of (79) is clearly small compared with the first term. If $\mathbf{v}$ were zero, (80) shows that $\rho_e$ would decay exponentially, in most conductors with a very short decay time. We may say in a rough way that the residual value of $\rho_e$ will be of order $\varepsilon Bv$, and the last term on the right in (79) will be of order $\varepsilon\mu v^2$ or $v^2/c^2$ compared with the first, and may therefore be neglected.[12] What it amounts to is the well-known result that the "displacement current" $\dot{\mathbf{D}}$ may be neglected except for high-frequency oscillations, and that, when this is neglected, although the electrostatic effects of charge accumulations may be important, their effects on current flow must, for consistency, also be neglected, and the term $\rho_e \mathbf{v}$ in (58) disregarded.

This amounts to dropping the second and third terms on the right in (79). The equation for **B**, to which we return later, then becomes

(81) $\qquad \partial \mathbf{B}/\partial t + \operatorname{curl}(\mathbf{B} \times \mathbf{v}) = -\eta \operatorname{curl} \operatorname{curl} \mathbf{B}.$

For a more detailed investigation of $\rho_e$ it is not sufficiently accurate to use (57) instead of (74). When (74) is used the charge density *in a local frame* is $-(1/c^2)\mathbf{H}\cdot\boldsymbol{\omega}$, where $\boldsymbol{\omega}$ is the vorticity. This holds accurately to our order of approximation if $\partial \rho_e/\partial t + \boldsymbol{\nabla}\cdot(\rho_e \mathbf{v}) = 0$, as in some simple examples; it still holds approximately otherwise unless the magnetic Reynolds number $UL/\eta$, where $U$ and $L$ are a typical macroscopic velocity and length of the system, respectively, is very small. Thus in a conducting medium moving with vorticity there is a space-charge density at any point where there is a magnetic field along the axis of the vorticity, even when observed in a local frame moving with the point. (See S. Goldstein, "Concerning a Continuum Theory of the Electrodynamics and Dynamics of Moving Media," *Proc. Symposium on Electromagnetics and Fluid Dynamics of Gaseous Plasma*, Polytechnic Institute of Brooklyn, 1961, pp. 65–80.) However the charge density is small in any conceivable terrestrial experiment.

[12] G. K. Batchelor, *Proc. Roy. Soc. (London)* **A201** (1950), pp. 405–416.

CHAPTER 4

# Inviscid Fluids

**4.1. Integration of the equation of motion. Conditions for steady motion. Bernoulli's equation. The equation for the velocity potential in the irrotational motion of a gas with contant entropy.**

For inviscid fluids, $\pi_{ij} = 0$. We also neglect heat conduction, so $\mathbf{q} = 0$, and we shall assume no change of energy by chemical action, or radiation, etc., so $R = 0$. With no electromagnetic effects, the equation of motion is

(82) $\quad \mathbf{f} = \partial \mathbf{v}/\partial t + \nabla(\tfrac{1}{2}\mathbf{v}^2) + \boldsymbol{\omega} \times \mathbf{v} = \mathbf{F} - \rho^{-1}\nabla p,$

and the equation of energy is

(83) $\quad\quad\quad\quad\quad DS/Dt = 0,$

which is immediately obvious, since there is no heat conduction, no viscous dissipation, no Joule heat, and no heat addition to any fluid element.

Let $\mathbf{F}$ be conservative, and equal to $-\nabla\Omega$, so that $\Omega$ is the potential energy per unit mass. (In the earth's gravitational field, if the axis of $x_3$ is vertically upwards, $\Omega = g\, x_3$ if variations in $g$ are neglected.)

Now from the thermodynamic relation

$$\rho^{-1}dp = dI - T\,dS,$$

if we set

(84) $\quad\quad\quad\quad I_0 = I + \tfrac{1}{2}\mathbf{v}^2,$
$\quad\quad\quad\quad \rho^{-1}\nabla p + \nabla(\tfrac{1}{2}\mathbf{v}^2) = \nabla I_0 - T\nabla S,$

so

(85) $\quad\quad \partial\mathbf{v}/\partial t + \boldsymbol{\omega} \times \mathbf{v} = -\nabla(I_0 + \Omega) + T\nabla S.$

For *steady motion*, it follows at once from (83) that $S$ is constant along each streamline. In (85) the term $\partial\mathbf{v}/\partial t$ disappears, $\boldsymbol{\omega} \times \mathbf{v}$

is at right angles to both **v** and **ω**, i.e., to the streamlines and vortex lines, and has no component along either, $\nabla S$ has no component along a streamline, and therefore neither has $\nabla(I_0+\Omega)$, so $I_0+\Omega$ is also constant along a streamline in steady flow.

If a gas starts from rest in a container, and is supposed maintained in mechanical and thermodynamic equilibrium there, $I_0 + \Omega$ and $S$ will be constant in the container, and therefore the same for all streamlines which start from the container. They will therefore be constant throughout the whole region occupied by streamlines which start from the container, except for the change of entropy in passing through a shock wave. Then $\omega \times \mathbf{v} = 0$, so, upstream of any shock waves that may appear, the motion is irrotational, or **ω** and **v** are parallel and the streamlines and vortex lines coincide. Vorticity, if it appears, will appear as vortex sheets, with vortex lines along streamlines. All this is for steady motion, with the space filled by streamlines from the container, and upstream of any shock waves. We shall see that in passing through a shock wave there is a definite increase of entropy, which depends on the inclination of the shock wave to the stream. Downstream of a shock wave which does not intersect the streamlines at a constant angle, the entropy will in general vary from streamline to streamline (though it will again be constant along each streamline). The motion will not be irrotational downstream of such a shock wave. For weak shock waves the entropy change is of the third order in the shock-wave strength, so neglect of the entropy change and of the consequent vorticity downstream does not lead to seriously incorrect results in this case. We shall return to a consideration of shock waves in due course; here, after this cautionary note, we do not mention them further.

$I_0 + \Omega$ is constant along a streamline; if we neglect the variation of $\Omega$ along a streamline, $I_0$ is constant, so $I_0$ is the value of the enthalpy at a stagnation point, $\mathbf{v} = 0$. It is called the stagnation enthalpy. $I_0 + \Omega$ will, in any case, be denoted by $\mathfrak{U}$. Note that irrotationality of a steady motion requires, in general, the constancy of *both* $\mathfrak{U}$ and $S$.

If the *entropy* is *constant* throughout the fluid,

(86) $$\partial \mathbf{v}/\partial t + \boldsymbol{\omega} \times \mathbf{v} = -\boldsymbol{\nabla}\mathfrak{U}.$$

Note that, in general, there is a thermodynamic equation of state connecting any three thermodynamic variables, and allowing us to express all of them in terms of any chosen two. (For gas mixtures, this assumes equilibrium concentrations; from this point of view, dissociated and ionized gases are to be counted as gas mixtures, as are gases with an appreciable time lag in any one of the possible molecular modes of motion. Some of these matters will be briefly mentioned later). If the entropy is constant, the pressure is a definite function of the density.

We may consider three cases together, in all of which the equation of motion takes the form (86).

(i) A gas with *constant entropy*, as above,

(87) $$\begin{aligned}\mathfrak{U} &= I_0 + \Omega \\ &= I + \tfrac{1}{2}\mathbf{v}^2 + \Omega.\end{aligned}$$

(ii) Any (artificial) case in which we suppose that it is sufficiently accurate to represent $p$ as *a function of $\rho$ only*. Then $\int \rho^{-1} dp$ is a definite function of $\rho$ (or $p$),

$$\rho^{-1}\boldsymbol{\nabla} p = \boldsymbol{\nabla} \int \rho^{-1} dp,$$

and

(88) $$\mathfrak{U} = \int \rho^{-1} dp + \tfrac{1}{2}\mathbf{v}^2 + \Omega.$$

(It is convenient to set out (i) and (ii) as separate cases. But if $S$ is constant, $p$ is a function of $\rho$ only, and

$$\rho^{-1} dp = dI, \quad \int \rho^{-1} dp = I.)$$

(iii) An incompressible fluid of *constant density*. In this case (86) clearly applies with

(89) $$\mathfrak{U} = p/\rho + \tfrac{1}{2}\mathbf{v}^2 + \Omega.$$

Note that for a perfect gas with constant specific heats $c_p$, $c_v$, such that $c_p = \gamma c_v$,

(90) $$I = a^2/(\gamma - 1),$$

where $a$ is the speed of sound.

From (86), for *steady motion* $\mathfrak{U}$ is constant along each streamline and each vortex line, but may in general take different values along different streamlines and different vortex lines. $\mathfrak{U}$ is constant throughout any region of irrotational flow, $\omega = 0$, and more generally throughout any region in which the streamlines and vortex lines coincide, $\omega$ and $\mathbf{v}$ being parallel. Conversely, if $\mathfrak{U}$ is constant throughout any region where there is motion ($\mathbf{v} \neq 0$), either the motion is irrotational or the streamlines and vortex lines coincide. For a region filled with streamlines, since $\mathfrak{U}$ is constant along each streamline, if at some station (for example, a long way upstream) we know that $\mathfrak{U}$ is the same for all streamlines, then $\mathfrak{U}$ is constant throughout the region. Hence, either the motion is irrotational, or, if vorticity has in some way been produced, the vortex lines coincide with the streamlines, throughout any region filled with streamlines which come from a station where $\mathfrak{U}$ is the same for all streamlines.

For a steady or unsteady *irrotational motion*, where $\mathbf{v} = \nabla \phi$, $\partial \mathbf{v}/\partial t = \nabla \partial \phi/\partial t$, $\omega = 0$,

(91) $$\mathfrak{U} + \partial \phi/\partial t = F(t)$$

throughout the region of irrotational motion for any one of the cases (i), (ii), (iii). If we can solve the equation for $\phi$ ($\nabla^2 \phi = 0$ for an incompressible fluid), with the boundary values of $\partial \phi/\partial \nu$ (and, if necessary, the correct conditions at infinity), we can calculate the pressure from (91), apart from an additive function of $t$ (a constant for steady motion); the pressure is then known everywhere if the value at any one point is given for all time.

For the *steady* flow of a perfect gas with constant specific heats, if

(92) $$\beta = 1/(\gamma - 1),$$

and the variation of $\Omega$ is neglected, $\beta a^2 + \tfrac{1}{2}\mathbf{v}^2 =$ constant along a streamline, and is also constant throughout any region of constant entropy when the motion is irrotational (or the streamlines and vortex lines coincide). According to this equation, there is a mathematically greatest value of $v$, say $v_G$, when $a = 0$ and therefore $T = 0$; if $a_0$ is the velocity of sound at a stagnation point, $v = 0$, and $v_s$ the velocity when the speed is just sonic, $v = a$, then

(93) $$\beta a^2 + \tfrac{1}{2}v^2 = \beta a_0^2 = \tfrac{1}{2}v_G^2 = (\beta + \tfrac{1}{2})v_s^2.$$

For air at ordinary temperatures and pressures, $\gamma = 1.4$, $\beta = 2.5$. The motion is subsonic ($v < a$), sonic ($v = a$), or supersonic ($v > a$) at any point according as $0 \leq v < v_s$, $v = v_s$, or $v_s < v < v_G$.

Quite generally, when the variation in $\Omega$ is neglected, constancy of $I + \tfrac{1}{2}v^2$ and $S$ give $I$ and $S$, and therefore $p$, $\rho$, and $T$ as functions of $v$ and the thermodynamic variables at a stagnation point. In particular, for a perfect gas with constant specific heats, when (93) applies, we have

(94) $$a^2/a_0^2 = T/T_0 = (\rho/\rho_0)^{\gamma-1} = (p/p_0)^{(\gamma-1)/\gamma} = 1 - v^2/v_G^2,$$

where $p_0$, $\rho_0$, and $T_0$ are the values at a stagnation point.

Turn now to the formula (15) for the acceleration $\mathbf{f}$ with moving axes. Since we have used $\Omega$ for the potential energy per unit mass, we shall now (temporarily) write $\boldsymbol{\Psi}$ for the angular velocity of the axes. We see at once that we have quite similar results for the cases corresponding with (87), (88), and (89).

(95) $$\partial \mathbf{v}/\partial t + \boldsymbol{\omega} \times \mathbf{v}' = -\nabla \mathfrak{U},$$

where

(96a) $$\mathfrak{U} = I + \Omega + \tfrac{1}{2}(\mathbf{v} - \mathbf{V})^2 + \boldsymbol{\Psi} \cdot (\mathbf{v} \times \mathbf{x})$$

or

(96b) $$\mathfrak{U} = \int \rho^{-1} dp + \Omega + \tfrac{1}{2}(\mathbf{v} - \mathbf{V})^2 + \boldsymbol{\Psi} \cdot (\mathbf{v} \times \mathbf{x})$$

or

(96c) $$\mathfrak{U} = p/\rho + \Omega + \tfrac{1}{2}(\mathbf{v} - \mathbf{V})^2 + \boldsymbol{\Psi} \cdot (\mathbf{v} \times \mathbf{x})$$

for cases (i), (ii), (iii), respectively. ($\mathbf{v}$ and $\boldsymbol{\omega}$ are used for $\mathbf{v}_s$ and $\boldsymbol{\omega}_s$, and $\mathbf{v}'$ for $\mathbf{v}_b$. From (15), the time-derivative $\partial \mathbf{v}/\partial t$ in (95) is taken with respect to the rotating axes; a motion is said to be steady if it is steady relative to the rotating axes.)

The deductions are similar, but note that, though we now take the actual vorticity and actual vortex lines, since the vector product is $\omega \times \mathbf{v}'$, $\mathfrak{U}$ is constant in steady motion if the actual vortex lines and the *relative* streamlines coincide, and conversely. (This has an application to flow past a screw propeller, or single-stage fan, just as the result for fixed axes has an application to flow past an airfoil). For irrotational motion, (91) holds with the altered value of $\mathfrak{U}$.

For the irrotational motion of an incompressible fluid, the equation for $\phi$ ($\nabla^2 \phi = 0$) comes immediately from the equation of continuity alone; Bernoulli's equation is used only to calculate $p$. But for the *irrotational motion of a gas with constant entropy*, $\rho$ must be eliminated between the equation of continuity and (91) to obtain the equation for $\phi$. The equation of continuity is

$$0 = (a^2/\rho)\{\nabla \cdot (\rho \mathbf{v}) + \partial \rho/\partial t\} = a^2 \nabla \cdot \mathbf{v} + (a^2/\rho)\mathbf{v} \cdot \nabla \rho + (a^2/\rho)\partial \rho/\partial t.$$

With $dS = 0$, $dI = dp/\rho = a^2 \, d\rho/\rho$, so

$$(a^2/\rho)\nabla \rho = \nabla I = -\nabla(\tfrac{1}{2}\mathbf{v}^2 + \Omega + \partial \phi/\partial t)$$

$$(a^2/\rho) \, \partial \rho/\partial t = \partial I/\partial t = -\partial(\tfrac{1}{2}\mathbf{v}^2)/\partial t - \partial^2 \phi/\partial t^2 + F'(t),$$

(since $\partial \Omega/\partial t = 0$), and the equation for $\phi$ is

(97) $\quad a^2 \nabla \cdot \mathbf{v} - \mathbf{v} \cdot \nabla(\tfrac{1}{2}\mathbf{v}^2 + \Omega + \partial \phi/\partial t) - \partial(\tfrac{1}{2}\mathbf{v}^2)/\partial t - \partial^2 \phi/\partial t^2 = 0,$

where $\mathbf{v} = \nabla \phi$, since the term $F'(t)$ may be absorbed by adding a function of $t$ to $\phi$; this does not affect the velocity, nor the determination of the pressure and density, which must in any case be completed from known values at one point. Also when $\phi$ is chosen so that the right-hand side of (91) is a constant, $a^2$ is expressed in terms of $\partial \phi/\partial t$ and $\mathbf{v}^2$ from (91), with $\mathfrak{U}$ as in (87) and the entropy constant. For a perfect gas with constant specific heats, (90) is used.

This derivation of equation (97) does not assume that the gas is a perfect gas.

Now

$$\mathbf{v}\cdot\boldsymbol{\nabla}\,(\partial\phi/\partial t) = \mathbf{v}\cdot(\partial\mathbf{v}/\partial t) = \partial(\tfrac{1}{2}\mathbf{v}^2)/\partial t,$$

and written out in full (97) (with $\boldsymbol{\nabla}\Omega$ neglected) is

(98)
$$\begin{aligned}
&(a^2 - v_1^2)\frac{\partial v_1}{\partial x_1} + (a^2 - v_2^2)\frac{\partial v_2}{\partial x_2} + (a^2 - v_3^2)\frac{\partial v_3}{\partial x_3} \\
&- v_1 v_2\left(\frac{\partial v_2}{\partial x_1} + \frac{\partial v_1}{\partial x_2}\right) - v_2 v_3\left(\frac{\partial v_3}{\partial x_2} + \frac{\partial v_2}{\partial x_3}\right) \\
&- v_3 v_1\left(\frac{\partial v_1}{\partial x_3} + \frac{\partial v_3}{\partial x_1}\right) - 2\left(v_1\frac{\partial v_1}{\partial t} + v_2\frac{\partial v_2}{\partial t} + v_3\frac{\partial v_3}{\partial t}\right) \\
&- \frac{\partial^2 \phi}{\partial t^2} = 0,
\end{aligned}$$

with

$$v_i = \partial\phi/\partial x_i,\ \partial v_i/\partial x_j + \partial v_j/\partial x_i = 2\partial^2\phi/\partial x_i \partial x_j,\ \text{etc.}$$

If we divide by $a^2$, and let $a \to \infty$, we recover the equation $\nabla^2\phi = 0$ for an incompressible fluid, as we should; $a$ is the velocity of wave propagation for an infinitesimal pressure change, and in an incompressible fluid this, together with the velocity of propagation for any finite change, is infinite. For an incompressible fluid, because of the infinite value of the velocity of wave propagation, the equation for $\phi$ takes the form $\nabla^2\phi = 0$ both for steady and unsteady motion. (Mathematically this is simply the equation of continuity, which does not contain $t$ explicitly for an incompressible fluid.)

The equation for steady flow in two dimensions, with $\partial/\partial x_3 = 0$, $v_3 = 0$, is easily seen to be elliptic where the flow is subsonic, and hyperbolic where the flow is supersonic.

Consider a small perturbation about a constant velocity $U$ along the axis of $x_3$, and take the velocity potential as $U(x_3 + \phi)$, so that $U\phi$ is the perturbation potential. If we neglect all squares and products of the derivatives of $\phi$, we obtain the linearized approximate equation

(99)
$$\frac{\partial^2\phi}{\partial x_1^2} + \frac{\partial^2\phi}{\partial x_2^2} + (1 - M^2)\frac{\partial^2\phi}{\partial x_3^2} = \frac{1}{a^2}\left(\frac{\partial^2\phi}{\partial t^2} + 2U\frac{\partial^2\phi}{\partial t\partial x_3}\right),$$

where

(100) $$M^2 = U^2/a^2,$$

and for the linearized equation the value of $a^2$ may be approximated by the value in the "undisturbed" stream. $M$ is then the Mach number of the undisturbed stream.

For the ordinary equation of acoustics, with sound waves of infinitesimal amplitude, we may simply put $U = 0$, to find the equation

$$\nabla^2 \phi = a^{-2} \partial^2 \phi / \partial t^2$$

It may easily be shown that the equation for the two-dimensional stream function introduced in Section 1.11, equation (30) for steady flow is

$$(a^2 - v_1^2)\frac{\partial^2 \psi}{\partial x_1^2} - 2v_1 v_2 \frac{\partial^2 \psi}{\partial x_1 \partial x_2} + (a^2 - v_2^2)\frac{\partial^2 \psi}{\partial x_2^2} = 0,$$

and the equation for Stokes's stream function for motion symmetrical about the axis of $x_3$ in Section 1.11 equation (36) is[13]

$$(a^2 - v_3^2)\frac{\partial^2 \psi}{\partial x_3^2} - 2v_3 v_r \frac{\partial^2 \psi}{\partial x_3 \partial r} + (a^2 - v_r^2)\frac{\partial^2 \psi}{\partial r^2} - \frac{a^2}{r}\frac{\partial \psi}{\partial r} = 0.$$

These equations also do not require the gas to be a perfect gas.

### 4.2. The theorems of Kelvin, Helmholtz, and Lagrange. Cauchy's vorticity equations

With $\mathbf{f}$ as in (82), $\mathbf{F}$ conservative and equal to $-\nabla\Omega$, and $\rho^{-1}\nabla p$ equal to the gradient of $I$, or $\int \rho^{-1} dp$, or $p/\rho$, as in cases (i), (ii), (iii) in Section 4.1, $\mathbf{f}$ is the gradient of a function which must be single-valued. From the formula for $DK/Dt$ in either (25) or (27),

$$DK/Dt = 0,$$

[13] For extensions of these equations, for a perfect gas with constant specific heats, when $I + \frac{1}{2}v^2$ is constant throughout the fluid, but $S$ varies from streamline to streamline, as in the flow behind a curved shock wave which does not intersect the streamlines at a constant angle, see the account of Crocco's stream function by L. Howarth in *Modern Developments; High Speed Flow*, Chapter II, Appendix.

and $K$ is constant. This is Kelvin's theorem. For a fluid of constant density, or a gas of constant entropy, or any gas in which $p$ is a function of $\rho$ only, with the body forces conservative, and viscosity and heat conduction (etc.) neglected, the circulation in any circuit moving with the fluid is constant for all time.

Consider now a surface $S$ which at any instant $t_0$ is composed wholly of vortex lines. (Such a surface may be a closed surface, or may reach to infinity, or may be bounded by curves lying on the boundaries of the fluid.) Draw on the surface a circuit $\gamma$ which is the boundary (or rim) of a continuous open surface which, lying wholly in the fluid, forms part of $S$. Since $\omega_\nu = 0$ on $S$, the circulation round $\gamma$ is zero by Stokes's theorem at time $t_0$. Now let $S$ and $\gamma$ move with the fluid. The circulation round $\gamma$ remains zero. This is true for any such circuit. Hence

$$\int \omega_\nu dS = 0$$

for all portions of $S$, and $\omega_\nu = 0$ at every point of $S$ at all subsequent times. Hence $S$ continues to consist of vortex lines. Moreover, at any instant a vortex line is the intersection of two surfaces such as $S$; at any point on the intersection of these surfaces, as they move with the fluid the vorticity continues to have no component normal to either of them, and therefore to be tangential to their intersection, which continues to be a vortex line. Hence the vortex lines move with the fluid, i.e., a line which is a vortex line at any instant, and then continues to pass through the same fluid particles, continues to be a vortex line.

Further, by Stokes's theorem and the definition of the strength of a vortex tube, the circulation round a small circuit embracing a vortex tube is the strength of the tube. Since the vortex lines move with the fluid, the vortex tube moves with the fluid, and the circuit, if it moves with the fluid, continues to embrace the vortex tube. Since the circulation round the circuit remains constant, the strength of the vortex tube remains constant. Hence:

*Any vortex tube moves with the fluid and its strength remains constant.*

These are Helmholtz's theorems. This proof, which is Kelvin's, holds if **v** is continuous, even if $\omega$ is discontinuous at certain surfaces.

To visualize the meaning of Helmholtz's theorems, consider a small cylinder of fluid which at any instant is part of a vortex tube, with its axis along a vortex line. As the cylinder moves, its axis will always be along a vortex line—i.e., the axis of the vorticity—and it will be part of a vortex tube; also, at any instant its cross-section will be inversely proportional to the magnitude of the vorticity.

If, in addition to our previous assumptions, the motion of a portion of the fluid is irrotational at any instant, then the circulation in any reducible circuit in that portion of fluid is initially zero, and stays zero as the circuit moves with the fluid. Hence the vorticity, initially zero, stays zero at any point of that portion of fluid. In particular, if the fluid is initially at rest, the vorticity is initially zero, and stays zero. Note that the theorem refers to portions of fluid, not to regions of space. However, if all the fluid is initially at rest, the vorticity is zero in every portion of fluid. Diffused vorticity cannot therefore occur. Since we have assumed **v** continuous, we have excluded vortex sheets and line vortices; hence the motion remains irrotational at every point. This is Lagrange's theorem.

Note that Kelvin's circulation theorem, by itself, does not preclude the appearance of vortex sheets. Only the assumption that **v** is continuous everywhere in the fluid (and not only in each of several regions into which the fluid may be supposed divided) does that. We may illustrate by considering the theory of two-dimensional flow round an airfoil in accelerated motion. See Fig. 7.

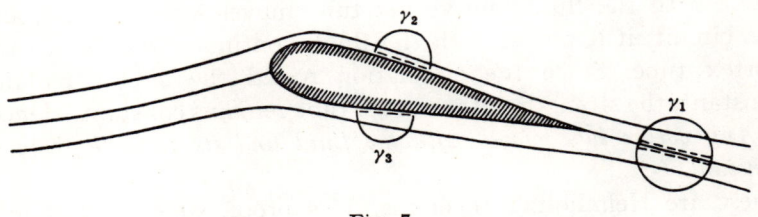

Fig. 7

INVISCID FLUIDS 73

Then a vortex sheet may spring from the trailing edge (or elsewhere, in fact, as far as we could assert with our present considerations). With the vortex lines normal to the plane shown, the circulation in $\gamma_1$ is then not zero. But initially $\gamma_1$ was not necessarily a circuit drawn in the fluid. The part of $\gamma_1$ above the assumed vortex sheet may have come from above the airfoil ($\gamma_2$), and the part below from below the airfoil ($\gamma_3$). In the theory of flow past an airfoil of finite span, there is also a trailing vortex sheet. If Fig. 7 shows a cross-section, the trailing vortex lines are parallel to the plane shown, and the circuits $\gamma_1$, $\gamma_2$, $\gamma_3$ must be taken normal to the plane. Similar remarks then apply. This kind of argument was put forward by Prandtl.

Helmholtz originally derived his theorems from his equations for the vorticity, which may be generalized to apply to a compressible fluid with the restrictions previously stated. (cf. Lamb, Section 146.). Since $\mathbf{f}$ is the gradient of a single-valued function,

$$(101) \qquad \operatorname{curl} \mathbf{f} = \boldsymbol{\omega}^* = \frac{\partial \boldsymbol{\omega}}{\partial t} + \operatorname{curl}(\boldsymbol{\omega} \times \mathbf{v}) = 0,$$

i.e.,

$$\frac{\partial \omega_i}{\partial t} + \mathbf{v} \cdot \boldsymbol{\nabla} \omega_i - \boldsymbol{\omega} \cdot \boldsymbol{\nabla} v_i + \omega_i \boldsymbol{\nabla} \cdot \mathbf{v} - v_i \boldsymbol{\nabla} \cdot \boldsymbol{\omega} = 0.$$

But $\boldsymbol{\nabla} \cdot \boldsymbol{\omega} = 0$, $\boldsymbol{\nabla} \cdot \mathbf{v} = -(1/\rho) D\rho/Dt$. Hence, dividing by $\rho$, we have

$$\frac{1}{\rho}\frac{D\omega_i}{Dt} - \frac{\omega_i}{\rho^2}\frac{D\rho}{Dt} = \frac{\boldsymbol{\omega}}{\rho} \cdot \boldsymbol{\nabla} v_i,$$

i.e.,

$$(102) \qquad \frac{D}{Dt}\left(\frac{\omega_i}{\rho}\right) = \frac{\boldsymbol{\omega}}{\rho} \cdot \boldsymbol{\nabla} v_i.$$

From these equations it easily follows that if at any instant we take a line element $\delta\mathbf{x}$ along a vortex line, such that $\delta\mathbf{x} - \varepsilon\boldsymbol{\omega}/\rho = 0$, where $\varepsilon$ is infinitesimal, then $D(\delta\mathbf{x} - \varepsilon\boldsymbol{\omega}/\rho)/Dt = 0$, from which, following Helmholtz, the conclusion is sometimes drawn that $\delta\mathbf{x} - \varepsilon\boldsymbol{\omega}/\rho$ remains zero. The result is known to be correct, but the proof is certainly incomplete. For in its simplest form it requires that a function $f(t)$, which is such that $f'(t) = 0$ when $f(t) = 0$, should be identically zero. $1 - \cos t$ is an immediate counter example.[14]

A rigorous proof may be constructed along these lines, since the components of $\delta \mathbf{x} - \varepsilon \omega/\rho$ satisfy the equations
$$D(\delta x_i - \varepsilon \omega_i/\rho)/Dt = (\partial v_i/\partial x_j)(\delta x_j - \varepsilon \omega_j/\rho),$$
but if we are to find a proof from equations such as (101), surely it would be best to integrate them. This is, in fact, easily achieved, for Helmholtz's equations, which express the vanishing of $\omega^*$, are exactly equivalent to $DK/Dt = 0$, as we saw in Section 1.9, and this equation immediately integrates to $K = $ constant. So the obvious thing to do is to express this last equation in terms of the vorticity. This leads to Cauchy's equations (which are found in quite a different fashion in the textbooks, and by Cauchy originally (see Lamb, Section 146)), from which a rigorous demonstration of Helmholtz's equations is easily given. These equations are, physically, so fundamental, that we here sketch this method of proof.

We transform the integral for $K$ at time $t$ into a surface integral in the **a**-space by the use of the formulae for the transformation of surface integrals. (A simpler, but longer, method is to transform the line integral for $K$ into a line integral in the **a**-space, then transform to a surface integral by Stokes's theorem, and use elementary calculus.) With the surface integrals over a surface spanning the circuit, as usual,

$$\begin{aligned} K &= \int \omega_\nu dS = \iint \{\omega_1 dx_2 dx_3 + \omega_2 dx_3 dx_1 + \omega_3 dx_1 dx_2\} \\ &= \sum \iint \omega_1 dx_2 dx_3 \\ &= \sum \iint \omega_1 \left\{ \frac{\partial(x_2, x_3)}{\partial(a_2, a_3)} da_2 da_3 + \frac{\partial(x_2, x_3)}{\partial(a_3, a_1)} da_3 da_1 \right. \\ &\quad + \left. \frac{\partial(x_2, x_3)}{\partial(a_1, a_2)} da_1 da_2 \right\}. \end{aligned}$$

[14] Larmor pointed out to Lamb that the same fallacy was contained in this as in certain defective proofs of Lagrange's theorem, and that Stokes had pointed out the fallacy in 1845. It is, however, still occasionally revived.

## INVISCID FLUIDS

Let $\xi_i$ be the initial values of $\omega_i$. Initially, $\partial x_\lambda/\partial a_\sigma = 1$ if $\lambda = \sigma$, and zero otherwise.

Consider first an area that transforms into an area in the $(a_2, a_3)$ plane. Equate the value of $K$ to its initial value.

$$\iint \left\{ \omega_1 \frac{\partial(x_2, x_3)}{\partial(a_2, a_3)} + \omega_2 \frac{\partial(x_3, x_1)}{\partial(a_2, a_3)} + \omega_3 \frac{\partial(x_1, x_2)}{\partial(a_2, a_3)} \right\} da_2 da_3$$

$$= \iint \xi_1 da_2 da_3.$$

Hence

$$\omega_1 \frac{\partial(x_2, x_3)}{\partial(a_2, a_3)} + \omega_2 \frac{\partial(x_3, x_1)}{\partial(a_2, a_3)} + \omega_3 \frac{\partial(x_1, x_2)}{\partial(a_2, a_3)} = \xi_1.$$

Similarly

$$\omega_1 \frac{\partial(x_2, x_3)}{\partial(a_3, a_1)} + \omega_2 \frac{\partial(x_3, x_1)}{\partial(a_3, a_1)} + \omega_3 \frac{\partial(x_1, x_2)}{\partial(a_3, a_1)} = \xi_2$$

$$\omega_1 \frac{\partial(x_2, x_3)}{\partial(a_1, a_2)} + \omega_2 \frac{\partial(x_3, x_1)}{\partial(a_1, a_2)} + \omega_3 \frac{\partial(x_1, x_2)}{\partial(a_1, a_2)} = \xi_3.$$

Solve for $\omega_1$ by multiplying the $i$th equation by $\partial x_1/\partial a_i$ and adding, and use the Lagrangian equation of continuity (equation (7)). Thus, if $\rho_0$ is the initial value of $\rho$,

(103) $$\frac{\omega_1}{\rho} = \frac{\xi_1}{\rho_0} \frac{\partial x_1}{\partial a_1} + \frac{\xi_2}{\rho_0} \frac{\partial x_1}{\partial a_2} + \frac{\xi_3}{\rho_0} \frac{\partial x_1}{\partial a_3} = \frac{\xi}{\rho_0} \cdot \nabla_a x_1.$$

Similarly

(103) $$\frac{\omega_2}{\rho} = \frac{\xi}{\rho_0} \cdot \nabla_a x_2,$$

$$\frac{\omega_3}{\rho} = \frac{\xi}{\rho_0} \cdot \nabla_a x_3,$$

where $\nabla_a$ is the gradient operator in the **a**-space. These are Cauchy's equations.

Lagrange's theorem follows at once, for if $\xi = 0$ for any fluid element (at time $t = 0$), then $\omega = 0$ for all time for that element.

More generally, at time $t = 0$ take a line element along a

vortex line, $\delta \mathbf{a} = \varepsilon \boldsymbol{\xi}/\rho_0$, where $\varepsilon$ is infinitesimal. Let the element move with the fluid. At time $t$ its projections on the axes are

$$(\partial x_i/\partial a_j)\delta a_j = \varepsilon \xi_j (\partial x_i/\partial a_j)/\rho_0$$
$$= \varepsilon(\boldsymbol{\xi} \cdot \boldsymbol{\nabla}_a x_i)/\rho_0 = \varepsilon \omega_i/\rho.$$

The element still lies along a vortex line. Its length, originally $\delta s_0 = \varepsilon \xi/\rho_0$, is now $\delta s = \varepsilon \omega/\rho$. If $\sigma$ is the normal cross-section of a narrow vortex tube, and $\delta s$ an element of its length, then, from the conservation of mass, $\rho \sigma \delta s = \text{constant} = \rho_0 \sigma_0 \delta s_0$. Hence $\sigma \omega = \sigma_0 \xi = \text{constant}$, and the strength of the vortex tube is constant. These are Helmholtz's theorems.

Note that for two-dimensional and axisymmetrical motions (without an azimuthal velocity component around the axis of symmetry) Helmholtz's equations (102) give

$$\frac{D}{Dt}\left(\frac{\omega}{\rho}\right) = 0, \quad \frac{D}{Dt}\left(\frac{\omega}{r\rho}\right) = 0,$$

respectively, the vortex lines being respectively perpendicular to the plane of the motion, and circles in planes normal to the axis of symmetry, with centers on the axis. The integrated forms of these come immediately from the deductions from Cauchy's equations as follows. $\omega/(\rho \delta s) = \xi/(\rho_0 \delta s_0) = \text{constant}$. So $\omega/\rho$ varies as $\delta s$. For the two-dimensional motion, with the vortex lines straight lines perpendicular to the plane of the motion, $\delta s$ is constant. So $\omega/\rho$ is constant for a fluid element. For the axisymmetrical case, with the vortex lines circles, and $\omega$ constant on each circle, $\delta s$ is proportional to the circumference of the circle, i.e., $\delta s \propto r$, $\omega/\rho \propto r$.

### 4.3. The $\omega$, B analogy.

If $\mathbf{B}$ is the magnetic induction, as in Section 3, then, since $\boldsymbol{\nabla} \cdot \mathbf{B} = 0$, there is a magnetic vector potential $\mathbf{A}$ such that

(104) $$\mathbf{B} = \operatorname{curl} \mathbf{A}, \quad \boldsymbol{\nabla} \cdot \mathbf{A} = 0.$$

Let $C$ be the circulation of $\mathbf{A}$ round a circuit, so

(105) $$C = \int \mathbf{A} \cdot d\mathbf{x} = \int B_\nu dS$$

is the flux of magnetic induction through the circuit. Since $\mathbf{\nabla} \cdot \mathbf{B} = 0$,

(106) $\qquad DC/Dt = \int B_\nu^* \, dS, \qquad \mathbf{B}^* = (\partial \mathbf{B}/\partial t) + \operatorname{curl}(\mathbf{B} \times \mathbf{v}).$

If the magnetic diffusivity $\eta$ is put equal to zero, i.e., if the electric conductivity $\sigma$ is effectively infinite, then from (81), $\mathbf{B}^* = 0$, and $DC/Dt = 0$.

There is a complete mathematical analogy between $C$, $\mathbf{B}$, $\mathbf{B}^*$ on the one hand and $K$, $\boldsymbol{\omega}$, $\boldsymbol{\omega}^*$ on the other, for a fluid of constant density, or a gas flow with constant entropy (or, in general, with $p$ a function of $\rho$ only), and with conservative body forces. (We shall see later that if $\eta \neq 0$, the analogy persists, with the vorticity taken in a fluid of constant viscosity $\mu$ and density $\rho$, and $\eta$ corresponding to the kinematic viscosity, $\nu = \mu/\rho$.) For the case we are now considering, $\eta = 0$, it follows at once that the lines of magnetic flux move with the fluid, that the strength $(B\sigma)$ of a magnetic tube remains constant as it moves with the fluid, that

$$D(\rho^{-1} B_i)/Dt = \rho^{-1} \mathbf{B} \cdot \mathbf{\nabla} v_i,$$

and that if $\mathbf{B}_{(0)}$ is the initial value of $\mathbf{B}$ (at time $t = 0$),

$$\rho^{-1} B_i = \rho_0^{-1} \mathbf{B}_{(0)} \cdot \mathbf{\nabla}_a x_i.$$

If in a two-dimensional flow the lines of magnetic induction are all perpendicular to the plane of the motion, $\rho^{-1} B$ is a constant at any fluid element as it moves. If in an axisymmetric flow the lines of induction are circles in planes normal to the axis of symmetry, with centers on the axis, $\rho^{-1} B \propto r$ at a fluid element.

When the lines of induction move with the fluid in this way, the magnetic field is said to be "frozen" into the material medium.

Some remarks on the irrotational motion of an incompressible fluid will be found in Appendix III. In Section 4.4, we pass to a consideration of shock waves; additional formulae for shock waves are set out for reference in Appendix IV. A centered (Prandtl-Meyer) expansion in a perfect and inviscid gas is considered in Appendix V. In the main text we pass to a consideration of viscosity and heat conduction, and suitable substitutions for $\pi_{ij}$ and $\mathbf{q}$. We return later (in Section 11) in the main text to further remarks on the dynamics of inviscid gases.

## 4.4. Shock waves. Remarks on plane sound waves of finite amplitude

### (a) Shock waves

At a surface of discontinuity $\rho v_\nu$ must, by physical continuity, be continuous, but in an inviscid gas $\rho$ and $v_\nu$ may each separately (as well as the tangential velocity component) be discontinuous. Let us examine the general conditions at such a discontinuity. We now use the subscripts 1 and 2 to denote conditions on the two sides of the surface of discontinuity. If $m$ is the rate of mass flow, per unit area, across the surface at any point

$$(107) \qquad \rho_1 v_{1\nu} = \rho_2 v_{2\nu} = m.$$

In time $dt$ the pressure difference $(p_1 - p_2)\boldsymbol{\nu}$ per unit area produces a change in the momentum of the fluid, per unit area, equal to $m(\mathbf{v}_2 - \mathbf{v}_1)dt$; hence the equation of linear momentum gives

$$(108) \qquad (p_1 - p_2)\boldsymbol{\nu} = m(\mathbf{v}_2 - \mathbf{v}_1).$$

Thus either $m = 0$, $p_1 = p_2$, or $\mathbf{v}_2$ is in the same plane as $\mathbf{v}_1$ and $\boldsymbol{\nu}$, and the tangential components of $\mathbf{v}_2$ and $\mathbf{v}_1$ are equal. In the former case there is no flow across the surface, $v_\nu = 0$ on both sides, and there is no discontinuity in pressure; the density and temperature may, however, be discontinuous, and we have what is called a contact discontinuity; the tangential velocity component may be discontinuous, and the contact discontinuity may also be a vortex sheet. We are here concerned with the second possibility,[15] in which case we have a shock wave.

If there is no loss of energy by radiation from the shock wave, the change in the energy (internal energy plus kinetic energy) of the fluid crossing per unit area is equal to the work done by the pressure forces in a time $dt$, so

$$m(\mathfrak{E}_2 + \tfrac{1}{2}\mathbf{v}_2^2 - \mathfrak{E}_1 - \tfrac{1}{2}\mathbf{v}_1^2) = p_1 v_{1\nu} - p_2 v_{2\nu} = m(p_1/\rho_1 - p_2/\rho_2),$$

i.e.,

$$(109) \qquad \mathfrak{E}_1 + p_1/\rho_1 + \tfrac{1}{2}\mathbf{v}_1^2 = \mathfrak{E}_2 + p_2/\rho_2 + \tfrac{1}{2}\mathbf{v}_2^2.$$

[15] For both shock waves and contact discontinuities, see Courant and Friedrichs, *Supersonic Flow and Shock Waves*.

If we introduce the enthalpy $I$,

$$(110) \qquad I_1 + \tfrac{1}{2}\mathbf{v}_1^2 = I_2 + \tfrac{1}{2}\mathbf{v}_2^2$$

and we may substitute $v_{1\nu}^2$, $v_{2\nu}^2$, for $\mathbf{v}_1^2$, $\mathbf{v}_2^2$, since the tangential components are continuous.

The entropy is also discontinuous through the shock wave, and under the present conditions, the entropy of a fluid element cannot decrease. If the gas passes from the side 1 to the side 2,

$$(111) \qquad S_1 \leqq S_2,$$

the sign of equality holding only when there is no discontinuity.

If we take normal components in (108), we have

$$(112) \qquad p_2 - p_1 = \dot{m}(v_{1\nu} - v_{2\nu}) = m^2(\rho_1^{-1} - \rho_2^{-1})$$

and the following relations are easily obtained.

$$(113) \quad (\rho_1^{-1} + \rho_2^{-1})(p_2 - p_1) = m(\rho_1^{-1} + \rho_2^{-1})(v_{1\nu} - v_{2\nu}) = v_{1\nu}^2 - v_{2\nu}^2,$$

$$(114) \qquad I_2 - I_1 = \tfrac{1}{2}(p_2 - p_1)(\rho_1^{-1} + \rho_2^{-1})$$

$$(115) \qquad \mathfrak{E}_2 - \mathfrak{E}_1 = \tfrac{1}{2}(p_2 + p_1)(\rho_1^{-1} - \rho_2^{-1})$$

$$(116) \qquad v_{1\nu} v_{2\nu} = (p_2 - p_1)/(\rho_2 - \rho_1)$$

$$(117) \quad (\rho_1^{-1} - \rho_2^{-1})(p_2 - p_1) = (v_{2\nu} - v_{1\nu})^2.$$

In Appendix IV, the relations for a shock in a perfect gas with constant specific heats are written down. We see that for a normal shock the flow is supersonic before it passes through the shock, and subsonic after; that, whether the shock be normal or oblique, the pressure and density must both rise across the shock; and that the increase of entropy across a shock is of the third order in the shock strength for a weak shock. These theorems may be proved for any gas. (For the mathematical conditions and the proof, see Courant and Friedrichs, Section 65.) Moreover, for a weak shock, the entropy increase affects the calculations of other quantities only when terms of at least the third order in the shock strength are retained.

So far we have assumed the shock wave to be at rest, and considered fluid velocities relative to it. Suppose now that we have a normal shock wave advancing with a uniform velocity

$V$ from a medium in which the pressure, density, and fluid velocity are $p_2$, $\rho_2$, $u_2$ into one in which they are $p_1$, $\rho_1$, $u_1$. Since all our equations are unaltered if we change the frame of reference to one moving with a uniform relative translation, we fall back on the previous case if we write $v_1 = V - u_1$, $v_2 = V - u_2$ (Fig. 8). The simplest and most important case is that in which the frame of reference is taken relative to the fluid in medium 1;

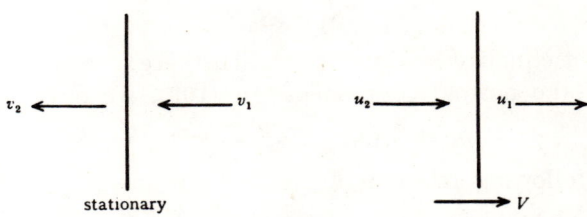

Fig. 8

$u_1 = 0$. In any case, relations which do not contain the velocities, such as (114), (115), and (A30) and the first of (A31) in Appendix IV, will be unaltered. In (117) we may substitute $(u_2 - u_1)^2$ for $(v_{2\nu} - v_{1\nu})^2$. Also, from (112), we find that

$$(118) \qquad \frac{1}{\rho_1} \left( \frac{p_2 - p_1}{1/\rho_1 - 1/\rho_2} \right)^{\frac{1}{2}} = V - u_1$$

However, we usually need to calculate $V$. If the pressure ratio $p_2/p_1$ is given, then for a perfect gas with constant specific heats, equation (A32) of Appendix IV allows us to calculate $V$ at once when the thermodynamic variables in medium 1 are known (and if $u_1$ is also known; here we simply put $u_1 = 0$). For $v_1 = V$, if $p_2/p_1 = 1 + x$

$$(119) \qquad \frac{V^2}{a_1^2} = 1 + \frac{\gamma + 1}{2\gamma} x.$$

The velocity, $u_2$, behind the shock wave is easily found from the second of equations (A31) of Appendix IV, since $u_2 = V - v_2$ ($= v_1 - v_2$); also $a_1/a_2 = (T_1/T_2)^{\frac{1}{2}}$, which is given by (A30).

The results are

(120) $$u_2 = \frac{a_1 x}{\gamma}\left(1 + \frac{\gamma + 1}{2\gamma} x\right)^{-\frac{1}{2}} = \frac{a_2 x}{\gamma}(1+x)^{-\frac{1}{2}}\left(1 + \frac{\gamma - 1}{2\gamma} x\right)^{-\frac{1}{2}}$$

If $u_2$ is given, instead of $p_2/p_1$, then $V$ is found from Eq. (A35) of Appendix IV, with the result

(121) $$V = \tfrac{1}{4}(\gamma + 1)u_2 + \{a_1^2 + [\tfrac{1}{4}(\gamma + 1)u_2]^2\}^{\frac{1}{2}};$$

other relations may be found as before. The inference from the Second Law of Thermodynamics is that, in general (with $u_1$ not necessarily zero), $u_2 > u_1$, and the wave is still compressive, $\rho_2 > \rho_1$, etc.

For a weak shock, in which $p_2/p_1 - 1 = \Delta p/p_1$ is small, $\rho_2/\rho_1 - 1$ and $T_2/T_1 - 1$, together with $M_1^2 - 1$ for a stationary normal shock, are all of order $\Delta p/p_1$, and for a normal shock propagating into still gas, the velocity of the shock differs from the velocity of sound ahead by a quantity of order $a_1 \Delta p/p_1$. For a perfect gas with constant specific heats, approximate formulae are easily written down. On the other hand, for a very strong shock, if we still assume such a gas, we see that $\rho_2/\rho_1$ tends to a finite limit as $p_2/p_1 \to \infty$, but $T_2/T_1$, together with $M_1^2$ for a stationary normal shock, and $V^2$ and $u_2^2$ for a propagating shock, all tend to infinity like a multiple of $p_2/p_1$. Since $T_2/T_1$ becomes large for a strong shock, it ceases to be a good physical approximation to consider the gas as a perfect gas with constant specific heats. For illustrative purposes we may note that the limiting value of $\rho_2/\rho_1$ in such a gas is $(\gamma + 1)/(\gamma - 1)$. From equation (A37) of Appendix IV we see that the Mach number behind a normal stationary shock cannot be less than $\{(\gamma - 1)/2\gamma\}^{\frac{1}{2}}$, which is the theoretical limiting value for an infinitely strong shock.

Formulae for the deflection and downstream Mach number for an oblique shock are also discussed in Appendix IV. The main purpose of the appendix is to set out the mathematical formulae and their immediate interpretation; applications are considered elsewhere (see the references in Section 11). Appendix IV therefore also contains the requisite formulae for applications in the case of weak shocks, and also for small deflections in "hypersonic" flow. Here we would point out that a uniform supersonic flow may be turned through an angle $\delta$ by an oblique shock (Fig. 9), provided $\delta$ does not exceed the maximum for the Mach number,

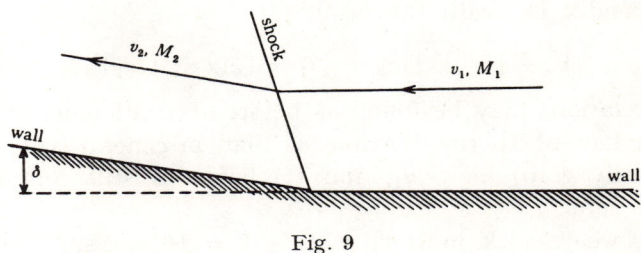

Fig. 9

$M_1$, of the incident flow; this turning is accomplished only for a concave turn, along a wall which "turns into the flow," and the effect of such a turn is compressive. For a deflection less than the permitted maximum, there are theoretically two shocks, a "weak" shock and a "strong" shock, which will accomplish the required deflection.

For a small deflection, $(p_2/p_1) - 1$ and $M_1^2 \cos^2 \alpha_1 - 1$ are small for the "weak" shock unless $M_1$ is large. If $M_1$ is large, we must use a different method of approximation — the hypersonic approximation. The values of $p_2/p_1 - 1$ and $M_1^2 \cos^2 \alpha_1 - 1$, when $\delta_1$ is small, then depend on the parameter $M_1 \delta$, which may be neither very large nor very small.

In Appendix V we consider a mathematical model which will turn a flow round a convex corner; the effect of such a flow will be expansive, and cannot be accomplished by a shock wave — it requires an *increase* in the velocity and a *decrease* of density, pressure, and temperature.

(b) *Plane waves of finite amplitude, and shock-wave formation*

We may consider briefly a simple example which illustrates the necessity of allowing shock-wave discontinuities in the flow of an inviscid gas. Illustrations are given in many places, so we consider the matter only in simple terms. We now temporarily write $u$ for $v_1$ and $x$ for $x_1$, and consider one-dimensional, irrotational, isentropic flow.

In the theory of plane sound waves of infinitesimal amplitude, with $a^2 = (\partial p/\partial \rho)_s$, the equations of motion and continuity give at once

$$\partial u/\partial t = -a_0^2 \partial s/\partial x, \quad \partial u/\partial x = -\partial s/\partial t,$$

where $\rho - \rho_0 = \rho_0 s$, and $\rho = \rho_0$, $a = a_0$, when $u = 0$. Hence $u_{tt} - a_0^2 u_{xx} = 0$, and

$$u = f_1(x - a_0 t) + f_2(x + a_0 t).$$

The first term represents a wave travelling in the positive direction, and the second a wave travelling in the negative direction, of the axis of $x$. For the former, $u = a_0 s$; for the latter $u = -a_0 s$. If we consider a positive wave, $u = (\partial p/\partial \rho)_s^{\frac{1}{2}}(\rho - \rho_0)/\rho_0$ is the condition that the wave motion should consist of a positive wave; if the condition is violated a wave will emerge travelling in the negative direction.

Now consider a finite disturbance, consisting of a plane progressive wave moving in the positive direction. We may assume that at any point the condition that a negative wave should not be generated depends only on the state of affairs in the immediate neighborhood of that point, which may be considered to be an infinitesimal disturbance relative to the state *at* the point, which becomes the reference state. The condition for the absence of negative waves becomes $\delta u = a\delta \rho/\rho$, or, in the limit, $du/d\rho = a/\rho$, so

$$u = \int_{\rho_0}^{\rho} a \frac{d\rho}{\rho},$$

and the value of $u$ is propagated with the velocity $a$ relative to the local moving frame through the point, i.e., with the velocity $(a + u)$ relative to fixed axes. Hence

(122) $\quad \partial u/\partial t = -(a + u)\partial u/\partial x$, and $u = f[x - (a + u)t]$.

This is Rayleigh's argument. More generally (Riemann), if we have both positive and negative waves (advancing and receding

waves), and put
$$\omega = \int_{\rho_0}^{\rho} a \frac{d\rho}{\rho},$$
the equations of motion and continuity,
$$\partial u/\partial t + u\, \partial u/\partial x = -a\, \partial \omega/\partial x,$$
$$\partial \omega/\partial t + u\, \partial \omega/\partial x = -a\, \partial u/\partial x,$$
may be added and subtracted to provide the equations
$$\partial r/\partial t + (a+u)\partial r/\partial x = 0, \quad \partial s/\partial t - (a-u)\partial s/\partial x = 0,$$
where $r = \tfrac{1}{2}(\omega + u)$, $s = \tfrac{1}{2}(\omega - u)$.

In an advancing wave, $s = 0$, $u = \omega$, $r = u$. The propagation velocity, $a + u$, is constant only if $a + \omega$ is constant, which leads to a pressure-density relation $p = A + B/\rho$, where $A$ and $B$ are constants. For a perfect gas with constant specific heats,
$$p/p_0 = (\rho/\rho_0)^\gamma, \quad \omega = 2(a - a_0)/(\gamma - 1),$$
i.e.,
$$a = a_0 + \tfrac{1}{2}(\gamma - 1)u,$$
and
$$a + u = a_0 + \tfrac{1}{2}(\gamma + 1)u$$
so
(123) $$u = f\{x - [a_0 + \tfrac{1}{2}(\gamma + 1)u]t\}.$$

If there is a continuous solution to a problem requiring a positive wave only, then it is of the form (123). Also $u$, $a$, $\rho$, and $p$ increase together. A given wave form does not keep its shape. For the wave form at time $t$ (the curve of $u$ as a function of $x$ at time $t$) is found by shifting each value of $u$ at time $t = 0$ through a positive distance $a_0 t + \tfrac{1}{2}(\gamma + 1)ut$ parallel to the $x$-axis. The points with the larger values of $u$ are shifted further. A wave in which $u$ is originally decreasing with $x$ becomes steeper and steeper. In such a wave, the pressure and density at any point increase during the passage of the wave, which is called a wave of compression. In the opposite case, the wave is said to be one of expansion or rarefaction. Such a wave becomes less and less steep. More generally, the wave may have both compressive and expansive phases. The former become steeper, the latter less steep, as time increases.

In a compression wave, a time is reached at which $\partial u/\partial x$ becomes infinite at some point. If

$$\xi = x - a_0 t - \tfrac{1}{2}(\gamma + 1)ut,$$

then $\xi$ is the abscissa of the point at which the value of $u$ at $x$ at time $t$ was to be found at time $t = 0$, and $\partial u/\partial x$ is infinite when

$$1 + \tfrac{1}{2}(\gamma + 1)tf'(\xi) = 0.$$

This occurs only when $f'(\xi)$ is negative, and the first appearance is at the point where $-f'(\xi)$ is a maximum, at time

$$t = 2\{(\gamma + 1)[-f'(\xi)]_{\max}\}^{-1}.$$

If, for example, $u = U \cos 2\pi x/\lambda$ at $t = 0$, $\partial u/\partial x$ becomes infinite when $t = \lambda/[(\gamma + 1)\pi U]$.

After $\partial u/\partial x$ becomes infinite, the continuous solution becomes meaningless, for there are regions in which $u$ is a triply-valued function of $x$. See Fig. 10. After the slope of the curve of $u$ against $x$ has become infinite, the required solution must be discontinuous; at the discontinuity there is an infinitely rapid change of $u$, $\rho$, and $p$, etc.

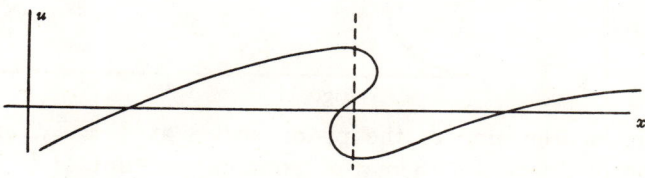

Fig. 10

As a second example, consider the wave produced by a piston in a straight tube, which begins, at time $t = 0$, to move towards a gas in the tube with a constant acceleration $f$, with all friction and heat conduction neglected. At time $T$, let the piston be at $x = X$ and moving with velocity $U$. Then $U = 0$ for $T \leq 0$, $U = fT$ for $T \geq 0$, $X = 0$ for $T \leq 0$, $X = \tfrac{1}{2}fT^2$ for $T \geq 0$; $u = U$ at $x = X$, $u = 0$ at $t = 0$. At any time $t \geq T$, $u = U$ at

$$x = X + [a_0 + \tfrac{1}{2}(\gamma + 1)U](t - T).$$

Hence $u = 0$ at $x = a_0(t - T)$ for $T \leq 0$, $t \geq T$, i.e., $u = 0$ for $x \geq a_0 t$. Also $u = fT$ at

$$x = \tfrac{1}{2}fT^2 + [a_0 + \tfrac{1}{2}(\gamma + 1)fT](t - T) \text{ for } 0 \leq T \leq t.$$

At time $t$ these values of $x$ cover the whole range of $x$ from the piston at $x = \tfrac{1}{2}ft^2$ to $x = a_0 t$. Eliminate $T$ to find the relation between $u$ and $x$.

$$x = a_0 t + u[\tfrac{1}{2}(\gamma + 1)t - a_0/f] - \tfrac{1}{2}\gamma u^2/f.$$

This is the equation of a parabola. $\partial u/\partial x$ is infinite at the vertex, for which

$$t = 2(a_0 + \gamma u)/[(\gamma + 1)f].$$

The infinity occurs first at $u = 0$, at $x = a_0 t$,

$$t = 2a_0/[(\gamma + 1)f],$$

when $U = 2a_0/(\gamma + 1)$. At later times the graph of $u$ against $x$ would be as shown in Fig. 11, with $u$ triply-valued in a certain region.

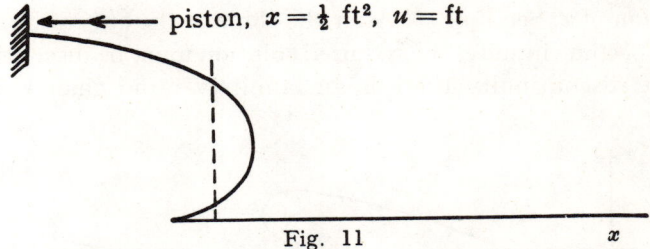

Fig. 11

If the acceleration of the piston ceases at $T = T_1$, and the velocity of the piston thereafter remains constant at $U_1 = fT_1$, the solution above remains for $t < T_1$ (if no discontinuity has been produced). For greater values of $t$ the parabola joins $u = 0$ to $u = U_1$; for values of $x$ between the piston and the abscissa where $u = U_1$ on the parabola, $u$ is constant and equal to $U_1$. The discontinuity is produced, as before, when $t = 2a_0/[(\gamma+1)f]$.

For the solution of such piston problems by characteristics, see Courant and Friedrichs, *Supersonic Flow and Shock Waves*. The characteristics are straight lines, which in a compressive wave intersect and form an envelope.

CHAPTER 5

# Newtonian Fluids and the Navier-Stokes Equations

## 5.1. Relation between $\pi_{ij}$ and $e_{ij}$. Equations of momentum

We return to the discussion of the effects of viscosity. For such a discussion it is necessary to connect the tensor $\pi_{ij}$ in Section 2, which is zero for a fluid at rest, with quantities arising from the motion of the fluid. Since we anticipate that no viscous stresses will arise in a rigid-body motion of a fluid, we assume that $\pi_{ij}$ depends on the rate-of-strain tensor. Independence of the vorticity may be argued, for the purely linear dependence which we shall assume, from the fact that whereas $\pi_{ij}$ and $e_{ij}$ are symmetrical, $\xi_{ij}$ is antisymmetrical. Moreover, a dependence of $\pi_{ij}$ on $\xi_{ij}$ would imply that an external moment is required to maintain a uniform rigid-body rotation of a fluid, and this is known not to be correct.[16]

We now assume that the stress components are linear functions of the rate-of-strain components. This is a generalization of a concept of Isaac Newton for a simple shearing motion (*Modern Developments*, Section 2), and this model is known as a Newtonian fluid. We further assume that the fluid is isotropic, its properties being the same in all directions, so the relations between the $\pi_{ij}$ and $e_{ij}$ must be unaltered by any rotation of the axes. The principal axes of stress and rate-of-strain must then coincide. For if we consider a small rectangular volume element with its edges parallel to the principal axes of stress, subject to normal pressures on its faces, the shear strain must present the same appearance from whichever side of each of the principal planes of stress it

---

[16] If we include electric and magnetic forces for a conducting fluid, and the second terms on the right in (74) in Section 3 are not neglected, $T_{ij}$ is not symmetrical. $\pi_{ij} + T_{ij}$ must still be symmetrical, so $\pi_{ij}$ is not symmetrical. The arguments and formulae of this section may then be taken to apply to the symmetrical part of $\pi_{ij}$.

is viewed. It follows at once that there is no shear. The result is also easily proved mathematically, by rotations of the axes through two right angles. If now $\pi_1$, $\pi_2$, $\pi_3$ are the principal viscous stresses, and $e_1$, $e_2$, $e_3$ twice the principal rates of extension, we may (using isotropy) write

(124) $$\pi_i = (\beta - \tfrac{2}{3}\mu)\varDelta + \mu e_i$$

for $i = 1, 2, 3$. Here

(125) $$\varDelta = \tfrac{1}{2}(e_1 + e_2 + e_3) = \mathbf{\nabla} \cdot \mathbf{v} = \tfrac{1}{2}e_{ii}$$

($\varDelta$ is invariant for change of axes, and the last two expressions in (125) hold for any axes); $\mu$ is the coefficient of shear viscosity, or simply the viscosity; $\beta$ is the coefficient of bulk viscosity, or simply the bulk viscosity.[17] (It is easily seen that $\pi_i = A\varDelta + Be_i$, where $A$ and $B$ are constants. The form chosen relates $\beta$ directly to the difference between the thermodynamic pressure and minus one third of the trace of the stress tensor; see equation (127) below.)

The formulae in (124) may be transformed from principal axes to any axes by the use of the tensor transformation formulae; the result is

(126) $$\pi_{ij} = \beta \varDelta \delta_{ij} + \mu(e_{ij} - \tfrac{2}{3}\varDelta \delta_{ij}).$$

Note that

(127) $$\tfrac{1}{3}\pi_{ii} = \beta\varDelta,$$

and, if $\varDelta \neq 0$, vanishes only if $\beta$ is taken as zero.

We now substitute (126) into the equations of motion (Section 2, Eq. (48)). In general $\mu$ and $\beta$ are not to be taken as absolute constants, but depend on the thermodynamic state of the fluid, and therefore may vary with the coordinates. The equations of motion become

(128) $$\rho f_i = \rho F_i - \partial p/\partial x_i + \mu \nabla^2 v_i + \tfrac{1}{3}\mu\, \partial\varDelta/\partial x_i \\ + e_{ij}\, \partial\mu/\partial x_j - \tfrac{2}{3}\varDelta\, \partial\mu/\partial x_i + \partial(\beta\varDelta)/\partial x_i.$$

[17] There are differences in notation. $\beta - 2\mu/3$ is called the "second" coefficient of viscosity, and denoted by $\lambda$ or $\mu'$; the bulk viscosity is often denoted by $\kappa$, but this symbol is required for thermometric conductivity.

NEWTONIAN FLUIDS AND NAVIER-STOKES EQUATIONS

In invariant vector form this equation is

(129) $\rho \left\{ \dfrac{\partial \mathbf{v}}{\partial t} + \boldsymbol{\nabla}(\tfrac{1}{2}\mathbf{v}^2) + \boldsymbol{\omega} \times \mathbf{v} \right\} = \rho \mathbf{F} - \boldsymbol{\nabla} p -$ curl curl $\mu \mathbf{v}$
$- \mathbf{v}\nabla^2 \mu + \boldsymbol{\nabla}\mu \times \boldsymbol{\omega} + \boldsymbol{\nabla}(\beta \Delta) + \tfrac{4}{3}\boldsymbol{\nabla}(\mu \Delta) + \boldsymbol{\nabla}(\mathbf{v}\cdot\boldsymbol{\nabla}\mu) - \Delta\boldsymbol{\nabla}\mu,$

or the last three terms may be put into the alternative form

$$\tfrac{1}{3}\mu\boldsymbol{\nabla}\Delta - \tfrac{2}{3}\Delta\boldsymbol{\nabla}\mu + \boldsymbol{\nabla}\boldsymbol{\nabla}\cdot(\mu\mathbf{v}).$$

The simplifications if the variation of $\mu$ is neglected, and if $\beta$ is taken as zero, are clear. For an incompressible fluid ($\Delta = 0$) of assumed constant properties ($\rho$ and $\mu$ constant), the equation becomes

(130) $$f_i = F_i - \dfrac{1}{\rho}\dfrac{\partial p}{\partial x_i} + \nu \nabla^2 v_i,$$

i.e.,

(131) $\partial \mathbf{v}/\partial t + \boldsymbol{\omega} \times \mathbf{v} = \mathbf{F} - \boldsymbol{\nabla}(p/\rho + \tfrac{1}{2}\mathbf{v}^2) - \nu$ curl $\boldsymbol{\omega}$,

where

(132) $$\nu = \mu/\rho,$$

and is called the kinematic viscosity.

### 5.2. Dissipation of energy

In the equation of energy (Section 2, equation (50a)), the expression for the dissipation function becomes

(133)
$$\begin{aligned}\Phi &= \tfrac{1}{2}\pi_{ij}e_{ij} = \tfrac{1}{2}\mu e_{ij}e_{ij} - \tfrac{2}{3}\mu\Delta^2 + \beta\Delta^2 \\ &= (\mu/6)[(e_{11} - e_{22})^2 + (e_{22} - e_{33})^2 + (e_{33} - e_{11})^2] \\ &\quad + \mu(e_{12}^2 + e_{23}^2 + e_{32}^2) + \beta\Delta^2.\end{aligned}$$

$\Phi$ is seen to be non-negative. If $\beta = 0$, it vanishes if, and only if,

$$e_{11} = e_{22} = e_{33}, \; e_{12} = e_{23} = e_{31} = 0,$$

i.e., in a spherically symmetrical expansion or contraction. Conversely, $\Phi$ vanishes in a spherically symmetrical expansion or contraction if, and only if, $\beta = 0$; if $\beta \neq 0$, there is dissipation of mechanical energy into heat in any spherically symmetrical expansion or contraction, including one which returns the gas to its original volume.

The expression for $\Phi$ in terms of the rate-of-strain components is valid in general orthogonal coordinates. Expressions in terms of derivatives of velocity components are usually found, if required, by tensor manipulation, but there is an expression in invariant vector form:

(134) $\quad \Phi = \mu[\mathbf{\nabla} \cdot \mathbf{\nabla}\mathbf{v}^2 + 2\mathbf{\nabla} \cdot (\boldsymbol{\omega} \times \mathbf{v}) - 2\mathbf{v} \cdot \mathbf{\nabla}\varDelta + \omega^2 - \tfrac{2}{3}\varDelta^2] + \beta\varDelta^2.$

In the equation of energy we also now substitute the usual relation for the heat flux vector,

(135) $\qquad\qquad\qquad \mathbf{q} = k\mathbf{\nabla}T,$

where $k$ is the thermal conductivity. (With regard to the sign, see the remark towards the bottom of p. 40.) The various forms of the equation of energy are easily expressed in vector form, and in orthogonal coordinates other than cartesian.

We now have six equations — the equation of continuity, three equations of momentum, the equation of energy, and a thermodynamic equation of state — for six dependent variables, three velocity components, and three thermodynamic state variables, say pressure, density, and temperature (or enthalpy, or entropy). For an incompressible fluid of assumed constant properties, we have four equations — three of momentum and one of continuity — for the three velocity components and the pressure.

For an incompressible fluid of constant properties, an expression is easily derived for the rate of dissipation of energy in any finite volume. From (134) we have at once, with $\varDelta = 0$, that the integral of $\Phi$ throughout any volume is

(136)
$$\int \Phi d\tau = \mu \int \mathbf{\nabla} \cdot (\mathbf{\nabla}\mathbf{v}^2 + 2\boldsymbol{\omega} \times \mathbf{v}) d\tau + \mu \int \omega^2 d\tau$$
$$= \mu \int \frac{\partial \mathbf{v}^2}{\partial \nu} dS + 2\mu \int (\boldsymbol{\omega} \times \mathbf{v})_\nu dS + \mu \int \omega^2 d\tau,$$

where the surface integrals are over the boundary. The integrand of the second term is the determinant

# NEWTONIAN FLUIDS AND NAVIER-STOKES EQUATIONS

$$\begin{vmatrix} v_1 & v_2 & v_3 \\ \omega_1 & \omega_2 & \omega_3 \\ v_1 & v_2 & v_3 \end{vmatrix}$$

where $\omega_i$ are the vorticity components.

Since $\partial \mathbf{v}^2/\partial \nu = 2v_i\, \partial v_i/\partial \nu$, and since in a viscous fluid $\partial v_i/\partial \nu$ is bounded (except at particular instants in certain circumstances), the contributions to the surface integrals vanish over the surface of any solid body at rest, on which $\mathbf{v} = 0$. Further, if, in an unbounded fluid, we take the average value of the dissipation per unit volume for a very large volume, in the limit this average value becomes $\mu\overline{\omega^2}$. For a motion which is nearly irrotational, the last two terms of (136) may be neglected.

Since $\Phi = \tfrac{1}{2}\pi_{ij}e_{ij}$, and the $\pi_{ij}$ are taken as linear functions of the $e_{ij}$, $\Phi$ is a homogeneous quadratic function of the $e_{ij}$. Also $\Phi$ is invariant for rotation of the axes. The $e_{ij}$ are the coefficients in the equation for the rate-of-strain quadric. $\Phi$ must therefore be the sum of multiples of the two independent quadratic invariants of the discriminating cubic. The multipliers may be found by considering simple cases, for example a simple shearing motion and a spherically symmetrical contraction or expansion. $\Phi$ may also be easily found in this way in terms of the $\pi_{ij}$, if required.

## 5.3. Rate of change of circulation and equations for the vorticity

From (130), and equation (25) of Section 1, we see at once that in an incompressible fluid of constant properties, with conservative body forces, the rate of change of circulation in a circuit moving with the fluid is

$$\frac{DK}{Dt} = -\oint \nu\, \text{curl}\, \boldsymbol{\omega} \cdot d\mathbf{x}. \tag{137}$$

In a viscous compressible fluid the expression for $DK/Dt$ is necessarily much more complicated. $\nu$ is now variable round the circuit, and

$$-\oint \frac{dp}{\rho} = \oint (T\,dS - dI) = \oint T\,dS,$$

which is not, in general, zero. Also terms arise from variations, round the circuit, of the viscosity and the dilatation $\Delta$. The result (apart from the effect of bulk viscosity) will be found in *Modern Developments*: *High Speed Flow*, Section II 6.

The equation for the vorticity $\omega$ is found by dividing (129) by $\rho$ and taking the curl of both sides. For an incompressible fluid of constant properties, with conservative body forces,

(138) $\qquad \partial \omega/\partial t + \text{curl } (\omega \times \mathbf{v}) = - \nu \text{ curl curl } \omega,$

in agreement with (137) (*cf*. (27) and the discussion immediately preceding it.) Hence

(139) $\qquad D\omega_i/Dt = \omega \cdot \boldsymbol{\nabla} v_i + \nu \nabla^2 \omega_i.$

The first term on the right expresses the rate at which the $\omega_i$ vary for a fluid element when $\nu$ is zero, in which case the vortex lines move with the fluid and the strength of a vortex tube remains constant; in this sense, the vorticity is convected with the fluid. The additional variation of the $\omega_i$, due to viscosity, expressed by the last term, follows the same law as the variation of temperature in the conduction of heat, the kinematic viscosity $\nu$ taking the place of the thermometric conductivity (or diffusivity of heat). Thus vorticity is convected with the fluid, but at the same time diffuses like heat. Therefore in an incompressible fluid of constant properties, even if it is viscous, vorticity cannot originate in the interior of the fluid, but must arise from the boundaries.

By comparing (138) with (81), we may carry further the ($\omega$, **B**) analogy. The equations are identical, with **B** for $\omega$, and the magnetic diffusivity $\eta$ taking the place of the kinematic viscosity.

### 5.4. Results from the kinetic theory of gases

For a general account of the application of the kinetic theory of gases, see Chapman and Cowling, *The Mathematical Theory of Non-uniform Gases*. A brief account of the mechanism of viscosity and other diffusion effects is given by Lighthill, "Viscosity Effects in Sound Waves of Finite Amplitude," in *Surveys in Mechanics*, The G. I. Taylor 70th. Anniversary Volume (ed. by

G. K. Batchelor and R. M. Davies; Cambridge Univ. Press, 1956). This work also contains discussions of relaxation effects in the attenuation of sound and in shock waves. Reference should also be made to Liepmann and Roshko, *Elements of Gas Dynamics*; Chap. 14.

From the kinetic theory of gases (for the present without relaxation effects), density, pressure, internal energy per unit mass, temperature, and other thermodynamic variables, may be defined so that the equations of equilibrium thermodynamics apply, the general equations of Section 2 for the flux of matter, momentum, and energy, may be derived, and the stress tensor $p_{ij}$ and heat flux vector $\mathbf{q}$ may be related to the molecular properties — the mass and motion of the molecules. (Deviations from the perfect gas laws, and the influence of van der Waals forces, must be treated separately.) The considerations by which this is accomplished are very general, and do not depend on the velocity-distribution function for the molecular velocities, nor on a model for the molecules and the nature of molecular encounters. For gases in motion the distribution function differs from that in equilibrium conditions, which is the Maxwellian distribution. If this difference is ignored, the equations of an inviscid gas without heat conduction result. The usual method is then to set up Boltzmann's equation for the velocity-distribution function, and proceed by successive approximation, with

$$(140) \qquad \begin{aligned} p_{ij} &= p_{ij}^{(0)} + p_{ij}^{(1)} + \cdots \\ q_i &= q_i^{(0)} + q_i^{(1)} + \cdots \end{aligned}$$

where $p_{ij}^{(0)} = -p\delta_{ij}$, $q_i^{(0)} = 0$. With all relaxation effects neglected in the molecule,

$$p_{ij}^{(1)} = -\tfrac{2}{3}\mu\Delta\delta_{ij} + \mu\left(\frac{\partial v_i}{\partial x_j} + \frac{\partial v_j}{\partial x_i}\right), \quad q_i^{(1)} = k\frac{\partial T}{\partial x_i},$$

so this approximation gives the theory of the Newtonian fluid without bulk viscosity. The models used all agree on this form of the stress tensor and heat conduction vector, and with them $\mu$ is independent of density and is a function of temperature only.

The simpler models all agree in showing that $\mu c_v/k$ is a constant. The simplest model of all — smooth, spherically symmetrical molecules — gives $\mu \propto T^{\frac{1}{2}}$ and $\mu c_v/k = \frac{2}{5}$. For this model the kinetic energy of translation is the only communicable energy, and the gas behaves as a perfect monatomic gas, with $c_v = 1.5\Re$, $c_p = 2.5\Re$, $\gamma = \frac{5}{3}$ (where $p/\rho T = \Re$). The increase of $\mu$ with $T$ is, in fact, more rapid than $T^{\frac{1}{2}}$. Sutherland's formula, based on spherically symmetrical molecules surrounded by fields of attractive force, gives $\mu \propto T^{\frac{3}{2}}/(A+T)$, where $A$ is a constant. By appropriate choice of $A$ this result may be made to represent closely the experimental values for a good many gases, (for example, nitrogen) over fairly wide ranges of temperature. The results for air, while not as good as for nitrogen alone, are not bad. But for purposes of calculation a simpler formula is needed, and if we are dealing with a restricted temperature range, it is usual to take a simple power law $\mu \propto T^n$. Between 90°K and 300°K, Cope and Hartree suggest $n = \frac{8}{9}$; between 200°K and 400°K, $n = \frac{3}{4}$; Crocco has suggested $n = 0.54$ at 3000°K. (We choose the most important reference temperature $T_1$ in the range considered, and the corresponding viscosity $\mu_1$, and write $\mu/\mu_1 = (T/T_1)^n$.) The dependence of $\mu$ (and $k$) on $T$ only, and their independence from $\rho$, is not intended to apply at very high, or very low, values of $\rho$.

For a monatomic gas the internal energy is made up of the translational kinetic energy of the atoms, the translation energy of the ions and electrons (if any), and the energy of the bound electronic energy states. For the first, if $p/\rho = \Re T$,

$$\mathfrak{E} = 1.5\Re T, \quad c_v = 1.5\Re, \quad c_p = 2.5\Re.$$

All except the first are negligible (except at "high" rarefactions) at temperatures below 7,000°K–10,000°K for argon, atomic oxygen and atomic nitrogen.

For a gas with diatomic molecules, we must introduce the rotational and vibratory energy. If the state of a portion of any gas is altered, the translational energy of a molecule is (statistically) independent of its original value and adjusted to the new state after a few collisions (say, four or five), and the rotational energy is so adjusted after a very few more collisions. But a much larger number of collisions may be necessary to adjust the energy in the vibrational degrees of freedom (and also in any excited atomic states), and gases in which this energy is a significant fraction of the total energy will then exhibit "heat-capacity lag," which may have important effects on macroscopic processes of sufficiently high frequencies. When calculating the specific heats for a diatomic gas, we would, therefore, almost always include the contribution of the rotational energy, which, except at low temperatures ($< 50°K$) may be taken as $\Re T$. (Hydrogen is exceptional: *cf.* Fowler and Guggenheim, *Statistical Thermodynamics*.) But whether the vibratory energy is to be included or not will depend on the rapidity of the changes considered. For changes that take place in a time of the order of one second, this contribution is to be included, and we have what is called the "equilibrium" value of the specific heats; for changes that take place in the time of oscillation of a high-frequency sound wave, it is to be omitted. If it is omitted, $c_v = 2.5\Re$, $c_p = 3.5\Re$, $\gamma = \frac{7}{5}$. The equilibrium values of $c_p/\Re$ for air (as a mixture of oxygen, nitrogen, and argon; dry air without any allowance for departure from the perfect gas laws) vary only from 3.4906 at $T = 150°K$ to 3.5176 at $T = 373.1°K$. In general, the assumption that $c_v$, $c_p$, and $\gamma$ are independent of $T$ is quite a fair approximation at moderate temperatures. (The presence of moisture in the air reduces the value of $\gamma$ and makes it much more variable with change of state, particularly at low temperatures.)

We return to $\mu c_v/k$. In gas dynamics, it is usual to use instead $\mu c_p/k$, called the Prandtl number, and denoted by $Pr$. For monatomic gases, with $\mu c_v/k = \frac{2}{5}$, $Pr = \frac{2}{3}$. For polyatomic gases, if vibrational energy is negligible, Eucken's formula gives $Pr = 4\gamma/(9\gamma - 5)$; the derivation is open to some criticism, and there are theoretical arguments for a somewhat smaller value. With $\gamma = \frac{7}{5}$, the formula gives $Pr = 0.737$; for air the experimental evidence is that $Pr$ is fairly constant, and equal to about 0.715. We may note that, just as the effect of viscosity is determined by the value of $\nu$ (the ratio of the viscosity to the density), so the effect of heat conductivity is determined by the value of $\kappa = k/(\rho c_p)$, since $k$

units of heat raise the temperature of unit volume by $k/\rho c_p$ degrees. $\kappa$ is called the thermometric conductivity, or diffusivity of heat, and $Pr = \nu/\kappa$.

For the sake of completeness, we may add that if $D$ is the coefficient of self-diffusion, $D/\nu$ appears experimentally to be between 1.2 and 1.5, and is usually about 1.4. Note that $\mu$, $k$, $D$ represent respectively the results of molecular transfer of momentum, heat energy, and mass.

The kinetic theory of gases, as summarized above, has had nothing to say about the bulk viscosity, $\beta$. In fact, for spherically symmetrical molecules, $\beta$ is effectively zero (except for very condensed gases), $\beta/\mu$ being of the order of the square of the ratio of the effective "volume" of the molecules to the volume of the gas; and for non-condensed monatomic gases, $\beta$ is, in fact, zero. Otherwise, $\beta$ is a parameter which may be used to describe one or more relaxation processes, or types of "lag," associated with the molecular constitution of a fluid, and capable of being otherwise represented as thermodynamically irreversible processes. (It must also be possible to represent the results of shear viscosity and heat conduction in this way, since the effects produced arise from the "lag" — of only a few collisions, it is true — with which a molecule adjusts, statistically, its translational energy.) We may say that when a coefficient of viscosity, or other similar coefficient, is used, the phenomenon is regarded more as a diffusion or transport phenomenon, and that, when it is legitimate to do so, this is an alternative to considering a thermodynamically irreversible process. It is legitimate to regard a phenomenon in this way when the time constant of the macroscopic processes with which we are concerned is large compared with the time constant of the relaxation process or lag. Thus, in a given motion, some relaxation processes may be replaced, for example, by the introduction of a coefficient of viscosity, but for others we may have to consider the relaxation effects in more detail. Some care must now be taken in the definition of the pressure and the temperature; when relaxation processes are explicitly taken into account, the pressure can no longer simply be defined as a thermodynamic pressure such that equilibrium thermodynamics can be applied; in general, the definition of pressure should proceed through a suitable definition of temperature and through the equation of state; if all relaxation processes are sufficiently taken into account explicitly, by the use of equations of irreversible thermodynamics using more than two independent variables of state, we should take $\beta = 0$, and the pressure is then minus one-third of the trace of the stress tensor $-\frac{1}{3}p_{ii}$. The definition and use of bulk viscosity is therefore a matter of some subtlety, but its use, when applicable, is well justified by the fact that calculations become simpler.

Note also that the bulk viscosity is important only when the dilatation, $\Delta$, is sufficiently large. In practice, this has so far restricted its importance to considerations of such matters as sound absorption and acoustic streaming, and shock waves (see Rosenhead *et al.*, *Proc. Roy. Soc.* (*London*) **A**226 (1954) pp. 1–69; Lighthill, *op. cit.* For irreversible thermodynamics, see I. Prigogine, *Introduction to the Thermodynamics of Irreversible Processes*, and the references in Rosenhead *et al.*; *op. cit.*).

It was previously mentioned that the translational energy of a molecule is adjusted to a new state after a very few collisions, and the rotational energy is so adjusted after a very few more collisions. The latter is an example of a "lag," with a very short relaxation time-constant, where the process can usually be well represented by a bulk viscosity. For if the gas expands or contracts, the lag of the rotational energy produces a lag in the adjustment of the total internal energy, and therefore a difference of the "thermodynamic" pressure from $-\frac{1}{3}p_{ii}$, the value it would have if all the energy were translational. This difference is taken care of by the term $\beta\Delta\delta_{ij}$. In this case, we should expect the process to be well represented by a bulk viscosity except in processes that are extremely rapid. On the other hand, the lag of the vibrational degrees of freedom is much greater, and in cases where it is important, we shall usually have to consider the relaxation processes explicitly.

To sum up we may say: (i) it would be possible to use the

"Eulerian" equations (equations with viscosity and heat conduction neglected), and express all the effects of "lag" in molecular processes by the use of equations of irreversible thermodynamics (though this has not, in fact, been considered); it is, however, mathematically much more convenient for further calculation to keep reversible, equilibrium, thermodynamics and introduce viscosity and heat conduction. (ii) For a monatomic gas, or for any gas in which every "lag" may be neglected except the lag of a molecule in taking up (statistically) the translational energy appropriate to its location, equilibrium thermodynamics may be used when $\mu$ and $k$ are introduced. (iii) When further molecular relaxation processes are to be considered, the use of equilibrium thermodynamics may sometimes be saved by the introduction of a bulk viscosity; this is particularly so for the "lag" of the molecule in taking up the energy of rotational degrees of freedom behind the taking up of the translation energy. (iv) In more "extreme" circumstances (rapid macroscopic processes and/or long relaxation times), the use of equilibrium thermodynamics cannot be completely saved; in such circumstances, however, it is still most convenient to introduce $\mu$, $\beta$, and $k$ in order to complicate the thermodynamics as little as possible. (v) If we preserve linear isotropic relations between the $\pi_{ij}$ and $e_{ij}$, and $\mathbf{q}$ and $\nabla T$, we can never do better, in order to save complication in the thermodynamics, than the introduction of $\mu$, $\beta$, and $k$ (of course, with the corresponding coefficients of diffusion when varying concentrations are to be considered).

The variation of $\beta/\mu$ with temperature, which must depend on the variation of the ratio of the relaxation frequency to the collision frequency, has not been carefully studied, and the best we can do at present is to take $\beta/\mu$ as independent of temperature. But this assumption is probably incorrect. $\beta$ probably varies more slowly with $T$ than $\mu$ does, at any rate in some gases. Such values as are known for gases indicate values of $\beta/\mu$ between 0.5 and 0.8.

We return to (140), and remark that the next approximation ($p_{ij}^{(2)}$, $q_i^{(2)}$) has been worked out, and leads to equations which are known as the Burnett equations, or, as a generalization, the

# NEWTONIAN FLUIDS AND NAVIER-STOKES EQUATIONS

thirteen moment equations of Grad. As the ratio of the length, $\lambda$, of a molecular mean free path to a characteristic length, $L$, of the flow field increases, it might have been expected that some such modification of the Navier-Stokes equations would be superior to those equations themselves. But all the present evidence indicates that that is not so; in fact, where the Navier-Stokes equations are themselves perhaps not completely adequate, the higher-order equations may even be inferior. Perhaps the expansion in (140) is asymptotic; when the first two terms give a very good approximation, the third may give an even better approximation; but when the first two terms do not provide so good an approximation, taking another term may make matters worse; this is a known behavior in asymptotic series.

On the other hand, a modification of the boundary conditions at a solid surface, when $\lambda/L$ (which is known as the Knudsen number) is not very small, to allow for slip and a temperature jump, does appear to improve matters a little. [S. A. Schaaf and P. L. Chambré, Section H of Vol. III of *High Speed Aerodynamics and Jet Propulsion*, Princeton University Press. The formulae cited are given by Kennard.] For the usual boundary conditions at a solid impermeable boundary at rest, $\mathbf{v} = 0$ and $T = T_w$, where $T_w$ is the temperature at the wall. With slip and temperature jump, if $u$ is a tangential velocity component at the surface

$$(141) \qquad u = \zeta \, \partial u / \partial v + \zeta' \, \partial T / \partial s, \quad T - T_w = \zeta_T \, \partial T / \partial v,$$

where the gradients on the right are taken at the wall, in $\partial T/\partial s$ the gradient is in the direction of $u$, and

$$(142) \qquad \zeta = \frac{2-\sigma}{\sigma} \lambda, \quad \zeta' = \frac{3}{4} \frac{\mu}{\rho T}, \quad \zeta_T = \frac{2-\alpha}{\alpha} \frac{2\gamma}{\gamma+1} \frac{\lambda}{Pr}$$

$\sigma$ is the reflection coefficient, i.e., the fraction of diffusely reflected molecules (whose average tangential velocity is zero), and $1 - \sigma$ is the fraction of specularly reflected molecules (whose average tangential velocity is the same as that of the molecules incident on the wall from a layer at a distance $\lambda$). $\alpha$ is known as a thermal accomodation coefficient. There are some disagreements about the form of $\zeta_T$, but for macroscopic calculations these are all absorbed in the use of values of $\alpha$ obtained from macroscopic experiments. In many cases, $\sigma$ is not far from 1 (about 0.8 to 1.0); for air on various surfaces $\alpha$ is in the neighborhood of 0.9. For a perfect gas, $\lambda$ is about $1.26\,\gamma^{\frac{1}{2}}\nu/a$, so $\lambda$ may be eliminated, with

$$\zeta = CM\nu/U, \qquad C = 1.26\gamma^{\frac{1}{2}}(2-\sigma)/\sigma,$$
$$\zeta_T = C_T M\nu/U, \quad C_T = 1.26\gamma^{\frac{1}{2}}[(2-\alpha)/\alpha][2\gamma/(\gamma+1)]Pr^{-1}.$$

$U$ is a reference velocity, and $M$ the Mach number $U/a$. $a$ and $\nu$ are values at the wall. (The appearance of $a$ and $M$ is a convenient way of introducing the temperature, and does not imply that the equations cannot be used when compressibility is not taken into account.)

The flow resulting from the term $\zeta'\partial T/\partial s$ is known as "thermal creep."

Higher-order correction terms to (141) may be found, but probably, like the Burnett equations, will not improve matters. When the Navier-Stokes equations, with the boundary conditions (141), are not adequate because $\lambda/L$ is too large, there is probably no alternative at present to struggling with a molecular treatment and the use of Boltzmann's or an equivalent equation. When $\lambda/L$ is sufficiently large, the régime of "free-molecule" flow is reached, and theoretical calculations become easier, since collisions of molecules which are re-emitted from the surface of a body with molecules in the incident stream are neglected. (See Schaaf and Chambré, *op. cit.*, and references there given.)

Note that (i) the physical order of magnitude of $\lambda/L$ may vary from one part to another of one and the same flow field, (ii) $L$ should be locally defined. It appears that a suitable definition is the distance over which a macroscopic variable, such as a velocity component, varies by an appreciable fraction of itself. (So $\lambda/L$ would be small in a flow which is sufficiently nearly uniform.) If, for example, we consider flow past an obstacle, a linear dimension of the obstacle is not a suitable definition of $L$ in *all* parts of the flow field.

Finally we remark that experimental values for $\mu$, $\rho$, $\nu$, $k$, $\kappa$, and $Pr$ may be found in books of reference, and that for liquids we rely on these entirely, making no attempt here to review the results of theories of the molecular constitution of liquids. In liquids, such as either water or lubricating oils, the viscosity $\mu$ diminishes fairly rapidly as the temperature rises; $\nu$ falls with increase of temperature in water, and rises in a lubricating oil. $Pr$ is not independent of temperature in a liquid; it falls as the temperature rises. There is a very wide range of values for $Pr$ in liquids; for example, its value at about 20°C is between 0.02 and 0.03 for mercury and between 9,000 and 10,000 for a lubricating oil.

### 5.5. Dimensionless groups. Dynamical similarity

Consider first the motion of an incompressible fluid of constant properties. The arguments used are quite general, but to fix ideas consider some definite flow, such as the steady flow of an otherwise unbounded fluid past a fixed obstacle, of a given shape at a given orientation to the stream, the flow being assumed uniform apart from the disturbance due to the obstacle. Body forces such as gravity are eliminated by replacing the actual stresses by their differences from the hydrostatic pressure; in speaking of the force on a solid body immersed in a moving fluid, the static buoyancy force is omitted.

Let $d$ be a representative length and $U$ a representative velocity (for example, the undisturbed velocity in flow past an obstacle) of the system. The system is fixed if $U$, $d$, $\rho$, and $\nu$ are fixed. If $v$ is any velocity component, $p$ a pressure, $\tau$ a shearing stress, then $v/U$, $p/(\rho U^2)$, $\tau/(\rho U^2)$ are non-dimensional; they are functions of $\mathbf{x}/d$, but for the same value of $\mathbf{x}/d$ in all geometrically similar systems their values will depend only on a non-dimensional combination of $U$, $d$, $\rho$, and $\nu$. The only possible non-dimensional combination is the Reynolds number, $R = Ud/\nu$ (or some function of $R$). Thus $v/U$, $p/(\rho U^2)$, $\tau/(\rho U^2)$ are functions of $\mathbf{x}/d$ and $R$ only. If $R$ is the same for two geometrically similar systems, the systems are also dynamically similar. If $F$ is a component of the force on the obstacle due to the stresses over its surface, and $S$

is a representative area associated with the obstacle, then $F/(\tfrac{1}{2}\rho U^2 S)$ is called a force coefficient. If $F$ is the drag force, $D$, in the direction of the undisturbed stream, this coefficient is the drag coefficient $C_D$; if $L$ is the "lift" force at right angles to this direction, the coefficient is the lift coefficient $C_L$. Force coefficients are functions of $R$ only.

If
$$y_i = x_i/d, \ u_i = v_i/U, \ p' = p/(\rho U^2), \ t' = Ut/d,$$
the equations of motion and continuity become

(143) $\quad \partial u_i/\partial t' + u_j \partial u_i/\partial y_j = -\partial p'/\partial y_i + R^{-1} \nabla^2 u_i, \quad \nabla \cdot \mathbf{u} = 0,$

where the operators $\nabla^2$ and $\nabla$ refer to the $y$-coordinates; boundary conditions become independent of $U$ and $d$. (If the problem is one of unsteady flow, the initial conditions are also assumed to become independent of $U$ and $d$ under the stated transformation.) The formula for the stresses becomes

$$p_{ij} = \rho U^2(-p' \delta_{ij} + R^{-1} e'_{ij}),$$

where the $e'_{ij}$ are formed from the $u_i$ and $y_i$ in the same way as $e_{ij}$ from the $v_i$ and $x_i$; the force on any immersed solid in the direction of the axis of $x_i$ is

$$\int p_{\nu i}\, dS = \int \nu_j p_{ji}\, dS,$$

with the integral over the surface of the solid. The statements previously made on the basis of dimensional reasoning follow at once.

For the applications of these results in experiment, and the necessary precautions in applying them, see *Modern Developments*, Section 5.

When the system includes a free surface, or a surface of separation of one fluid from another (such as water from air), as in the study of surface waves, then the actual stresses enter into the boundary conditions. When the body force is gravity, the quantities depend on $g$, as well as on $U$, $d$, $\rho$, and $\nu$, and another nondimensional parameter, the Froude number, equal to $U^2/gd$, enters. If surface tension, $\tau$, is important, a nondimensional parameter

$\tau/(\rho d U^2)$ enters. In meteorological and tidal problems, we must often take into account the angular velocity of rotation of the earth. For bodies with geometrically similar roughnesses a ratio of two lengths may be used as a roughness parameter. Enough has been said to show how the ideas may be adapted to particular circumstances.

For an inviscid gas without heat conduction it is well known that a relevant non-dimensional parameter is the Mach number, $M_0 = U/a_0$, of the undisturbed stream. Clearly, however, if we are to have dynamical similarity for the flow of two different gases in geometrically similar systems, there are necessary conditions on the equations of state, in addition to the condition that the Mach numbers must be the same. For perfect gases, $\gamma$, the ratio of the specific heats, must be the same; in general, nondimensional coefficients will depend on $M_0$ and $\gamma$. (Of course the extent of the variation with $\gamma$ is a matter for further investigation, just as it is for any other non-dimensional parameter.) (See *Modern Developments: High Speed Flow*, Section II 9.) Body forces may usually be neglected; the effect of gravity, for example, if important, will entail dependence on $gh/a^2$, where $h$ is a measure of the vertical extent of the system.

When viscosity and heat conduction are taken into account (and they should be taken into account together) in a compressible fluid, matters are necessarily much more complicated. To a considerable extent this is also the case when compressibility may be neglected, if large temperature differences are imposed; in both cases the physical properties of the fluid are variable. When compressibility may be neglected, and the imposed temperature differences are not large, either the whole process may be considered as isothermal, or the variation of physical properties may be neglected for a first approximation, and then the flow problem and any problem of heat transfer that may be involved may be separated. But in the high speed motion of viscous, heat-conducting gases, the problems of flow and heat transfer are interwoven. Consequently we have to consider here non-dimensional heat transfer coefficients.

We can see immediately that (with body forces neglected) our non-dimensional coefficients will now, in general, depend on $R$ (with some selected value of $\nu$, often, but not always necessarily, the value in the undisturbed stream, when there is one), on $M_0$, on $\gamma$, and on the ratio $Pr$ of the kinematic viscosity to the thermometric conductivity. If $T_0$ is a characteristic temperature, and $T_1 - T_0$ a characteristic temperature difference, then, associated with the variation of physical properties with temperature, we also introduce $(T_1 - T_0)/T_0$ as a non-dimensional parameter.

Now on the simplest view heat is convected, diffused by conduction, and produced by dissipation and by compression of the fluid. Not only do temperature changes change the physical properties of the fluid, but, under the influence of gravity, the density changes produce buoyancy forces which give rise to convection currents: this is known as free convection. Convection of heat by a stream otherwise produced is called forced convection. Unless the velocities in such a stream are small, or the imposed temperature differences very high, free convection may usually be ignored in comparison with the forced convection. But if there is no forced stream, free convection is certainly important. In some situations, with slow forced flows and large temperature differences, both must be taken into account at once.

Let $b$ be the coefficient of expansion of the fluid with temperature, equal to
$$\rho(\partial(1/\rho)/\partial T)_p = -(1/\rho)(\partial \rho/\partial T)_p.$$
For liquids $b$ is usually small (about $1.5 \times 10^{-4}$ per °C for water at 0°C), and it is usually sufficiently accurate to write
$$(\rho - \rho_0)/\rho_0 = -b(T - T_0).$$
For perfect gases, $b = T^{-1}$ from the equation of state.

For free convection, we need a parameter to take the place of the Reynolds number. This is provided by the Grashof number,
$$Gr = d^3 b g (T_1 - T_0)/\nu^2,$$
or the Rayleigh number
$$Ra = Pr\, Gr = d^3 b g (T_1 - T_0)/\nu\kappa.$$

Temperature differences produced by dissipation and compression are usually of order $U^2/c_p$; the ratio of this to the temperature difference $T_1 - T_0$ may be introduced as another parameter. (It is called by Schlichting the Eckert number.) In a perfect gas,

$$U^2/c_p = (\gamma - 1)M_0^2 T_0,$$

so the ratio is the combination, $(\gamma - 1)M_0^2 T_0/(T_1 - T_0)$, of the Mach number, $\gamma$, and $(T_1 - T_0)/T_0$. We may denote it by $Ec$. For a perfect gas, it is not, however, a new additional parameter.

If $Q$ is the quantity of heat transferred in unit time across a surface $S$, the usual heat transfer coefficient is the Nusselt heat transfer coefficient, $Nu$, for which

$$Q = Nu\, kS(T_1 - T_0)/d.$$

($k$ is the thermal conductivity.) For many problems of forced convection, there is an analogy between heat transfer and skin-friction, which is more clearly displayed by use of the Stanton heat-transfer coefficient, $St$, for which

$$Q = St\, \rho c_p SU(T_1 - T_0).$$

The ratio

$$Nu/St = Ud\rho c_p/k = Ud/\kappa = R\, Pr.$$

This ratio is called the Péclet number. For any finite area $S$, the above definitions give the mean Nusselt and Stanton numbers for the area in question. If we let the area shrink up to a point, and take the limit of $Q/S$, we obtain the local values of these numbers. Taking the heat transfer per unit time per unit area at a point as $-k\partial T/\partial v$ (the minus sign has been inserted to give the heat transferred *from* the surface), we see that the local values are given by

$$Nu = \frac{d\partial T/\partial v}{T_1 - T_0}, \quad St = \frac{k\partial T/\partial v}{\rho c_p U(T_1 - T_0)},$$

the value of $\partial T/\partial v$ being taken on the surface. For further discussion, see Squire in *Modern Developments: High Speed Flow*. Chap. 14; Schlichting, *Boundary Layer Theory*, Chap. 14.

CHAPTER 6

# Exact Solutions of the Navier-Stokes Equations

## 6.1. Incompressible fluids of constant properties

What is to be said under this heading is contained in the textbooks (see Lamb's *Hydrodynamics*; Schlichting's *Boundary Layer Theory*; *Modern Developments*) and will be stated briefly. In the main, only those matters are discussed to which reference will be made later.

We put on record that, if $(x, y, z)$ and $(u, v, w)$ are written in place of $x_i$ and $v_i$, the equation of continuity is

$$(144) \qquad \frac{\partial u}{\partial x} + \frac{\partial v}{\partial y} + \frac{\partial w}{\partial z} = 0,$$

and the equations of motion are

$$(145) \qquad \frac{\partial u}{\partial t} + u\frac{\partial u}{\partial x} + v\frac{\partial u}{\partial y} + w\frac{\partial u}{\partial z} = -\frac{1}{\rho}\frac{\partial p}{\partial x} + \nu\nabla^2 u,$$

and two similar equations.

For a two-dimensional motion, if $\psi$ is the stream function (with $u = \psi_y$, $v = -\psi_x$) the equation for $\psi$ is

$$(146) \qquad \frac{\partial}{\partial t}(\nabla^2\psi) - \frac{\partial(\psi, \nabla^2\psi)}{\partial(x, y)} = \nu\nabla^4\psi.$$

(For the equation for Stokes's stream function for motion symmetrical about an axis, and transformations to general orthogonal coordinates, see *Modern Developments*, Chap. 3.)

There are a number of simple, but physically important, solutions for which the non-linear terms in the equations of motion are identically zero. First consider steady motions. The most important is Poiseuille flow — steady, laminar flow through a circular tube under the influence of a pressure gradient, sufficiently far from the ends of the tube for the velocity distribution across a section to be the same at all sections. The velocity distribution is then parabolic; if $P$ is the pressure gradient, $a$ the radius of the tube, and $r$ distance from the axis of the tube, the velocity $u = P(a^2 - r^2)/4\mu$. The average velocity, $u_m$, over a cross-section is half the maximum, and each is proportional to the pressure gradient. The tangential stress, $\tau_0$, at the wall is $\frac{1}{2}Pa = 4\mu u_m/a$; if $R$ is the Reynolds number $u_m d/\nu$, where $d$ is the tube diameter, the skin-friction coefficient $c_f$, defined as $\tau_0/(\frac{1}{2}\rho u_m^2)$, is $16/R$.

If the Reynolds number is increased beyond a certain value, the motion ceases to be of the rectilinear type considered here, and becomes turbulent.

$P$ is the gradient of the difference of the pressure from the hydrostatic pressure. In a tube inclined to the horizontal, such a gradient is provided by gravity.

The above solution is for zero slip at the wall. With slip (equation 141), $u = P(a^2 + 2a\zeta - r^2)/4\mu$. If squares of $\zeta$ are neglected, the results are the same as for flow through a tube of radius $a + \zeta$.

For simple shearing flow (Couette flow) between the two parallel planes $y = 0$ and $y = h$, with the former at rest and the second moving with velocity $U$ parallel to the axis of $x$, the steady state solution is (without slip) simply $u/U = y/h$ (with $v = w = 0$). The arrangement is experimentally unrealizable; it may be approximated by the flow between two coaxial circular cylinders; if end effects may be neglected, so that the flow is the same in any section normal to the axis, the steady flow between rotating coaxial cylinders is easily found.

Problems of unsteady motion, with the quadratic terms zero, are essentially problems in the diffusion of vorticity, for which the same mathematical methods may be used as for other linear diffusion problems, as in the solution of the heat-conduction equation. Note that in the solutions for the vorticity, and for the part of the velocity which does not depend on pressure gradients, $\nu$ and $t$ enter only in the combination $\nu t$.

We refer to the texts for solutions of more complicated steady-state problems, such as two-dimensional flow between nonparallel plane walls, and the flow due to a rotating disk; and also for the results for a two-dimensional flow and an axi-symmetrical flow against a plane athwart the stream. These give the steady flow in a boundary layer at the forward stagnation point of a blunt-nosed cylinder and a blunt-nosed body of revolution, respectively; they are, however, solutions of the complete Navier-Stokes equations. The two-dimensional motion with $y = 0$ as the wall, is an adaptation to a viscous fluid of the potential solution

$$w = \phi + i\psi = \tfrac{1}{2}c(x + iy)^2, \quad \psi = cxy, \quad u = cx, \quad v = -cy$$

in an ideal fluid. In the solution $y = 0$ remains a streamline, and as $y \to \infty$, $u/cx$ and $-v/cy \to 1$; in the viscous solution there is no slip on $y = 0$, i.e., $u = 0$ on $y = 0$. The axi-symmetrical solution is a similar adaptation of the potential solution $\psi = -cr^2x$ (which satisfies $D^2\psi = 0$, Eq. (39)). The notation of equation (39) has been used here and in that notation the plate is now $x = 0$.

For the two-dimensional flow, we set

$$\eta = (c/\nu)^{\tfrac{1}{2}} y, \quad \psi = (\nu c)^{\tfrac{1}{2}} x f(\eta),$$

so

$$u = cx f'(\eta), \quad v = -(\nu c)^{\tfrac{1}{2}} f(\eta).$$

The equation for $f$ is

(147) $$f''' + f f'' - f'^2 + 1 = 0,$$

with the boundary conditions $f(0) = f'(0) = 0$, $f'(\eta) \to 1$ as $\eta \to \infty$. The pressure is given by

$$p/\rho = \text{constant} - \tfrac{1}{2}(c^2 x^2 + v^2) + \nu \partial v/\partial y.$$

It is easily checked that these equations provide an exact solution of the Navier-Stokes equations.

For the axi-symmetrical case, $v_r = v_x = 0$ on $x = 0$, and

$$v_r/cr \to 1, \quad -v_x/2cx \to 1$$

as $x \to \infty$. With

$$rv_r = -\partial\psi/\partial x, \quad rv_x = \partial\psi/\partial r,$$

as in (37), we set

so
$$\eta = (c/\nu)^{\frac{1}{2}}x, \quad \psi = -(c\nu)^{\frac{1}{2}}r^2 f(\eta),$$

$$v_r = crf'(\eta), \quad v_x = -2(c\nu)^{\frac{1}{2}}f(\eta).$$

The equation for $f$ is now

(148)  $\qquad f''' + 2ff'' - f'^2 + 1 = 0,$

with the boundary conditions $f(0) = f'(0) = 0$, $f'(\infty) = 1$. The equation for the pressure is

$$p/\rho = \text{constant} - \tfrac{1}{2}(c^2 r^2 + v_x^2) + \nu \partial v_x/\partial x.$$

These solutions are further discussed in Section 7(d).

In addition to exact solutions, approximations are to be considered for large and for small Reynolds numbers. Approximations for large Reynolds numbers (boundary-layer theory) will be discussed later. At low Reynolds numbers (the approximations of Stokes and Oseen) reference may be made to the cited texts, to papers by Lagerstrom and Kaplun, and by Kaplun, at the 9th International Congress of Applied Mechanics, Brussels, 1956, and to Proudman and Pearson, *J. Fluid Mechs.* 2 (1957), pp. 237–262. (The papers of Lagerstrom and Kaplun, and Kaplun, have been published in the *J. Math. and Mechs.* 6 (1957), pp. 585 *et seq.*

### 6.2. Compressible fluids

Probably the first discussion of exact solutions of the Navier-Stokes equations for compressible fluids (other than consideration of flow through a shock wave) was given by C. R. Illingworth, *Proc. Cambridge Phil. Soc.* **46** (1950) pp. 469–78. The simplest, and perhaps the most instructive, example considered by Illingworth was the generalization to compressible fluids of the simple Couette shearing flow. A solution may be obtained under very general conditions. Illingworth took the Prandtl number as constant, but it has lately been pointed out by A. J. A. Morgan (*J. Ae. Sci.* **24** (1957) pp. 315, 316) that even this is not necessary, the analysis being easily generalized to allow variations of $Pr$. The results of these calculations repay careful study. An account of this problem, with new applications, has been given by Liepmann and Roshko in their book, *Elements of Gasdynamics*, Wiley, New York, 1957. The fundamental mathematics is straightforward and easy. As before, the flow is assumed to take place between the two parallel planes, $y = 0$, and $y = h$, with the former at rest and the latter moving with velocity $U$ parallel to the axis of $x$. Then $v = 0$, $w = 0$ at both plates. We are here considering steady motion only, and we may begin by assuming that all quantities are functions of $y$ only. From the equation of continuity, and the third equation of momentum, together with the boundary conditions, it then follows that $v$ and $w$ are identically zero. We either neglect the effect of gravity, or assume that the walls are horizontal. In either case the first equation of momentum and the energy equation are

$$\frac{d}{dy}\left(\mu \frac{du}{dy}\right) = 0, \quad \frac{d}{dy}\left(k \frac{dT}{dy}\right) + \mu \left(\frac{du}{dy}\right)^2 = 0.$$

These have the immediate first integrals

$$\mu \frac{du}{dy} = C_1, \quad k \frac{dT}{dy} + C_1 u = C_2.$$

Hence

$$\frac{k}{\mu} \frac{dT}{du} + u = \frac{C_2}{C_1}.$$

In general, $k$ and $\mu$ are functions of $T$. Let

$$\int \frac{k}{\mu} dT = F(T).$$

Since $k/\mu = c_p/Pr$, and $c_p dT = dI$ for a perfect gas, $F(T) = I/Pr$ for a perfect gas with a constant Prandtl number. (This does not assume constant specific heats.) In any case, there is an integral

$$F = C_3 + \frac{C_2}{C_1} u - \tfrac{1}{2} u^2,$$

whence $T$ is found as a function of $u$. Hence $\mu$ (and also $k$) are found as functions of $u$. Then $y$ is found as a function of $u$ from

$$C_1 y = \int_0 \mu \, du,$$

where no slip has been assumed at $y = 0$ and the boundary condition $u = 0$ at $y = 0$ has been applied. Three other boundary conditions determine $C_1$, $C_2$, $C_3$. One of these is (without slip) $u = U$ at $y = h$. There are two boundary conditions on the temperature; at each wall either the temperature is given, or, if the wall is insulated, $dT/dy = 0$.

If gravity is neglected, the pressure is constant. If the walls are horizontal, the second equation of momentum gives

$$dp/dy + g\rho = 0,$$

i.e.,

$$C_1 \, dp/du = -g\mu\rho.$$

$\mu$ is a known function of $u$, and from the equation of state $\rho$ is a known function of $p$ and $T$, and therefore of $p$ and $u$. For a perfect gas, $\rho = p/\Re T$ and

$$\frac{C_1}{p}\frac{dp}{du} = -\frac{g}{\Re}\frac{\mu}{T}.$$

Hence $p$ is also found as a function of $u$. The variation of pressure is small if $gh$ is small compared with the square of the velocity of sound at the moving wall.

For further details, see the works cited. Illingworth gives other examples of generalizations, to compressible fluids, of solutions for incompressible fluids.

Mention must also be made here of the discussion of flow through a shock wave. Discussions of solutions are to be found in *Modern Developments*: *High Speed Flow*, Chap. 4 (C. R. Illingworth) and in Liepmann and Roshko's book. Reference should also be made to Lighthill's paper on Viscosity Effects in Sound Waves of Finite Amplitude in the G. I. *Taylor Anniversary Volume*, previously cited, to Gilbarg and Paolucci, *J. Rat. Mech. Anal.* **2** (1953) pp. 617–642, and to Grad, *Comm. Pure Appl. Math.* **5** (1952), pp. 257–300.

Steady flow through a stationary normal shock is considered; there is only one component of velocity, $u$, and $u$ and the thermodynamic variables are functions of $x$ only. The equations of continuity, momentum, and energy are, therefore,

$$\frac{d}{dx}(\rho u) = 0, \quad \rho u \frac{du}{dx} = \frac{d}{dx}(\rho u^2) = \frac{d}{dx}\left(-p + \mu' \frac{du}{dx}\right),$$
(149)
$$\rho u \frac{d}{dx}(I + \tfrac{1}{2}u^2) = \frac{d}{dx}\left(k \frac{dT}{dx}\right) + \frac{d}{dx}\left(\mu' u \frac{du}{dx}\right),$$

where

(150) $$\mu' = \tfrac{4}{3}\mu + \beta.$$

These equations have immediate first integrals

$$\rho u = m, \quad p + mu - \mu' \, du/dx = C_1$$
(151)
$$m(I + \tfrac{1}{2}u^2) - k\,dT/dx - \mu' u \, du/dx = C_2.$$

Any two sets of values ($\rho_1$, $u_1$, $p_1$, $I_1$, $T_1$, and the same with subscript 2) for which $\mu' du/dx$ and $k\,dT/dx$ vanish, are immediately seen to be connected by the Rankine-Hugoniot relations for a normal shock, as previously obtained. These relations connect the initial state upstream (subscript 1) with the final state downstream (subscript 2). Details in the transition zone are to be found by the integration of (151), with use of the equation of state. Note that, if the gas is a perfect gas, so that $dI = c_p dT$, and if

$$\frac{\mu' c_p}{k} = \tfrac{4}{3}\mu \frac{c_p}{k}\left(1 + \frac{3\beta}{4\mu}\right) = \tfrac{4}{3} Pr\left(1 + \frac{3\beta}{4\mu}\right) = 1,$$

then

$$k\frac{dT}{dx} + \mu' \frac{d}{dx}(\tfrac{1}{2}u^2) = \frac{k}{c_p}\frac{d}{dx}[I + \tfrac{1}{2}u^2] = m[I + \tfrac{1}{2}u^2] - C_2,$$

and the only permissible solution is $I + \tfrac{1}{2}u^2 = $ constant. The equations may then be fairly easily integrated. Unfortunately, the solution cannot be regarded as physically realistic. For although $Pr = \tfrac{3}{4}$ may be regarded as a fairly satisfactory value for a diatomic gas, it is unrealistic to take $\beta = 0$ for such a gas; for a monatomic gas, for which $\beta = 0$, $\tfrac{3}{4}$ is too high a value for $Pr$, which is about $\tfrac{2}{3}$. For further details of the integration, see the works cited.

The energy equation may also be written

$$m\frac{dS}{dx} = \frac{\mu'}{T}\left(\frac{du}{dx}\right)^2 + \frac{1}{T}\frac{d}{dx}\left(k\frac{dT}{dx}\right)$$

$$= \frac{\mu'}{T}\left(\frac{du}{dx}\right)^2 + \frac{k}{T^2}\left(\frac{dT}{dx}\right)^2 + \frac{d}{dx}\left(\frac{k}{T}\frac{dT}{dx}\right).$$

The increase of entropy through the shock is therefore given by

(152) $$m(S_2 - S_1) = \int_1^2 \left\{\frac{\mu'}{T}\left(\frac{du}{dx}\right)^2 + \frac{k}{T^2}\left(\frac{dT}{dx}\right)^2\right\} dx.$$

Lighthill (*op. cit.*) has used this expression to find a lower bound to the shock "thickness," which he suggests may be a good approximation to its value.

For comparisons of the calculated structure of shock waves with experimental results in helium, air, and argon, see F. S. Sherman, *N. A. C. A. Tech. Note* 3298 (1955) and L. Talbot and F. S. Sherman, *Univ. Calif. (Berkeley) Inst. Eng. Res., Tech. Rep.* HE-150-137 (1956). See also the references to Greene, Cowan, and Hornig, given by Sherman. The evidence is that a continuum theory, based on the Navier-Stokes equations with use of a bulk viscosity for polyatomic gases, and a satisfactory variation of viscosity with temperature, is at least as adequate in predicting the structure of shocks which are not too strong as any other theory whose consequences have so far been worked out. The most serious discrepancy was on the upstream side of a shock at $M_1 = 3.7$ in air. The temperature being lower on the upstream side, the mean free path is larger than on the downstream side, so it may be that the Navier-Stokes equations must fail. But somewhat different variations of viscosity with temperature, and particularly an increase in the values of the bulk viscosity upstream and a decrease in its variation with temperature, may improve the agreement with experiment without adversely affecting that for the weaker shocks. As previously mentioned, very little is known with certainty about the bulk viscosity and its variation with temperature; for fairly strong shocks in air, these matters are clearly important.

CHAPTER 7

# Boundary-Layer Theory for Incompressible Fluids

In this section attention is restricted to fluids of constant properties. Attention is also restricted to (1) certain physical and mathematical fundamentals; (2) some questions that seem promising for further mathematical elucidation; (3) material which is required for discussions in subsequent sections. Discussion of the following matters must unfortunately be omitted: (i) three-dimensional boundary layers; (ii) boundary-layer control, and the effects of suction; (iii) approximate and numerical methods of solution of the boundary-layer equations; (iv) boundary layers in unsteady flow; (v) the application of the theory to wakes and jets. Accounts of some of these matters may be found in the literature previously cited; for (i), in addition to the account in Schlichting's book, reference may be made, to begin with, to articles by W. R. Sears in *Applied Mechanics Reviews* **7** (1954), pp. 281–285, and by F. K. Moore in *Advances in Applied Mechanics*, Vol. 4, Academic Press, 1956, pp. 159–228. (Reference may now also be made to *Laminar Boundary Layers* (ed. by L. Rosenhead), Oxford University Press, 1963.)

Consider a steady two-dimensional flow, past a cylindrical obstacle, of a fluid of small viscosity, with no slip at the solid surface. If the viscosity is completely neglected, and the flow calculated as a potential flow without discontinuities, it is impossible to annul the slip. The condition of zero slip can be satisfied only by assuming a vortex sheet coincident with the solid surface, through which the relative tangential velocity falls to zero. In a fluid of small viscosity, the vorticity diffuses, and is convected with the stream. The result is to produce a layer of large vorticity adjacent to the solid surface. Through this layer the tangential velocity falls from its value in the main stream to zero at the surface, and in this layer the rate-of-shear, as well as the vorticity, is large. Suppose, for example, that a flat plate (theoretically of zero thickness) is placed in a stream of velocity

$U$, parallel to the stream; then the time required for the fluid to traverse a distance $x$ along the plate is of order $x/U$, and in this time the vorticity has become appreciable in a layer whose thickness is of order $(\nu x/U)^{\frac{1}{2}}$. This is the order of the thickness of the boundary layer. Outside this layer, diffusion will produce only a vorticity which is exponentially small.

Also, if a solid body is started from rest, the initial motion is irrotational, without circulation, and with a vortex sheet coincident with the solid surface. This vorticity immediately begins to diffuse, and after a short time $t$ there is a boundary layer whose thickness is of order $(\nu t)^{\frac{1}{2}}$. Initially the diffusion far outweighs the convection (and also the influence of a pressure gradient), and calculations of boundary-layer growth have been made by successive approximation, starting from this consideration. Such calculations are valid for only limited times, but further consideration of boundary layers in unsteady flow is still rather limited. (Compare Sears, *J. Ae, Sci.* **23** (1956), pp. 490–499; see also the discussion there of the steady-state circulation.)

The existence of boundary layers, as layers of large vorticity and large rate-of-shear, is well substantiated by experiment.

### 7.1. Two-dimensional boundary layer equations

We now begin to translate these ideas into mathematical form. (Compare *Modern Developments*, Section **44** *et seq.*) We consider a two-dimensional motion, in the $(x, y)$ plane, and remark that the velocity $u$ parallel to a wall will, at a given distance from the wall, vary along the wall as a result of the retarding action of the tangential stresses, and continuity requires that there should be a velocity component $v$ perpendicular to the wall. To begin with, let the wall be flat, with the axis of $x$ along the wall. The equations of continuity and momentum are the equations (144) and (145), with $w = 0$, and all quantities independent of $z$. The viscous terms must, in a boundary layer, be of the same order of magnitude for small $\nu$ as the inertia terms, for if the viscous terms are neglected, the equations relate to the theory of inviscid fluids; whereas if the inertia terms are omitted, after the manner of Stokes (or partly taken into account, after the manner of Oseen) we have the theory of approximations for small Reynolds numbers. And in the boundary layer, where the vorticity and rate-of-shear are large, $\partial u/\partial y$ is large; $u$ changes from zero at the wall to a value sensibly equal to that in the main stream in a length of the order of a boundary-layer thickness. Hence $\nu(\partial^2 u/\partial y^2)$ must be of the same order as the inertia terms, which we take as $O(1)$. The boundary-layer thickness may therefore be said to be of order $\nu^{\frac{1}{2}}$. From the equation of continuity, $v = O(\nu^{\frac{1}{2}})$, and the second equation of momentum gives $\partial p/\partial y = O(\nu^{\frac{1}{2}})$. The change of pressure, along a normal to a wall, through the boundary layer is $O(\nu)$, and is neglected. Hence $p$ and $\partial p/\partial x$ are taken as independent of $y$, and the boundary-layer equations become

$$\partial u/\partial t + u\partial u/\partial x + v\partial u/\partial y = -\rho^{-1}\partial p/\partial x + \nu\partial^2 u/\partial y^2,$$
(153) $$\partial u/\partial x + \partial v/\partial y = 0.$$

We may, as before, set $u = \psi_y$, $v = -\psi_x$. The equation for $\psi$ now becomes a third-order parabolic equation, in place of the previous fourth-order elliptic equation (146). On the other hand, $\partial p/\partial x$, which has, throughout the boundary layer, the value just outside it, in the main stream, must be supposed known.

Flow along a curved wall may be similarly considered by using a coordinate net consisting of the wall and parallel curves, together with normals to the wall. If $\kappa$ is the curvature of the wall, $\nu^{\frac{1}{2}}\kappa$ and $\nu\partial\kappa/\partial s$ are assumed small, where $\partial\kappa/\partial s$ is the rate of change of $\kappa$ along the wall. $(\partial p/\partial y)/\rho$ is now equal to $\kappa u^2$, since the pressure gradient must balance the centrifugal force, but the change of pressure through the boundary layer, along a normal to the wall, may still be neglected, being $O(\nu^{\frac{1}{2}})$. With these assumptions, the equations (153) remain unaltered if $x$ is distance along the (curved) wall and $y$ distance normal to it.

To express the theory in non-dimensional form we use the symbolism of equations (143), but with the non-dimensional coordinates represented by $(x_1, y_1)$ and the non-dimensional velocity components by $(u_1, v_1)$, and the primes dropped from the non-dimensional time $t'$ and the non-dimensional pressure $p'$.

We note that, according to our previous results, as $R \to \infty$ ($\nu \to 0$), $\partial u_1/\partial y_1 \to \infty$ with $R^{\frac{1}{2}}$, and $v_1 \to 0$ with $R^{-\frac{1}{2}}$. We therefore use the following manner of taking the limit of the equations as $R \to \infty$. Write $y' = R^{\frac{1}{2}} y_1$, $v' = R^{\frac{1}{2}} v_1$, leaving $x_1$, $u_1$, $p$, $t$ unaltered. After these transformations are made, assume that all derivatives that occur remain finite, and let $R \to \infty$. The result is

$$\partial u_1/\partial t + u_1 \partial u_1/\partial x_1 + v' \partial u_1/\partial y' = -\partial p/\partial x_1 + \partial^2 u_1/\partial y'^2$$
$$\partial p/\partial y' = 0$$
(154) $$\partial u_1/\partial x_1 + \partial v'/\partial y' = 0.$$

The same equations are obtained for a curved wall as $R \to \infty$, if $\kappa$, $\partial \kappa/\partial s$ are everywhere finite.

Note that (with dimensional quantities), the vorticity in the boundary layer is nearly $-\partial u/\partial y$, and the shearing stress nearly $\mu \partial u/\partial y$. The rate of dissipation of energy per unit time per unit volume is approximately $\mu (\partial u/\partial y)^2$; this has a finite limit as $R \to \infty$, but because the thickness of the boundary layer is of order $R^{-\frac{1}{2}}$, the total rate of dissipation of energy per unit time across the layer will $\to 0$ like $R^{-\frac{1}{2}}$ as $R \to \infty$. In steady motion the skin-friction will therefore also $\to 0$ like $R^{-\frac{1}{2}}$.

We leave aside for the present the specification of initial boundary conditions at a given value of $x$, and, for unsteady motions, at a given value of $t$. At a solid impermeable wall ($y = 0$) at rest, $u = 0$ and $v = 0$. Outside the boundary layer the vorticity, and also the difference between $u$ and its value $U_e$ in the main stream, tend to zero, and, in general, become exponentially small. The coordinate $y$ does not tend to infinity in the boundary layer, and the condition of matching with the main inviscid flow outside the boundary layer is best stated in terms of $y'$ if there is a geometrical representative length of the system. In any case we may state the condition in terms of $Y = y/\nu^{\frac{1}{2}}$. Clearly $Y \to \infty$ as $\nu \to 0$ for any non-zero $y$, no matter how small. We are concerned with an asymptotic approximation as $\nu \to 0$ to a solution of the full Navier-Stokes equations, so without further discussion here simply state that a further boundary condition is $u/U_e - 1 \to 0$ (in general, exponentially) as $y' \to \infty$ or $Y \to \infty$.

Since $u/U_e - 1$ does not become identically zero, but only exponentially small, outside the boundary layer, there is no satisfactory mathematical definition of the thickness of the layer. We may, however, define a definite thickness in the following way. Because the flow is retarded near the wall, and continuity requires a constant flux between a given streamline and the wall, a streamline at a large distance from the wall will be displaced through a distance $\delta_1$. This distance is called the displacement thickness of the boundary layer, and, with $d$ as a representative length of the system as on p. 101,

$$(155) \quad \delta_1 = \nu^{\frac{1}{2}} \int_0^\infty \left(1 - \frac{u}{U_e}\right) dY = R^{-\frac{1}{2}} d \int_0^\infty \left(1 - \frac{u}{U_e}\right) dy'.$$

Since the pressure is nearly constant across every section of the boundary layer, an adverse pressure gradient in the main stream is operative right through the layer. A thin stratum of fluid adjacent to the wall is retarded by friction at the wall, pulled forward by the stream above it through the action of viscosity, and retarded by the adverse pressure gradient. With a favorable pressure gradient, it would continue its forward motion; but, because of the action of the wall, its forward velocity, energy, and momentum are small, and may be insufficient for it to force its way for very long against an adverse pressure gradient. It is then brought to rest, and farther on, next to the wall, a slow back-flow in the direction of the pressure gradient may set in. The forward stream then leaves the surface. A thin layer of vorticity thus leaves the wall and makes its appearance in the interior of the fluid, where it corresponds, for a real fluid, to the vortex sheets previously considered in an inviscid one. The separation of the forward flow from the surface begins where $(\partial u/\partial y)_{y=0} = 0$. The position of this point can be calculated if the pressure distribution outside the boundary layer is given; there is at present no satisfactory method of calculating this pressure gradient when separation occurs. If the pressure distribution were to stay unchanged with change of Reynolds number, the position of the point of separation would also be unchanged according to the boundary-layer approximation; in fact, that theory gives the limiting position as $R \to \infty$, when a vortex sheet leaves the surface. (We may here interpolate that for the calculations made for unsteady boundary layers in the motion of a solid body started from rest, when separation begins at a point on the surface after a time from the commencement of the motion, the position and time are also independent of the Reynolds number.)

If there is no separation, for a first approximation the pressure may be taken to be that given by potential-flow theory. This is certainly the case for flow past a semi-infinite flat plate along the stream. It is also the case (although the approximation is not so accurate) for flow past a finite flat plate, or past a symmetrical cylinder of airfoil section with a cusped trailing edge, at zero incidence, and carefully "streamlined" towards the rear, to avoid flow separation. In such cases the boundary layers on the two sides join together at the trailing edge to form a rather narrow wake; for a first approximation to the pressure distribution the effect of this narrow wake, as well as of the boundary layers, is neglected. But when separation occurs, there is a broad wake, and we do not know how to calculate the pressure distribution. We may (and do) use experimental results for the pressure, and use boundary-layer theory to calculate the skin-friction, but we cannot construct a rational and complete theory of the boundary layer unless we know the limit of the solution of the Navier-Stokes equations as $R \to \infty$. This is still an unsettled question. Experimentally, the limiting motions are turbulent, and we do not wish to construct the limit of steady "laminar" motions for their own sake, but to begin a theory of asymptotic solutions of the Navier-Stokes equations for sufficiently large $R$. We return to this matter later.

We conclude this section by considering the equation (146) for $\psi$. We express this equation in non-dimensional form, using $\psi_1$ for the non-dimensional stream function, and $x_1$, $y_1$ for non-dimensional coordinates, with $\nu$ in (146) replaced by $R^{-1}$. We consider only steady motion. Then, $R^{-1}$ being small, the term $(R^{-1})\nabla^4 \psi_1$ is a perturbation of the non-viscous equation, containing a small coefficient but higher derivatives than occur in the unperturbed equation, and enabling more boundary conditions to be satisfied (no slip at a solid surface). We are therefore concerned with the theory of singular perturbations of differential equations. Here we have a linear perturbation of a non-linear equation. The non-linear part, by itself, may be integrated at once to provide the simple linear Laplace equation when viscosity is neglected, but the non-linearity makes itself strongly felt as soon as the viscous perturbation is considered. For the unperturbed equation, $\nabla^2 \psi$ is a function of $\psi$, and we have the classical theorem that, with no vorticity in the undisturbed flow at infinity upstream, $\nabla^2 \psi = 0$ in any region occupied by streamlines coming from infinity upstream. With viscosity included, this will remain correct within an exponentially small error for streamlines which have not entered a region where the perturbation terms are important.

According to the ideas of boundary-layer theory, we now set

$$y' = R^{\frac{1}{2}} y_1, \quad \chi = R^{\frac{1}{2}} \psi_1,$$

so

$$u_1 = u/U = \partial \chi / \partial y', \quad v' = R^{\frac{1}{2}} v/U = -\partial \chi / \partial x_1,$$

and we assume that after this transformation is made the derivatives that occur are bounded as $R \to \infty$, so the boundary-layer equation is simply obtained by putting $R^{-1} = 0$ in the resulting equation, which is (temporarily dropping the subscript 1 on $x$ and the prime on $y$)

(156) $$\frac{\partial}{\partial y} \{ \chi_{yyy} + \chi_x \chi_{yy} - \chi_y \chi_{xy} \} + R^{-1} \{ 2 \chi_{xxyy} + \chi_x \chi_{xxy} - \chi_y \chi_{xxx} \} + R^{-2} \chi_{xxxx} = 0$$

The boundary-layer equation is therefore

(157) $$\chi_{yyy} + \chi_x \chi_{yy} - \chi_y \chi_{xy} = P(x),$$

where

(158) $$P(x) = - \lim_{y \to \infty} \chi_y \chi_{xy} = - \lim u_1 u_{1x},$$

since $\chi_{yyy}$ and $\chi_{yy}$ must certainly $\to 0$ as $y \to \infty$. Hence, from Bernoulli's equation, the function $P(x)$, arising from the integration with respect to $y$, is simply the pressure gradient in the main stream outside the boundary layer.

### 7.2. Motion symmetrical about an axis

Consider the flow of a fluid of small viscosity past a body of revolution, with its axis of symmetry parallel to the undisturbed stream. Then, if $x$ is distance from the forward stagnation point, measured along a curve of intersection of the body surface with a meridian plane, $y$ is normal distance from the surface, and $u$ and $v$ are the corresponding components of velocity, it may be shown that we arrive again at the first of equations (153) — the equation of momentum — but the equation of continuity is now

$$\partial(ru)/\partial x + \partial(rv)/\partial y = 0,$$

where $r$ is distance from the axis of revolution; it may be replaced by $r_0$, the distance from the axis of revolution of the point on the same normal but on the body surface, so that $r_0$ is a given function of $x$.

The conditions on the curvature of a meridian section are exactly the same as those on the curvature of a section in two-dimensional motion. There is, however, a condition on the transverse curvature, that the boundary-layer thickness or, more precisely, the displacement thickness should be small compared with $r_0$. Near the forward stagnation point on a blunt-nosed body of revolution, it may be shown that the equations above may be allowed to remain. However, if the boundary-layer thickness grows along the surface of the body (as on a cylinder without pressure gradient, for example) the theory ceases to have validity when this thickness becomes too large. This will happen in flow along a long slender cylinder, or any long slender body of revolution. This transverse curvature effect has been considered by a number of authors. A few references are given below; others will be found in the papers cited. Note that some of the papers listed deal also with the flow of compressible fluids; we shall not return to this question when we consider boundary layers in compressible fluids.

See R. A. Seban and R. Bond, *J. Ae. Sci.* **18** (1951), pp. 671–675 (see also H. R. Kelly, *J. Ae. Sci.* **21** (1954), p. 634); M. B. Glauert and M. J. Lighthill, *Proc. Roy. Soc. (London)* **A230** (1955), pp. 188–203; K. Stewartson, *Quart. Appl. Math.* **13** (1955), pp. 113–122; R. F. Probstein and D. Elliott, *J. Ae. Sci.* **23** (1956), pp. 208–224, 236; S. I. Pai, *J. Ae. Sci.* **23** (1956), pp. 795, 796.

Mangler's transformation of an axisymmetrical problem into

a two-dimensional one applies to the boundary-layer flow of compressible as well as incompressible fluids, and will be set out in Section 9.2.

### 7.3. Flow along a flat plate

Consider the steady flow, with undisturbed velocity $U$, of an otherwise unbounded fluid of small viscosity past a semi-infinite flat plate parallel to the stream, lying on $y = 0$, $x \geq 0$. For the first boundary-layer approximation, there is no pressure gradient in this problem, $u$ is symmetrical, and $\psi$ and $v$ are antisymmetrical, about $y = 0$, and we solve for $y \geq 0$. There is no geometrical representative length of the system. Inside the boundary layer we have seen (p. 121, where $R \propto U/\nu$) that $u/U$ is a function of $y' = (U/\nu)^{\frac{1}{2}}y$. Since no scale length can be defined, the velocity distributions for different values of $x$ must be similar, i.e., there is a $g(x)$ such that $u/U$ is a function of $g(x)y/\nu^{\frac{1}{2}}$ only, which requires that $\psi g(x)/(\nu^{\frac{1}{2}}U)$ should be a function of $g(x)y/\nu^{\frac{1}{2}}$ only. Substitution into the equation shows that $g(x)$ is a constant multiple of $x^{-\frac{1}{2}}$. (The same result is obtained by a somewhat different form of argument on p. 135.) The solution now proceeds as follows. With

(159) $$\begin{cases} \eta_0 = \tfrac{1}{2}y'/x^{\frac{1}{2}} = \tfrac{1}{2}(U/\nu x)^{\frac{1}{2}}y, \\ \psi = (U\nu x)^{\frac{1}{2}}f(\eta_0), \quad \text{i.e. } \chi = \psi/(U\nu)^{\frac{1}{2}} = x^{\frac{1}{2}}f(\eta_0), \end{cases}$$

(160) $$\begin{cases} u_1 = u/U = \tfrac{1}{2}f'(\eta_0), \\ v_1 = v/U = \tfrac{1}{2}(\nu/Ux)^{\frac{1}{2}}\{\eta_0 f'(\eta_0) - f(\eta_0)\}, \end{cases}$$

and the equation for $\chi$ (essentially equation (157)) reduces to

(161) $$f''' + ff'' = 0.$$

(The factor $\tfrac{1}{2}$ in $\eta_0$ is inserted to facilitate later comparison with parabolic coordinates (p. 125).)

From the boundary conditions at the plate, $f(0) = f'(0) = 0$. We also impose the boundary condition $f' \to 2$ as $\eta_0 \to \infty$, which ensures that $u/U \to 1$ as $y' \to \infty$ (and also as $x \to 0$ for $y' \neq 0$). We cannot impose any other boundary conditions. If now $F(\eta_0)$

is any solution satisfying the conditions at $\eta_0 = 0$, so also is $cF(c\eta_0)$, where $c$ is any constant. With any selected value of $F''(0)$, the equation may be integrated numerically, series solutions obtained for sufficiently small $\eta_0$, and also asymptotic expressions for large $\eta_0$; the solution satisfying $f' \to 2$ as $\eta_0 \to \infty$ is then found by taking
$$c = \{2/\lim_{\eta_0 \to \infty} F'(\eta_0)\}^{\frac{1}{2}}.$$
It is thus found that
(162) $\qquad f = \alpha \eta_0^2/2! - \alpha^2 \eta_0^5/5! + 11\alpha^3 \eta_0^8/8! - \ldots$
for sufficiently small $\eta_0$, and that as $\eta_0 \to \infty$,
(163) $\qquad f \sim 2\eta_0 - \beta, \; f'' \sim \gamma e^{-(\eta_0 - \frac{1}{2}\beta)^2},$
where $\alpha, \beta, \gamma$ are constants ($\alpha = 1.328, \beta = 1.7208$).

If $\tau$ is the skin-friction,
(164) $\qquad \tau_1 = 4\tau/(\rho U^2) = \alpha(\nu/Ux)^{\frac{1}{2}},$
and the (non-dimensional) displacement thickness is
(165) $\quad \delta_1 = 2\left(\dfrac{\nu x}{U}\right)^{\frac{1}{2}} \int_0^\infty (1 - u_1) \, d\eta_0 = \left(\dfrac{\nu x}{U}\right)^{\frac{1}{2}} \lim_{\eta_0 \to \infty} [2\eta_0 - f(\eta_0)] =$
$\qquad = \beta\left(\dfrac{\nu x}{U}\right)^{\frac{1}{2}}.$

As $\eta_0 \to \infty$, $v \to V$, where
(166) $\qquad\qquad\qquad V = \frac{1}{2}\beta(\nu U/x)^{\frac{1}{2}}.$
($V = U \, d\delta_1/dx$, as it must from the conservation of mass.) The solution is invalid at, and near, the leading edge. (This was to be expected on physical grounds, since the Reynolds number governing the validity of the approximation at any section must, in fact, be the Reynolds number $R_x = Ux/\nu$, based on $x$.) The singularity in $\tau_1$ at $x = 0$ is integrable. But $v$ has a singularity all along the line $x = 0$; to order $\nu^{\frac{1}{2}}$, $V \neq 0$, and the boundary-layer flow does not join smoothly on to the potential flow, for which the stream function is $U\Psi = Uy$ (or on to any potential flow). Although terms of order $\nu$ only have been neglected in (157), there is an error of order $\nu^{\frac{1}{2}}$ in the merging of the boundary-layer solution and the potential-flow solution. The streamlines in the potential flow are deflected through a distance $\delta_1$, and to obtain a

result correct to order $v^{\frac{1}{2}}$, this effect must be taken into account, so as to bring into account the velocity $V$. The potential flow is a solution of the Eulerian equations for an inviscid fluid; to find the correction to the potential flow, the necessary boundary condition is applied at the plate. The stream function of the potential flow is taken to be $U[\Psi_0 + (v/U)^{\frac{1}{2}}\Psi_1]$, where $\Psi_0 = y$ in this case, and the boundary condition on $\Psi_1$ is

$$(v/U)^{\frac{1}{2}}\Psi_1 = -\delta_1 = -\beta(vx/U)^{\frac{1}{2}}$$

at $y = 0$, $x > 0$, with $\Psi_1$ continuous in the fluid. Then $P(x)$ is to be calculated for this new potential flow, and substituted into the equation for $\chi$ (essentially equation (157)); $\chi$ is corrected by taking $\chi = \chi_0 + v^{\frac{1}{2}}\chi_1$, where $\chi_0$ is given by (159).

The potential problem is immediately solved by the use of parabolic coordinates $(\xi_1, \eta_1)$ for which

(167) $\qquad z = x + iy = (\xi_1 + i\eta_1)^2 = \zeta_1^2.$

This transformation will also be used in later sections. We take $\arg z = 0$ on the upper side of the plate, and $\arg z = 2\pi$ on the lower side, so $0 \leq \arg z \leq 2\pi$, with $0 \leq \arg \zeta_1 \leq \pi$; hence the upper half of the $\zeta_1$-plane is the map of the $z$-plane cut along the positive real axis. $\eta_1 = 0$ on the plate; $\xi_1 = 0$ on the negative real axis in the $z$-plane, with $\xi_1 > 0$ on the upper half plane, and $\xi_1 < 0$ on the lower half plane. On the plate, $\xi_1 = \pm x^{\frac{1}{2}}$; near the plate, for positive $y$, and $y/x$ small, $\eta_1$ is approximately $y/(2x^{\frac{1}{2}})$, so $(U/v)^{\frac{1}{2}}\eta_1$ is approximately equal to $\eta_0$, as defined in (159). (This explains why, contrary to the usage in certain recent accounts, the factor 2 has been kept in the denominator in the definition of $\eta_0$.)

The solution for $\Psi_1$ is easily found to be $-\beta\xi_1$; hence for the potential flow, to order $v^{\frac{1}{2}}$,

(168) $\qquad \Psi = y - (v/U)^{\frac{1}{2}}\beta\xi_1 = y - \beta(v/2U)^{\frac{1}{2}}(r + x)^{\frac{1}{2}}$

(in the upper half plane), where $r = (x^2 + y^2)^{\frac{1}{2}}$. (The solution has a singularity at the leading edge.)

From (168), $P(x) = 0$ on the plate to order $v^{\frac{1}{2}}$, so $\chi_1 = 0$. Away from $x = 0$ the boundary-layer solution is correct to order $v^{\frac{1}{2}}$, and only the potential flow needed correction. (This is the result for a semi-infinite plate only. For a plate of finite length, $d$, the wake produces an effect of order $(Ud/\nu)^{-\frac{1}{2}}$ on the pressure gradient and the boundary-layer solution. An approximate calculation of this effect has been made by Kuo, *J. Math. Phys.* **32** (1952), pp. 83–100. Kuo's approximation is simply to take $\delta_1$ constant, $V$ zero, for $x/d > 1$ on $y = 0$ in calculating $\Psi_1$.)

A more satisfactory presentation of the theory of boundary-layer flow past a semi-infinite flat plate is obtained by using parabolic coordinates throughout. The equation for $\psi_1 = \psi/U$ is transformed to parabolic coordinates. It is

(169a)
$$\frac{\nu}{U}\left\{(\xi_1^2+\eta_1^2)\left(\frac{\partial^2 \Delta}{\partial \xi_1^2}+\frac{\partial^2 \Delta}{\partial \eta_1^2}\right)+4\left(\Delta-\xi_1\frac{\partial \Delta}{\partial \xi_1}-\eta_1\frac{\partial \Delta}{\partial \eta_1}\right)\right\}+$$
$$+(\xi_1^2+\eta_1^2)\frac{\partial(\psi_1,\Delta)}{\partial(\xi_1,\eta_1)}+2\left(\xi_1\frac{\partial \psi_1}{\partial \eta_1}-\eta_1\frac{\partial \psi_1}{\partial \xi_1}\right)\Delta=0,$$

where
(169b) $$\Delta = \partial^2\psi_1/\partial\xi_1^2 + \partial^2\psi_1/\partial\eta_1^2.$$

The substitutions
(170) $$\eta = (U/\nu)^{\frac{1}{2}}\eta_1, \quad \chi = (U/\nu)^{\frac{1}{2}}\psi_1 = \psi/(U\nu)^{\frac{1}{2}},$$
are then made, and all derivatives of $\chi$ with respect to $\eta$ and $\xi_1$ are assumed bounded as $\nu \to 0$. The boundary-layer equation is now obtained by writing $\nu = 0$ in the equation for $\chi$ in terms of $\xi_1$ and $\eta$. Note that the leading edge is now inside the boundary layer, and that the potential flow is approached everywhere when $\eta \to \infty$. The boundary-layer equation is found to be

(171) $$\xi_1^2\left(\frac{\partial^4\chi}{\partial\eta^4}+\frac{\partial\chi}{\partial\xi_1}\frac{\partial^3\chi}{\partial\eta^3}-\frac{\partial\chi}{\partial\eta}\frac{\partial^3\chi}{\partial\xi_1\partial\eta^2}\right)+2\xi_1\frac{\partial\chi}{\partial\eta}\frac{\partial^2\chi}{\partial\eta^2}=0,$$

with an immediate first integral

(172) $$\xi_1\left(\frac{\partial^3\chi}{\partial\eta^3}+\frac{\partial\chi}{\partial\xi_1}\frac{\partial^2\chi}{\partial\eta^2}-\frac{\partial\chi}{\partial\eta}\frac{\partial^2\chi}{\partial\xi_1\partial\eta}\right)+\left(\frac{\partial\chi}{\partial\eta}\right)^2 = P(\xi_1)$$

For the present problem, we take $P(\xi_1) = 0$, and find that there is a solution
(173) $$\chi = \xi_1 f(\eta),$$
where $f(\eta)$ is the function defined in (161)–(163). This function is called the Blasius function. The velocity is zero at the plate. As regards conditions for large $\eta$, we see that
(174) $$\chi \sim 2n\xi_1 - \beta\xi_1, \quad \psi_1 \sim y - \beta(\nu/U)^{\frac{1}{2}}\xi_1,$$

so in parabolic coordinates the external potential flow is obtained as the asymptotic value of the boundary-layer flow correctly to order $v^{\frac{1}{2}}$. On the plate, $\xi_1 = x^{\frac{1}{2}}$, and

(175) $$\tau_1 = \alpha(v/Ux)^{\frac{1}{2}},$$

as before. Away from the neighborhood of the leading edge, the singularity on $x = 0$ has disappeared.

These results (apart from consideration of the singularity at $x = 0$) are a particular case of a theorem proved by Kaplun, Z. Angew. Math. Phys. **5** (1954), pp. 111–135, for any flow without a wake. Kaplun shows that (i) there are always coordinates for which the external potential flow is included (asymptotically) in the boundary-layer solution correctly to order $v^{\frac{1}{2}}$; these coordinates (not necessarily or usually orthogonal) are, in the present notation, of the form $F_1(\Psi_1), \Psi_0 F_2(\Psi_1)$; (ii) if $\chi = F(x, y')$ is a boundary-layer solution in any coordinates $(x, y')$, and $x$ and $y'$ are expressed in terms of any other coordinates $\xi_1$ and $\eta$, then the boundary-layer solution in the coordinates $\xi_1$ and $\eta$ is

$$\chi = F[x(\xi_1, 0), \eta(\partial y'/\partial \eta)_{\eta=0}];$$

(iii) the skinfriction is unaltered. In the present case,

$$\Psi_0 = y = 2\xi_1 \eta_1, \qquad \Psi_1 = -\beta \xi_1;$$

for $F_1(\Psi_1)$ we may take $\xi_1$, and for $F_2$ we may take $1/(2\xi_1)$. The coordinates become $\xi_1$ and $\eta_1$. Also $x(\xi_1, 0) = \xi_1^2$, $(\partial y'/\partial \eta)_{\eta=0} = 2\xi_1$, and

$$\chi = x^{\frac{1}{2}} f[y'/(2x^{\frac{1}{2}})]$$

becomes

$$\chi = \xi_1 f[2\xi_1 \eta/2\xi_1] = \xi_1 f(\eta),$$

as above.

## 7.4. Solutions with similar velocity distributions. Flow near a stagnation point

In this section, attention is restricted to the first boundary-layer approximation for steady flow, i.e., to solutions of (153) with $\partial u/\partial t = 0$, and with $-\rho^{-1} \partial p/\partial x = U dU/dx$, where $U$ is the velocity outside the boundary layer, supposed given as a function of $x$; the boundary conditions to be imposed are $u = 0$,

$v = 0$ on $y = 0$, $u/U \to 1$ as $y \to \infty$ (strictly as $yv^{-\frac{1}{2}} \to \infty$, but to the approximation here considered, this is of no importance; in this form, however, it shows clearly that $u/U \to 1$ for any non-zero $y$ as $v \to 0$.)

In considering flow along a flat plate, we remarked that the velocity distributions for different values of $x$ must be similar, i.e., that there must be a $g(x)$ such that $u/U$ is a function of $g(x)y/v^{\frac{1}{2}}$ only. We may now generalize, and ask in what circumstances are there "similar" solutions when $U$ is a function of $x$, i.e., when is $u/U$ a function of $g(x)y/v^{\frac{1}{2}}$ only for some $g(x)$. The answer is immediately found by substitution into the equations (Goldstein, *Proc. Cambridge Phil. Soc.* **35** (1939), pp. 338–340); $U$ must be of the form $cx^m$ or $ce^{kx}$, where $c$, $m$ and $k$ are constants ($c > 0$), with $g \propto x^{\frac{1}{2}(m-1)}$ and $e^{\frac{1}{2}kx}$, respectively. (For cases in which $U$ is negative and the fluid has been flowing along a solid impermeable surface over an infinite distance, see Goldstein, *J. Fluid Mechanics* **21** (1965), pp. 33–45.)

Consider first $U = cx^m$ (Falkner and Skan, *Rep. Mem. Aero Res. Comm., London*, **1314** (1930); Hartree, *Proc. Cambridge Phil. Soc.* **33** (1937), pp. 223–239.) With

(176a)
$$\eta = c^{\frac{1}{2}}x^{\frac{1}{2}(m-1)}v^{-\frac{1}{2}}y = (U/vx)^{\frac{1}{2}}y,$$
$$\psi = c^{\frac{1}{2}}x^{\frac{1}{2}(m+1)}v^{\frac{1}{2}}f(\eta) = (Uvx)^{\frac{1}{2}}f(\eta),$$

(176b)
$$u/U = f'(\eta),$$

and

(176c)
$$f''' + \tfrac{1}{2}(m+1)ff'' - mf'^2 + m = 0.$$

(When $m = 0$, this equation becomes (161) if $\tfrac{1}{2}\eta$ is taken as the independent variable instead of $\eta$. When $m = 1$, the equation is equation (147).) The boundary conditions are $f(0) = f'(0) = 0$, $f'(\infty) = 1$.

With $m > -1$, put

(177)
$$Y = [\tfrac{1}{2}(m+1)]^{\frac{1}{2}}\eta, \ F(Y) = [\tfrac{1}{2}(m+1)]^{\frac{1}{2}}f(\eta),$$
$$\beta = 2m/(m+1);$$

the equation becomes

(178)
$$F''' + FF'' - \beta(F'^2 - 1) = 0,$$

with $F(0) = F'(0) = 0$, $F'(\infty) = 1$. When $U = ce^{kx}$ put

(179a) $\quad\quad\quad \eta = (ck/2\nu)^{\frac{1}{2}} y e^{\frac{1}{2}kx} = (kU/2\nu)^{\frac{1}{2}} y,$

(179b) $\quad\quad\quad \psi = (2\nu c/k)^{\frac{1}{2}} e^{\frac{1}{2}kx} F(\eta) = (2\nu U/k)^{\frac{1}{2}} F(\eta),$

(179c) $\quad\quad\quad u/U = F'(\eta),$

for $k > 0$. When $k < 0$, substitute $(-k)^{\frac{1}{2}}$ for $k^{\frac{1}{2}}$ (except in the exponentials and in $U$). For $k > 0$,

(180) $\quad\quad\quad F''' + FF'' - 2(F'^2 - 1) = 0,$

which is (178) with $\beta = 2$. The boundary conditions are $F(0) = F'(0) = 0$, $F'(\infty) = 1$, as before. Thus (180) is the limiting case of (175) as $m \to \infty$. For $k < 0$, the equation for $F$ is

$$F''' - FF'' + 2(F'^2 - 1) = 0,$$

with the same boundary conditions as before. In this case there is no solution. (It has been proved by G. H. Hardy (*Proc. Cambridge Phil. Soc.* **35** (1939), pp. 652, 653) that if $F$ satisfies this equation and is regular for large positive $\eta$, and $F' \to 1$ as $\eta \to \infty$, then $F = \eta + b$, where $b$ is a constant.)

Solutions of (178) with the stated boundary conditions (including the case $\beta = 2$) have been studied, and tabulated for several values of $\beta$, by Hartree (*op. cit.*). Existence proofs for $\beta \geqq 0$ have been given by H. Weyl, *Annals of Math.* **43** (1942), pp. 381–407. For $\beta < 0$, a mathematical theory of the equation has not been given. The circumstances seem to be as follows. Hartree found numerically that there is not a unique solution when $\beta < 0$. Physically, we are entitled to make the boundary condition at infinity more restrictive. The vorticity must tend to zero exponentially (not algebraically, like a negative power of $Y$), and $u/U$ must asymptote 1 with an exponentially small error. Also we here reject solutions in which the velocity in the boundary layer becomes greater than $U$ for some finite value of $Y$, and then decreases, so we require $1 - u/U \geqq 0$ for all positive $Y$. (See, however, Stewartson, *Proc. Cambridge Phil. Soc.* **50** (1954), pp. 454–465). With these conditions, Hartree found that a unique solution without reversed flow could be found for $0 > \beta > \beta_0$, where $\beta_0$ is about $-0.1988$. When $\beta = \beta_0$, $F''(0) = 0$; near $\beta_0$, $F''(0) = O[(\beta - \beta_0)^{\frac{1}{2}}]$. For $\beta < \beta_0$, Hartree found that there is no solution for which the above conditions are satisfied.

Plausible arguments may be given to show that there is a unique solution of (178) for which $F(0) = F'(0) = 0$ and $1 - F'(Y) \to 0$, if, for $\beta < 0$, $1 - F'(Y) \to 0$ exponentially. But for $\beta < \beta_0$ these solutions are no longer real. (The special case $\beta = -1$ may, in fact, be integrated in terms of parabolic cylinder functions.) There are real solutions for $\beta < \beta_0$ for which $1 - F'(Y)$ is not everywhere positive, and for which $1 - F'(Y) \to 0$ like a multiple of $Y^{2\beta}$; these solutions are not unique.

When $m \leqq -1$, it appears that there are no solutions at all of (176c) satisfying the boundary conditions.

The mathematical equations require further elucidation, with proofs of the results.

The equations for flow near a stagnation point, considered in Section 6.1, and for which the governing differential equations are (147) and (148), are particular cases of (178). We have noted that (147) is (176c) with $m = 1$, and is therefore transformed into (178) with $\beta = 1$. The equation (148) is transformed into (176c) with $m = \frac{1}{3}$ by the transformation $f^* = \sqrt{3}f$, $\eta^* = \sqrt{3}\eta$, applied to (148); hence it may be transformed into (178) with $\beta = \frac{1}{2}$. This is a particular case of Mangler's transformation (Section 9.2) of an axisymmetrical problem into a two-dimensional one.

We may, in fact, consider similar solutions for axisymmetrical boundary layers in the same way as for two-dimensional ones. In particular, if, outside the boundary layer, for radial flow outwards over a plane disc, $v_r = V = cr^n$ (cf. Mills, Thesis, Calif. Inst. Tech., 1935), then the solution is given by

$$\eta = (V/\nu r)^{\frac{1}{2}}x, \quad \psi = -(\nu V r^3)^{\frac{1}{2}}f(\eta), \quad v_r = Vf'(\eta),$$

where

$$f''' + \tfrac{1}{2}(n+3)ff'' - nf'^2 + n = 0,$$

$f(0) = f'(0) = 0$, $f'(\infty) = 1$. The transformation to (176c), with $m = n/3$, is again by $f^* = \sqrt{3}f$, $\eta^* = \sqrt{3}\eta$, and when transformed to (178), $\beta = 2n/(n+3)$.

CHAPTER 8

# Extension of Boundary-Layer Theory for Incompressible Fluids

## 8.1. Introduction

The main difficulties encountered in constructing solutions of the Navier-Stokes equations for steady flow arise from the non-linear terms. The known exact solutions, though physically important, are, mathematically, either trivial (in the sense that the non-linear terms are identically zero) or accidental (in the sense that the flow exhibits "similarity" and the solution depends on solving an ordinary differential equation — or, what is the same thing, a small number of simultaneous equations — not of high order). We should like to obtain a deeper mathematical, and mathematical-physical, understanding of the nature of solutions in more general cases, and in particular for flow past an obstacle. We are content, to begin with, to study two-dimensional motion. Construction of solutions which would show clearly both the mathematical and physical nature of the results, and allow fairly easy approximate computations, is the aim we have in mind. We do not expect to be able to find exact solutions of this nature; we therefore consider methods of approximation. Only two classes of such methods are known — those based on "low-Reynolds-number" approximations (such as the methods of Stokes and Oseen), and those based on "high-Reynolds-number" approximations, i.e., on boundary-layer theory. It is believed (though without proof) that in flow past a cylindrical obstacle, for example, a Reynolds number of, say, 20 would be a "high" Reynolds number for this purpose, and calculation would be best approached by attempting to find an approximation starting from the "high-Reynolds-number" end. Of course, at such Reynolds numbers as 20, the flow field may be computed numerically, especially if the use of high-speed computing machinery is available. As the Reynolds number increases, there are considerable difficulties associated even with the numerical computation; if sufficient effort were available, these might by no means prove insuperable. At present, the number of numerical solutions available is still regrettably small. Such numerical solutions would not, by themselves, give us the mathematical insight we should like into the nature of the information contained in the Navier-Stokes equations; but it would be helpful to have a large number of such solutions, especially if they were presented in an easily assimilable form, say graphically and with the aid of cinematograph film. (Even for steady two-dimensional motion, the results are functions of three independent variables or parameters, $x$, $y$, and $R$.) Such solutions in large numbers would be helpful even though in some ways the method of obtaining them must have by-passed fundamental questions such as existence, uniqueness, and (for an unbounded fluid) the correct formulation of conditions at infinity.

When we attempt to construct asymptotic approximations, based on "large" Reynolds numbers, to solutions of the Navier-Stokes equations, we must start from the large-Reynolds-number end, and can derive no help from experiment. For at sufficiently large Reynolds numbers steady flows do not exist in practice; actual flows are eddying or turbulent above a certain Reynolds number. (For descriptions of the flow patterns, and data for particular obstacles, reference may be made to *Modern Developments*.) We are optimistic enough to believe that solutions still exist of the equations of steady motion, but that the flow is unstable; this is, of course, one of the questions we should like definitely to settle. Moreover, apart from the desire to understand the mathematical nature of the information contained in the Navier-Stokes equations, we need to find approximations at very large Reynolds numbers in order to begin the study of asymptotic expansions that will apply at moderate, or even fairly small, Reynolds numbers. Also, if we could establish the existence of such flows at high Reynolds numbers, understand them, and learn to compute them easily, we might eventually learn something more about the question of stability.

Clearly, to start such an investigation we must know the limit of the steady solutions of the equations as the Reynolds number $R \to \infty$. For flows past cylindrical obstacles, with separation, this limit is not known. Classical potential theory is incorrect, but it is sometimes held that the "free-streamline" theory, mentioned in Appendix III(c), may give the correct limit. In general, a "free-streamline" solution is not unique; but the position of separation of the forward flow, as calculated by boundary-layer theory with the flow outside the boundary layer given by the free-streamline theory, must coincide with the position of separation as assumed for each free streamline; and although the evidence is not yet perfectly satisfactory, there is little doubt that the correct solution is that for which the free streamlines have finite curvature everywhere. (H. B. Squire, *Phil. Mag.* (7), **17** (1934), pp. 1150–1160; I. Imai, *J. Phys. Soc. Japan* **8** (1953) pp. 399–402; M. Kawaguti, *ibid.*, pp. 403–406). However, there are considerable difficulties in accepting the free-streamline result as the correct limit. (The question is a purely theoretical one at this stage, and is not that of adjusting the theory to fit more closely to experiment, which is concerned with turbulent, not steady, motion.) If we start from this limit for infinite $R$, and seek the correction for a large finite $R$, we know that the vortex sheets which are the free streamlines will diffuse, but this would still leave, at infinity downstream, a region of finite difference from the undisturbed flow, and it is not known how a perturbation for large finite $R$ could be carried out so that at infinity downstream the flow would return to the undisturbed flow. Zero difference from the undisturbed flow is the condition usually assumed at infinity.

If we consider a motion started from rest in a viscous fluid, it is known that $\lim_{t \to \infty} \lim_{\nu \to 0}$ and $\lim_{\nu \to 0} \lim_{t \to \infty}$ are different. Because of the action of viscosity in diffusing the vorticity, diffused vorticity cannot occur in the former, although vortex sheets may; but regions of diffused vorticity may occur in the second limit. A simple example is the flow between two parallel planes, when one is held stationary and the other started moving with constant velocity in its own plane. The first limit gives zero velocity everywhere between the planes, and a vortex sheet at the moving plane (exactly as at $t = 0$); the second gives a linear distribution of velocity and constant vorticity between the planes. (In a limiting two-dimensional steady flow, in any finite region, the diffused vorticity must be constant in the region, or in each of two or more parts of it; see Batchelor, *J. Fluid Mechs.* 1 (1956), pp. 177–190.) Now what we require is the second limit; it appears likely that what free-streamline theory gives is, in fact, the first limit. Batchelor has lately suggested (*J. Fluid Mechs.* 1 (1956), pp. 388–398) that the correct limit may well be a finite (and therefore necessarily cusped) wake, with regions of constant vorticity. (For details see the paper cited.) We cannot yet begin to construct an asymptotic theory for flows with separation.

If we consider the limit as $R \to \infty$ of the steady viscous flow past a finite flat plate ($y_1 = 0$, $0 < x_1 < 1$), we obtain $u_1 = 1$ everywhere except on $y_1 = 0$, $x_1 \geq 0$. Vortex sheets are present on both sides of the plate, through which $u_1$ drops from 1 to 0. There is also a singular surface on $y_1 = 0$, $x_1 > 1$, along which $u_1$ increases from 0 to 1 as $x_1$ increases from 1 to $\infty$. This singular surface may be regarded as the confluence of two vortex sheets. According to a calculation of the flow in the wake on boundary-layer theory (*Modern Developments*, Section 248), the limiting value of $u_1$ when $R \to \infty$ ($\nu \to 0$) is given by

$$u_1 = 1 - \alpha/(2\pi^{\frac{1}{2}} x_1^{\frac{1}{2}}) + \ldots$$

along this line for large $x_1$ (see Section 7.3 for $\alpha$), and by

$$u_1 = 1.23(x_1 - 1)^{\frac{1}{2}} - 1.18(x_1 - 1)^{\frac{3}{2}} + \ldots$$

for $x_1 - 1$ sufficiently small but away from the singularity at $x_1 = 1$. Thus even for this simple case, "classical" potential-flow theory (even with vortex sheets at the solid surface) does not give the correct limit as $R \to \infty$. A similar singular surface will arise in the limiting flow past any cusped, streamlined cylinder (airfoil) for which separation of the forward flow does not take place in the boundary layer.

Even for this simple case an attempt at a construction of an accurate asymptotic expansion for large $R$ is difficult.

We therefore restrict our considerations at present to flows without wakes. The simplest example seems to be that of flow along a semi-infinite flat plate, infinitely thin, parallel to the stream, i.e., a flow that has no wake and has zero pressure gradient in the limit when the boundary layer is infinitely thin. The first approximation was considered on pp. 123–126. Because of the "similarity" property of the solution, we have then to deal with a sequence of ordinary (not partial) differential equations. What we are now seeking is an asymptotic expansion as $\nu \to 0$ of a solution of the Navier-Stokes equations. There is no representative geometrical length of the system; the only length that enters is $\nu/U$; the nondimensional Cartesian coordinates that enter are $x_2$ and $y_2$, where $x_2 = Ux/\nu$, $y_2 = Uy/\nu$; and $u/U$ must be a function of $x_2$ and $y_2$ only. (This conclusion is easily checked from the Navier-Stokes equations and the boundary conditions on the assumption that a solution exists.) But inside the boundary layer $u/U$ is a function of $y' = (U/\nu)^{\frac{1}{2}}y$ at each value of $x$ or $x_2$, and must therefore be a function of $y_2/x_2^{\frac{1}{2}}$, or $\eta_0$, where $\eta_0$ is the variable introduced on p. 123, $\eta_0 = y_2/(2x_2^{\frac{1}{2}}) = \frac{1}{2}(U/\nu x)^{\frac{1}{2}}y$, at each value of $x_2$, and must be a function of $\eta_0$ and $x_2$ only. In order that $u/U$ should $\to 1$ as $\eta_0 \to \infty$, the first approximation must be a function of $\eta_0$ only, as on p. 123. If we now postulate an asymptotic expansion for small values of $\nu$, this becomes an asymptotic expansion for large values of $x_2$. If, for example, we postulate an expansion in powers of $\nu^{\frac{1}{2}}$, the expansion of the stream function would be of the form

$$\nu x_2^{\frac{1}{2}}[F_0(\eta_0) + x_2^{-\frac{1}{2}}F_1(\eta_0) + x_2^{-1}F_2(\eta_0) + \ldots],$$

where $F_0$ is the function $f$ defined on p. 123. The expansion in the boundary layer, on each side of the plate, must merge smoothly with the flow outside the boundary layer, which, apart from an exponentially small error, is a potential flow. This potential flow outside the boundary layer will have to be found as an asymptotic expansion for large values of $r_2 = (x_2^2 + y_2^2)^{\frac{1}{2}}$, and must be a continuous potential flow over the whole region far from any point of the plate, including the region upstream of the plate. In other words, we are seeking an asymptotic expansion of a solution of the Navier-Stokes equations for large $r_2$; the two expansions, inside and outside the boundary layer, must be expansions of the same full solution, and must match or merge smoothly.

The first approximation, $F_0(\eta_0)$, inside the boundary layer is determinate, but it turns out that an expansion in powers of $\nu^{\frac{1}{2}}$ or $x_2^{-\frac{1}{2}}$ is not sufficiently general. $F_1$ will be found to be zero, but succeeding terms will depend on $\log \nu$ or $\log x_2$, the next one linearly, and after that the next but one quadratically, and so on. Connected mathematically with the appearance of logarithms, and more importantly, undetermined constants appear, which can be found only when the whole flow field is determined sufficiently closely. The appearance of this indeterminacy is not surprising. Consider flow past a plate of finite thickness with a rounded leading edge, or past an infinitely wide and infinitely long cylinder with a rounded leading edge, symmetrically placed in a stream of velocity $U$. If $R_0$ is the radius of curvature at the leading edge, there is a boundary layer at the leading edge and in its neighborhood if $UR_0/\nu$ is large enough, but if it is small there is not. However, in any case, when the distance $x$ of a point on the surface far from the leading edge is large enough, the relevant Reynolds number is $x_2 = Ux/\nu$, and there will be a boundary layer sufficiently far back. If the pressure gradient there were zero for an infinitely thin boundary layer, we would have the same conditions as for an infinitely thin plate. So we are attempting to find an asymptotic expansion of a solution of an elliptic equation in an infinite region without specifying fully the boundary conditions when $r_2$ is not large. We could attempt to find an expansion of a completely different kind starting from small values of $r_2$, but there is now no question of merging or matching, as there was for the two different forms of the asymptotic expansion for large $r_2$. Theoretically only a sufficient knowledge of the full solution would settle the question; however, numerical computation might give approximate numerical values of the undetermined constants in the asymptotic expansion. For the flat plate there can be no high-Reynolds-number or boundary-layer approximation in the neighborhood of the leading edge, and any expansion for small $r_2$ would have to start from the low-Reynolds-number end (Stokes for "creeping" flows); if $UR_0/\nu$ is large enough there is a boundary layer solution near the leading edge, with an expansion of a type completely different from the asymptotic expansion here considered. The appearance of the logarithms and the undetermined constants has nothing to do with the singularity at the leading edge.

Something similar must happen with all flows which have a "similarity" property, as for certain flows past infinite wedges (which lead to (176) with $m < 1$ for the first approximation), and also for flows in laminar wakes and jets, though the stage in the expansion at which upstream influences make themselves felt may vary from case to case.

(For references see Goldstein, *Developments in Mechanics, Proc. 8th Midwestern Mech. Conf.*, Pergamon Press, 1965, pp. 205 *et seq*. Reference for this, and for second-order effects in general in boundary-layer theory, may also be made to Van Dyke, *Perturbation Methods in Fluid Mechanics*, Academic Press, 1964. See also Van Dyke, *Higher-Order Boundary-Layer Theory*, in *Annual Reviews of Fluid Mechanics*, Vol. **1** (1969).)

For the semi-infinite flat plate the discussion is best carried out in parabolic coordinates. (See p. 126.)

## 8.2. Asymptotic solution of the Navier-Stokes equations for flow past a flat plate

Use parabolic coordinates, as in (167), (169). Define $\eta$ as in (170), and define $\xi$ similarly, so

(181) $$\xi = (U/\nu)^{\frac{1}{2}}\xi_1, \quad \eta = (U/\nu)^{\frac{1}{2}}\eta_1.$$

With $\zeta_1 = \xi_1 + i\eta_1$, the transformation $(\zeta_1, z)$ was described for our purposes after equation (167). If $\zeta = \xi + i\eta$, we now have

(182) $$(U/\nu)z = (U/\nu)(x + iy) = \zeta^2 = (\xi + i\eta)^2.$$

Change the meaning of the symbol $\psi$ from its meaning in Section 7, and now let the stream function be $\nu\psi$, so that if $\chi$ is as defined in (170) and as it occurs on p. 126,

(183) $$\psi = (U/\nu)^{\frac{1}{2}}\chi.$$

We remark, to begin with, that it has been suggested that for flow past any semi-infinite body an asymptotic expansion of a solution of the Navier-Stokes equations for $\chi$ may (inside the boundary layer) be expected to proceed in powers of $v^{\frac{1}{2}}$.

Because of the similarity of the solutions, it may be shown that for the flat plate this must lead to an expansion for (the new) $\psi$ of the type

(184) $\quad \psi = \xi f_0(\eta) + \epsilon f_1(\eta) + f_2(\eta)/\xi + \epsilon f_3(\eta)/\xi^2 + \cdots,$

where $\epsilon = $ sign of $\xi$. It is now clear that (184) (if it is correct) is an asymptotic solution for large $\xi$.

Since $\psi$ must be odd in $\xi$, and since it appears from the equation for $\chi$ that only integral powers of $v$ may be needed in its expansion, it has been suggested that only odd powers of $\xi$ need be included in (184):

(185) $\quad \psi = \xi f_0(\eta) + \xi^{-1} f_2(\eta) + \xi^{-3} f_4(\eta) + \cdots.$

However, an asymptotic expansion need not be analytic, so perhaps (184) should be retained. If we substitute (184) into the Navier-Stokes equation, and take $f_0$ as the Blasius function (equations (161)–(163) and (173)), we find that

(186a) $\quad L_n(f_n) = F_n(\eta),$

where

(186b) $\quad L_n = d^4/d\eta^4 + f_0 d^3/d\eta^3 + (n+1)f_0' d^2/d\eta^2 + f_0'' d/d\eta \\ \qquad - (n-1)f_0''',$

(186c) $\quad F_1 = 0, \ F_2 = 2\eta f_0''(\eta f_0' - f_0) = d(\eta f_0' - f_0)^2/d\eta,$

and, in general, $F_n$ involves the $f$ up to $f_{n-1}$.

$f_n$ must have a double zero at $\eta = 0$. At infinity, either the boundary condition

(187a) $\quad f_n \to 0 \text{ as } \eta \to \infty$

or

(187b) $\quad f_n/\eta \to 0 \text{ as } \eta \to \infty$

has been taken, (187a) by Alden, *J. Math. Phys.* **27** (1948), pp. 91–104, (187b) by Kochin, *Sobranie Sochinenii* **2** (1949), *Akad. Nauk. S.S.S.R.*, and by Goldstein, *Proc. Intern. Cong. Mathematicians*, Amsterdam, 1954. It appears that each of these is unnecessarily restrictive. All that we may require is that the solution should merge smoothly into a potential flow, which, expressed in the original "unstretched" variables, $\psi_1$, $\xi_1$, and $\eta_1$,

## EXTENSION OF BOUNDARY-LAYER THEORY

together with $U/\nu$, should itself give the original undisturbed flow as $\eta_1 \to \infty$. This condition may be translated into terms of the stretched, similarity, variables, $\psi$, $\xi$, and $\eta$ (when $U/\nu$ does not appear). If, in the boundary layer, $\psi$ were to be expressed in the form (184) or (185), it may be shown that the procedure to be adopted is as follows. In the potential flow $\psi$ is given asymptotically by a series of harmonic functions. This series must be rearranged by expanding each of the harmonic functions in descending powers of $\xi$, i.e., for $|\xi| > \eta$; when $\psi$ in the potential flow is thus expressed as a descending series in $\xi$, the coefficient of $\xi^{-(2n-1)}$ (or $\varepsilon\xi^{-(2n-2)}$), which turns out to be a polynomial in $\eta$, gives the requisite asymptotic expression for $f_{2n}$ (or $f_{2n-1}$). These expressions must be asymptoted with exponentially small errors. Of course the expression for $\psi$ in the potential flow as a series of harmonic functions must be such that, as $\eta \to \infty$, the original uniform undisturbed flow is recovered, i.e.

$$\psi/(2\xi\eta) \to 1.$$

We return to this matter.

Without a close investigation of the existence of solutions of (186a), with double zeros at the origin and under the most general conditions that we may allow as $\eta \to \infty$, we certainly cannot accept the validity of (184), or the more restricted form (185). In fact, it turns out that these forms are not sufficiently general. First of all, we explain why this is so, explaining at the same time something of previous investigations.

(186a) may be integrated once to give

(188a) $$K_n(f_n) = A_n + G_n,$$

where

(188b) $$K_n = d^3/d\eta^3 + f_0 d^2/d\eta^2 + nf_0' d/d\eta - (n-1)f_0'',$$

$A_n$ is a constant, and

(188c) $$G_n(\eta) = \int_0^\eta F_n(\eta)d\eta, \quad G_1 = 0, \quad G_2 = (\eta f_0' - f_0)^2.$$

(186a) has four complementary functions, which for sufficiently small $\eta$ have series expansions

$$y_n^{(0)} = 1 + (n-1)\,\alpha\eta^3/3! + \ldots$$
$$y_n^{(1)} = \alpha\eta - \alpha^2\eta^4/4! + \ldots = f_0'$$
$$y_n^{(2)} = \eta^2/2! - (n+2)\,\alpha\eta^5/5! + \ldots$$
(189)
$$y_n^{(3)} = \eta^3/3! - 2(n+2)\,\alpha\eta^6/6! + \ldots.$$

$y^{(0)}$, $y^{(1)}$, $y^{(2)}$ are complementary functions of (188a); $y^{(3)}$ is a particular integral of (188a) with $A_n = 1$ and $G_n$ neglected.

With

(190) $$\lambda = \eta - \tfrac{1}{2}\beta,$$

three asymptotic approximations to complementary functions of (188a) are (i) a constant, (ii) $E_n$, (iii) $H_n$, where

(191) $$E_n \backsim \frac{1}{\lambda^{n-1}}\left\{1 + \frac{(n-1)n}{4\lambda^2} + \ldots\right\}$$

$$H_n = e^{-\lambda^2}\lambda^{n-2}\left\{1 - \frac{(n-2)(n-3)}{4\lambda^2} + \ldots\right\}.$$

A fourth complementary function of (186a), which is a particular integral of (188a) with $A_n = 1$ and $G_n$ neglected, asymptotes $\eta/(2n)$.

Since $f_n(0) = f_n'(0) = 0$, $y^{(0)}$ and $y^{(1)}$ must be absent. For any $n$, there must be constants $a_n^{(2)}$, $b_n^{(2)}$, $c_n^{(2)}$, $a_n^{(3)}$, $b_n^{(3)}$, $c_n^{(3)}$, such that

(192)
$$y_n^{(2)} \backsim a_n^{(2)} + b_n^{(2)} E_n + c_n^{(2)} H_n$$
$$y_n^{(3)} \backsim \frac{\eta}{2n} + a_n^{(3)} + b_n^{(3)} E_n + c_n^{(3)} H_n.$$

First try $f_1 = 0$, and consider the equation for $f_2$. Since $G_2 \sim \beta^2$, there is a particular integral $y_2^{(4)}$ of (188a) (with $A_2 = 0$), with a double zero at the origin, for which

(193) $$y_2^{(4)} \backsim \frac{\beta^2}{4}\eta + a_2^{(4)} + b_2^{(4)} E_2 + c_2^{(4)} H_2.$$

Any multiples of $y_2^{(2)}$ and $y_2^{(3)}$ may be added to $y_2^{(4)}$ to give the solution for $f_2$. The multiples may be chosen to annul the term in $\eta$ and the constant in the asymptotic expansion, and so to satisfy (187a) (Alden). Or we may annul the term in $\eta$ only, satisfying (187b), and leaving an arbitrary multiple of $y_2^{(2)}$. The solution is then indeterminate at this stage. It is made determinate only by joining the asymptotic solution for large $\xi$ to a permissible solution near the leading edge (Goldstein). In either case, a multiple of $E_2$ is left in the asymptotic expansion. But this asymptotic approximation must join smoothly, with an exponentially small error, to a potential flow. In the original unstretched variables, a constant term $A_1$ in the asymptotic expansion of $f_2$ is allowed for by a harmonic term $(\nu/U)^{\frac{1}{2}}A_1\xi_1/(\xi_1^2 + \eta_1^2)$ in $\psi_1$ in the potential flow, and the term in $E_2$ introduces an error of order $(\nu/U)^2$. But to proceed further this term must be annulled at the next stage, and this requires a potential function which behaves like $y^{-1}$ all along the positive real axis: it appears that there is no such function. So it is impossible to proceed further, and it was incorrect to leave a multiple of $E_2$ in the asymptotic expansion of $f_2$.

We may try to annul the terms in $\eta$ and $E_2$ in the asymptotic expansion by choosing the correct multiples of $y_2^{(2)}$ and $y_2^{(3)}$, and so satisfy (187b) and also remove $E_2$ (Kochin). But this fails, because $\eta f_0' - f_0$ is a complementary function of (188a), and $y_2^{(2)} = (\eta f_0' - f_0)/\alpha \backsim \beta/\alpha$, so $b_2^{(2)} = 0$. The multiple of $y_2^{(2)}$ affects neither the term in $\eta$ nor the term in $E_2$, and a multiple of $y_2^{(3)}$ cannot be found to annul both.

We cannot satisfy either (187a) or (187b) in this manner. However, we may remove the condition (187b), and seek a solution which merges into a potential flow, which itself gives the uniform undisturbed flow as $\eta_1 \to \infty$. To do this, we remove the term in $E_2$ from the asymptotic expansion of $f_2$, and leave both the constant and the multiple of $\eta$. To remove $E_2$, add the necessary multiple of $y_2^{(3)}$ to $y_2^{(4)}$. The solution is again indeterminate to the extent of an arbitrary multiple of $y_2^{(2)}$. Then $f_2 \backsim A_0 \eta + A_1$ (with $A_1$ indeterminate),

$$\psi \backsim 2\xi\eta - \beta\xi + A_0 \eta/\xi + A_1/\xi,$$

and this joins smoothly on to the potential flow

$$\psi = 2\xi\eta - \beta\xi + A_0 \tan^{-1}\frac{\eta}{\xi} + A_1 \frac{\xi}{\xi^2 + \eta^2}.$$

(The determination of the potential flow is unique.) To the accuracy involved, this solution is now a satisfactory extended boundary-layer flow on each side of the plate for large $\xi$. However, this is *not* sufficient. It must also be possible to continue and connect the potential flows into which the boundary-layer flows on the two sides merge, so as to obtain one continuous potential flow over the whole region far from any point of the plate, including the potential flow upstream. But $A_0 \tan^{-1} \eta/\xi$ introduces a discontinuity $\pi A_0$ in $\psi$ for any (sufficiently large) $\eta$ along $\xi = 0$, which is the upstream continuation of the plate.

We may put the matter in another way. Let $\psi = \text{Im}(w)$, where $w$ is a function of $\zeta$. If $\text{Im}(\log \zeta) = \arg \zeta$, then $\tan^{-1} \eta/\xi = \text{Im}(\log \zeta)$ for $\xi > 0$, but $= \text{Im}(\log \zeta) - \pi$ for $\xi < 0$. Then for $\xi > 0$ we find we can construct a boundary-layer flow of the form (185) which merges into the potential flow for which

$$w = \zeta^2 - i\beta\zeta + A_0 \log \zeta + \sum_{m=1} i^m A_m/\zeta^m.$$

But for $\xi < 0$, we have to take an extra term $-\pi A_0$ in $w$. This difference in the asymptotic representations of $w$ for large positive and for large negative $\xi$ is forced on us by the attempt to use (185). On the other hand, we may write

$$\zeta' = \zeta e^{-\frac{1}{2}\pi i} = \eta - i\xi,$$

and

$$w = -\zeta'^2 + \beta\zeta' + A_0 \log \zeta' + \sum_{m=1} A_m/\zeta'^m.$$

Now $\text{Im}(\log \zeta') = -\tan^{-1} \xi/\eta$ whether $\xi$ be positive or negative. To find the asymptotic expressions for large $\eta$ for the boundary-layer solution, we must re-arrange the expression for $\psi$ in the potential flow in descending powers of $\xi$, and therefore expand each term for $|\xi| > \eta$. We must therefore replace $\tan^{-1} \xi/\eta$ by $\frac{1}{2}\pi - \tan^{-1} \eta/\xi$ for $\xi$ positive, and by $-\frac{1}{2}\pi - \tan^{-1} \eta/\xi$ for $\xi$ negative. Hence we must try (184) instead of (185), and seek an $f_1$ with a double zero at the origin which asymptotes a non-zero constant $(-\frac{1}{2}\pi A_0)$ as $\eta \to \infty$. It is easily proved that there is no such solution for $f_1$.

Hence expansions of the above forms for $w$, and of the form (184) for $\psi$, are not sufficiently general. In particular, $w$ has a more complicated singularity at infinity. For the present, we consider only the possibility that $w$ may be given by a double series in $1/\zeta'$ and $(\log \zeta')/\zeta'$. There is not yet any proof that this will be sufficiently general, and that the singularity of $w$ at infinity is not even more complicated. However, at present it does seem

EXTENSION OF BOUNDARY-LAYER THEORY 143

that it is sufficiently general to allow an asymptotic solution to be constructed if we assume that

$$w = -\zeta'^2 + \beta\zeta' + \sum_{m=1} \frac{1}{\zeta'^m} \{b_{m,m}(\log \zeta')^m + b_{m,m-1}(\log \zeta')^{m-1}$$
(194)
$$+ \ldots + b_{m,1}\log \zeta' + b_{m,0}\}.$$

$\psi$ is antisymmetrical in $\xi$, and analytic on $\xi = 0$, $\eta > 0$. The $b$ are therefore taken as real. The resulting expression for $\psi$ is rearranged (by expanding each term for $|\xi| > \eta$) as a series in $1/\xi$ and $(\log \xi)/\xi$. In place of (184) we must assume, in the boundary layer, an asymptotic expression of the form

$$\psi = \xi f_0(\eta) + \xi^{-1}[f_2(\eta) + g_2(\eta) \log \xi]$$
(195)
$$+ \xi^{-2}[f_3(\eta) + g_3(\eta) \log \xi + h_3(\eta)(\log \xi)^2]$$
$$+ \xi^{-3}[f_4(\eta) + g_4(\eta) \log \xi + h_4(\eta)(\log \xi)^2 + k_4(\eta)(\log \xi)^3] + \ldots$$

The functions $f, g, h, k, \ldots$ must all have double zeros at the origin. When $\eta \to \infty$, comparison with the re-arranged series for $\psi$ in the potential flow shows that we require

$$f_2 \backsim b_{10}, \quad g_2 \backsim b_{11}$$
$$f_3 \backsim \tfrac{1}{2}\pi(b_{21} - b_{11}\eta), \quad g_3 \backsim \pi b_{22}, \quad h_3 \sim 0,$$
$$f_4 \backsim (\tfrac{1}{4}\pi^2 b_{32} - b_{30}) - \eta(\tfrac{1}{2}\pi^2 b_{22} + b_{21} - 2b_{20}) + \eta^2(\tfrac{3}{2}b_{11} - b_{10})$$
$$g_4 \backsim (\tfrac{3}{4}\pi^2 b_{33} - b_{31}) - 2\eta(b_{22} - b_{21}) - b_{11}\eta^2,$$
(196)
$$h_4 \backsim -b_{32} + 2b_{22}\eta, \quad k_4 \backsim -b_{33},$$

all with exponentially small errors. The equation for $g_2$ is $L_2(g_2) = 0$, and the required solution is $(b_{11}/\beta)(\eta f_0' - f_0)$, with $b_{11}$ undetermined at this stage. The equation for $f_2$ is now

$$L_2(f_2) = F_2 + (b_{11}/\beta) f_0''(f_0' - f_0^2)$$

(after substitution for $g_2$). If $y_2^{(5)}$ is a particular integral with $F_2$ omitted and $b_{11}/\beta = 1$, then $y_2^{(4)} + (b_{11}/\beta)y_2^{(5)}$ is now a particular integral for $f_2$. Multiples of $y_2^{(2)}$ and $y_2^{(3)}$ may be added to provide the solution for $f_2$. We find that we may choose $b_{11}/\beta$ and the multiple of $y_2^{(3)}$ so that the terms in $\eta$ and in $E_2$ in the asymptotic expansion of $f_2$ are both annulled. $f_2$ then asymptotes a constant. The multiple of $y_2^{(2)}$ is still left arbitrary, so the limiting value of $f_2$, namely $b_{10}$, is not determined.

It appears that $h_3$ must be zero, and that it follows that $g_3$ is also zero, so $b_{22} = 0$. Also $k_4$ is zero, so $b_{33} = 0$. Dr. J. D. Murray has investigated rather fully the equations for $f_3$, $g_3$, $h_3$, $f_4$, and $k_4$, and has tabulated the functions that occur (including those multiplying the constants that remain undetermined) in the part of the expansion written out in (195). (*J. Fluid Mechanics* **21** (1965), pp. 337–344.) Dr. Van Dyke had at an early stage made a rough numerical calculation of the value of $b_{11}/\beta$, and found it to be 1.66, where $\beta = 1.7208$ (see equation (163), p. 123). As regards the functions, $f_0$ is, of course, determinate. $g_3$, $h_3$, and $k_4$ are zero. $g_2$, $f_3$, and $h_4$ are determinate. However $f_2$ and $g_4$ involve an unknown constant, $A$, linearly; $f_4$ involves $A$ quadratically and also a second unknown constant linearly. The constants in (194) which are not determined are $b_{10}$ and $b_{30}$; $b_{10}$ involves $A$ linearly; $b_{30}$ involves $A$ quadratically and the second unknown constant linearly. $b_{31}$ also involves $A$ (and therefore $b_{10}$) linearly. $b_{11}$, $b_{21}$, and $b_{32}$ are determinate: $b_{11} = 2.8577$, $b_{21} = 2.267$, and $b_{32} = 5.95$. The validity of the procedure to this stage depended on $b_n^{(2)}$ (equation (192)) not being zero for $n = 3$ and $n = 4$; this was checked numerically. There is no strict mathematical proof, and no proof that the process may be continued without failure at any stage.

Moreover, reference should also be made to the series of characteristic solutions of Libby and Fox (*J. Fluid Mechanics* **17** (1963), pp. 433–449. See Van Dyke, *Higher-Order Boundary-Layer Theory*, in *Annual Reviews of Fluid Mechanics*, Vol. **1** (1969), p. 286.)

A summary interim report of this investigation was issued in December, 1956, from the University of California, Institute of Engineering Research, where the investigation was made in the summer of 1956. Acknowledgements are there made for the benefits derived from discussions with a number of mathematicians and workers in fluid mechanics, including Drs. Kaplun and van Dyke, Mr. R. T. Jones, Professors Lagerstrom, Schaaf, Schiffer, Lewy, and Carrier. Special mention was made of the contributions to Dr. Kaplun. Professor Isao Imai made a completely independent investigation, with exactly the same form for the terms in $1/\zeta'$, $(\log \zeta')/\zeta'$, $1/\xi$, $(\log \xi)/\xi$, and his numerical value agrees with that computed by Dr. Van Dyke and Dr. Murray. A short account was published in the *J. Ae. Sci.* **24** (1957), pp. 155, 156. Professor Imai also found an expression for the drag on a length $l$ of the plate, when $(Ul/\nu)^{\frac{1}{2}}$ is large, by considerations of momentum.

The form of the expansion near the leading edge of a flat plate for small values of $Ur/\nu$ (of course also involving unknown constants) (Carrier and Lin, *Quart. Appl. Math.* **6** (1948), pp. 63–68), corrected to make the stream function analytically anti-symmetrical about $\xi = 0$, $\eta > 0$, was provided by Prof. Carrier (private communication) and independently by Dr. Van Dyke (*Perturbation Methods in Fluid Mechanics*, Academic Press, 1964, p. 39). But it is not possible finally to confirm the validity of either this expansion or the asymptotic expansion and determine the unknown constants in them without a complete, or nearly complete, numerical solution of the whole problem. At this time published reports of valuable numerical work have appeared by A. I. Van de Vooren and D. Dijkstra, *J. Engineering Mathematics* **4** (1970), pp. 9–27, and by Akira Yoshizawa, *J. Physical Society of Japan* **28** (1970), pp. 776–779. (I am indebted to Dr. Van Dyke for these references.)

# CHAPTER 9

# Boundary Layers in Gases

Reference may be made to Chap. 10 (Boundary Layers, by A. D. Young) and also to Chap. 14 (Heat Transfer, by H. B. Squire) of *Modern Developments: High Speed Flow*.

In the boundary layers we shall now consider, because of the variations of viscosity and density with temperature, the effects of temperature variations must be taken into account in the determination of the velocity. Significant temperature differences may be imposed by the boundary conditions, or may arise in high-speed flow from the heat produced by dissipation of mechanical energy or by compression, or both. In flows in which, the effect of gravity being neglected, the stagnation enthalpy, $I + \tfrac{1}{2}\mathbf{v}^2$, may be taken as constant along a streamline, the temperature rise in a perfect gas with constant specific heats from a position where the velocity is $U$ to a stagnation point on the same streamline is $U^2/(2c_p)$. This expression, $U^2/(2c_p)$, as we shall see, may be taken generally as the order of magnitude of temperature differences arising, in high-speed flow, from the heat produced by compression and dissipation of energy in a perfect gas.

Boundary-layer theory in gases, with the fluid properties variable with temperature, is similar to, but more complicated than, the theory for fluids of constant properties. (i) In addition to the velocity boundary-layer, through which the velocity rises rapidly from zero at a wall to its value in the main stream, there is a temperature boundary-layer, through which the temperature changes rapidly from its value at the wall to its value in the main stream; the two boundary layers are of the same order of magnitude if the Prandtl number, $Pr$, is of order unity. (ii) The determination of the velocity and temperature are interconnected. (iii) If the main stream is supersonic, the boundary layer and the main stream may interact in a much more striking fashion than for a subsonic stream. (a) The assumptions of boundary-layer theory, especially that the rate of change of the velocity along the boundary layer is much smaller than that across it, break down at an intersection of a shock wave and a boundary layer, and special discussions become necessary (See *Modern Developments*: *High Speed Flow*, Chap. 10, Section 18). (b) In a region behind a curved shock the flow is rotational, and the rotational flow in the "mainstream" may have an important effect on the boundary layer. The presence of a boundary layer may also have an appreciable effect on the shock wave; it is then imperative to take the interaction between the boundary layer and the external stream into account. This is especially marked at high Mach numbers (hypersonic flow); even for flow past a flat plate along the stream, there is a shock wave of finite strength from the leading edge, or just before it, inclined at a small angle to the plate and curved, with rotational flow behind. For a discussion of hypersonic boundary layers, with references, see Lester Lees, *Hypersonic Flow*, 5th. Intern. Aero. Conf., Los Angeles, 1955, pp. 241–276, reprinted by the Inst. Ae. Sci. (Sherman M. Fairchild Fund.) (c) With a subsonic external flow any disturbance to the thickness of the boundary layer has only a small effect; and a thickening of the boundary layer, in a flow along an obstacle in a channel, for example, leads to an acceleration of the main stream and a consequent thinning of the boundary layer. However, with a supersonic main stream, the effects, in the first place, are concentrated along Mach lines or surfaces; and a thickening of the boundary layer, leading to a contraction of the stream tubes in the main flow, now produces a deceleration of that flow (see Section 11.1), and the effects may become widespread. (d) Separation with a supersonic main stream is a more complicated phenomenon than with a subsonic main stream. Since the velocity falls to zero through a boundary layer, there is necessarily a subsonic part of that layer (though that part may be very thin). Even with a supersonic main stream, effects may be transmitted upstream through the subsonic part of the layer, and this may lead to results quite different from those to be expected in purely supersonic flow. For example, the effect of a shock wave at the trailing edge of an airfoil may spread upstream, and separation may take place at a position where, theoretically, the pressure is decreasing in the direction of the flow.

## 9.1. Boundary-layer equations for two-dimensional flow

We neglect the effect of gravity.
With
$$\frac{D}{Dt} = \frac{\partial}{\partial t} + v_1 \frac{\partial}{\partial x_1} + v_2 \frac{\partial}{\partial x_2}, \quad \Delta = \frac{\partial v_1}{\partial x_1} + \frac{\partial v_2}{\partial x_2},$$
the equations of continuity and momentum for a two-dimensional flow are

(197a) $$\frac{\partial \rho}{\partial t} + \frac{\partial}{\partial x_1}(\rho v_1) + \frac{\partial}{\partial x_2}(\rho v_2) = 0,$$

(197b) $$\rho \frac{Dv_1}{Dt} = -\frac{\partial p}{\partial x_1} + \frac{\partial}{\partial x_2}\left[\mu\left(\frac{\partial v_1}{\partial x_2} + \frac{\partial v_2}{\partial x_1}\right)\right] + 2\frac{\partial}{\partial x_1}\left(\mu \frac{\partial v_1}{\partial x_1}\right)$$
$$+ \frac{\partial}{\partial x_1}[(\beta - \tfrac{2}{3}\mu)\Delta]$$

(197c) $$\rho \frac{Dv_2}{Dt} = -\frac{\partial p}{\partial x_2} + \frac{\partial}{\partial x_1}\left[\mu\left(\frac{\partial v_1}{\partial x_2} + \frac{\partial v_2}{\partial x_1}\right)\right] + 2\frac{\partial}{\partial x_2}\left(\mu \frac{\partial v_2}{\partial x_2}\right)$$
$$+ \frac{\partial}{\partial x_2}[(\beta - \tfrac{2}{3}\mu)\Delta].$$

We assume a perfect gas, with a constant Prandtl number, but, for the present, we allow for variation of the specific heats. The equation of energy is

(198a) $$\rho T \frac{DS}{Dt} = \rho \left[\frac{DI}{Dt} - \frac{1}{\rho}\frac{Dp}{Dt}\right] = \Phi + \frac{1}{Pr}\frac{\partial}{\partial x_1}\left(\mu \frac{\partial I}{\partial x_1}\right)$$
$$+ \frac{1}{Pr}\frac{\partial}{\partial x_2}\left(\mu \frac{\partial I}{\partial x_2}\right),$$

where

(198b) $$\Phi = \mu\left\{2\left(\frac{\partial v_1}{\partial x_1}\right)^2 + 2\left(\frac{\partial v_2}{\partial x_2}\right)^2 + \left(\frac{\partial v_1}{\partial x_2} + \frac{\partial v_2}{\partial x_1}\right)^2\right\} + (\beta - \tfrac{2}{3}\mu)\Delta^2.$$

For flow along a plane wall, we may now carry through an order-of-magnitude argument in the same way as for a fluid of constant properties. In exactly the same way, we find that the ratio of the thickness of the velocity boundary-layer to a characteristic length $d$ must be of order $R^{-\frac{1}{2}}$, where $R$ is a Reynolds number $Ud/\nu$; $U$ is the velocity just outside the boundary layer, in the main stream, and $\nu$ is the kinematic viscosity at any standard position. The first momentum equation in the boundary layer may be simplified, with similar approximations to those used for fluids of constant properties, to

$$(199) \qquad \rho \frac{Dv_1}{Dt} = -\frac{\partial p}{\partial x_1} + \frac{\partial}{\partial x_2}\left(\mu \frac{\partial v_1}{\partial x_2}\right).$$

The second momentum equation gives, as before, that $\partial p/\partial x_2$ is $O(R^{-\frac{1}{2}} \rho U^2/d)$, and $p$ may be taken as constant across a section of the boundary layer.

In the equation of energy, taking $I_e$ as of order unity (we use the subscript $e$ to denote conditions just outside the boundary layer in the main stream), and requiring that $(Pr)^{-1} \partial(\mu \partial I/\partial x_2)/\partial x_2$ should be of the same order as $\rho DI/Dt$, we find that the ratio of the thickness of the temperature boundary layer to the characteristic length $d$ must be of order $Pr^{-\frac{1}{2}} R^{-\frac{1}{2}}$, and that the equation of energy simplifies to

$$(200) \qquad \rho \frac{DI}{Dt} - \frac{Dp}{Dt} = \frac{1}{Pr} \frac{\partial}{\partial x_2}\left(\mu \frac{\partial I}{\partial x_2}\right) + \mu \left(\frac{\partial v_1}{\partial x_2}\right)^2$$

(For gases, $Pr$ is of order unity.)

To these equations we join the equation of continuity and the equation of state.

As regards boundary conditions, to the conditions on the velocity, that $v_1 = v_2 = 0$ at an impermeable wall, we add a condition on the temperature or enthalpy, that $T$ is given and equal to $T_w (I = I_w)$ at a wall, or that the rate of heat conduction is given, $\partial T/\partial x_2$ or $\partial I/\partial x_2$ given. For a heat-insulated boundary, $\partial T/\partial x_2 = \partial I/\partial x_2 = 0$. (With a given solid body of given heat conductivity, the full problem is a dual one, of fluid heat transfer and heat conduction in the solid.) We also require, for a first approximation, that $v_1$ and $T$ (or $I$) should go over smoothly and monotonically into their values in the main stream as $\nu^{-\frac{1}{2}} x_2 \to \infty$, with an exponentially small error for large $\nu^{-\frac{1}{2}} x_2$.

For a curved wall, we may follow the same procedure, with the same kind of result, as for a fluid of constant properties. We obtain the same equations (197a, 199) with $x_1$ measured along the wall and $x_2$ normal to it, if $\delta \kappa$, $\delta^2 \partial \kappa / \partial x_1$ are small, where $\kappa$ is the curvature of the wall, and $\delta$ a length of the order of either boundary-layer thickness. The pressure gradient normal to the wall must again balance the centrifugal force, but the pressure is still to be taken as constant across a cross-section of the boundary layer. The same approximation as before (200) is also obtained for the energy equation.

We shall consider only steady motions. Then, in the main stream, the entropy $S$ and the stagnation enthalpy $I + \frac{1}{2} \mathbf{v}^2$ are constant along each streamline. We take them as constant throughout the relevant part of the main stream, and we then have

(201) $$\frac{\partial p}{\partial x_1} = -\rho_e U \frac{dU}{dx_1} = \rho_e \frac{dI_e}{dx_1}.$$

In the equation of energy, for steady flow,

(202) $Dp/Dt = v_1 \partial p / \partial x_1 = -\rho_e v_1 U \, dU/dx_1 = \rho_e v_1 \, dI_e/dx_1.$

We now see under what conditions the dissipation term, and the term $Dp/Dt$, may be neglected in the equation of energy for steady flow. The dissipation term, $\mu(\partial v_1 / \partial x_2)^2$, and the pressure term, are both seen to be of order $(\rho_e U^3/d)$; they therefore produce temperature differences of the order of $U^2/c_p$; and if one may be neglected, the other may. If $|T_e - T_w|$ is a representative temperature difference imposed by the boundary conditions, the order of $\rho DI/Dt$ arising from this imposed temperature difference is $(\rho_e U c_p |T_e - T_w|/d)$. Both the dissipation and pressure terms may be neglected if $U^2/c_p \ll |T_e - T_w|$.

## 9.2. Boundary-layer equations for axisymmetrical flow. Mangler's transformation

For flow past a blunt-nosed body of revolution, with its axis parallel to the undisturbed stream, if the body is not too slender, the boundary-layer equations may be found, exactly as for a fluid of constant properties, to have the same form as for two-dimensional flow except for the equation of continuity. If $x_1$ and $x_2$ are distances along and normal to a meridian section, the equations of momentum and energy are as in (199) and (200), with $Dp/Dt$ as in (202). For steady flow, the equation of continuity at a point $P$ in the boundary layer may be taken as

$$\frac{\partial}{\partial x_1}(\rho r_0 v_1) + \frac{\partial}{\partial x_2}(\rho r_0 v_2) = 0,$$

where $r_0$ is the distance from the axis of symmetry of the point on the meridian section and on the same normal as $P$ (i.e., for the same $x_1$, but with $x_2 = 0$). This seemingly small change in the equation of continuity makes a big difference in calculations. For steady flows, however, the equations may be transformed into the equations for a two-dimensional flow by a transformation due to Mangler. (The same transformation may be applied, of course, for a fluid of constant properties.) Use primed quantities for the two-dimensional flow.

For the axisymmetrical flow, there is a stream function such that

$$\rho r_0 v_1 = \partial \psi / \partial x_2, \quad \rho r_0 v_2 = -\partial \psi / \partial x_1.$$

For a two-dimensional flow there is a stream function $\psi'$ such that

$$\rho' v_1' = \partial \psi' / \partial x_2', \quad \rho' v_2' = -\partial \psi' / \partial x_1'.$$

In the axisymmetrical flow, $r_0$ is a given function of $x_1$. The equations of the transformation are:-

$$x_1' = \int_0^{x_1} \frac{r_0^2}{d^2} dx_1, \quad x_2' = \frac{r_0}{d} x_2, \quad \psi'(x_1', x_2') = \frac{1}{d} \psi(x_1, x_2),$$

$$p' = p, \quad I' = I, \quad \rho' = \rho, \quad \mu' = \mu,$$

and $Pr$ is the same for both. In these equations, $d$ is any fixed reference length. Then

$$\rho v_1 = \frac{1}{r_0}\frac{\partial \psi}{\partial x_2} = \frac{\partial \psi'}{\partial x_2'} = \rho' v_1', \text{ i.e., } v_1 = v_1'.$$

$$\rho v_2 = -\frac{1}{r_0}\frac{\partial \psi}{\partial x_1} = -\frac{r_0}{d}\frac{\partial \psi'}{\partial x_1'} - \frac{x_2}{r_0}\frac{dr_0}{dx_1}\frac{\partial \psi'}{\partial x_2'},$$

i.e.,

$$v_2 = \frac{r_0}{d} v_2' - \frac{x_2}{r_0}\frac{dr_0}{dx_1} v_1',$$

or

$$v_2' = \frac{d}{r_0}\left\{v_2 + \frac{x_2}{r_0}\frac{dr_0}{dx_1} v_1\right\}.$$

Also

$$\rho\left\{v_1 \frac{\partial}{\partial x_1} + v_2 \frac{\partial}{\partial x_2}\right\} = \frac{r_0^2}{d^2}\rho\left\{v_1' \frac{\partial}{\partial x_1'} + v_2' \frac{\partial}{\partial x_2'}\right\},$$

and the axisymmetrical equations of continuity, momentum and energy are easily seen to transform into the corresponding two-dimensional ones.

The boundary conditions for the two-dimensional problem are (from $v_1 = v_2 = 0$ at $x_2 = 0$), $v_1' = v_2' = 0$ at $x_2' = 0$, and we have the same condition on $I'$ at $x_2' = 0$ as on $I$ at $x_2 = 0$ (if $\partial I/\partial x_2$ is given and not zero at $x_2 = 0$ the value of $\partial I'/\partial x_2'$ at $x_2' = 0$ must be suitably chosen); also we take $U' = U$, $I_e' = I_e$, and the boundary conditions on $v_1$ and $I$ at infinity are the same in the two cases. Note the correspondence of the stream functions at infinity.

If $\tau$ is the shear stress, $\mu(\partial v_1/\partial x_2)$, the skin frictions are the values at $x_2 = 0$, i.e.,

$$\tau(0) = \mu(0)(\partial v_1/\partial x_2)_{x_2=0},$$
$$\tau'(0) = \mu'(0)(\partial v_1'/\partial x_2')_{x_2'=0},$$

so

$$\tau'(0)/\tau(0) = dx_2/dx_2' = d/r_0.$$

Let $c_f$ be the corresponding local skin-friction coefficient, $= \tau(0)/(\frac{1}{2}\rho_e U^2)$, and $R$ the Reynolds number $Ux_1/\nu_e$. Then

$c_f'/c_f = \tau'(0)/\tau(0) = d/r_0$, and $R'/R = x_1'/x_1$.

It is convenient to compare values of $c_f R^{\frac{1}{2}}$. We have

$$\frac{c_f' R'^{\frac{1}{2}}}{c_f R^{\frac{1}{2}}} = \frac{d}{r_0}\left(\frac{x_1'}{x_1}\right)^{\frac{1}{2}} = \left[\frac{\int_0^{x_1} r_0^2 dx_1}{r_0^2 x_1}\right]^{\frac{1}{2}}.$$

Examples have previously been given for flows of fluids of constant properties. (See p. 130; $x_1$ and $x_2$ here are the coordinates there denoted by $r$ and $x$, respectively; $r_0$ is to be taken as $r$ (or $x_1$); the stream function on p. 130 is the stream function here divided by $\rho$ and with its sign changed.) Another example is provided by the flow past a circular cone at zero incidence, with a shock wave from the tip; behind the shock wave, the pressure, density, and velocity are constant along any radius vector through the tip and outside the boundary layer; and the flow is irrotational outside the boundary layer. Hence the corresponding two-dimensional flow (if both flows are laminar) is the flow in the boundary layer of a flat plate at zero incidence in a uniform stream, with the same pressure, density, and velocity as those just outside the boundary layer of the cone. If $\alpha$ is the semi-vertical angle of the cone, $r_0 = x_1 \sin \alpha$, so $c_f' R'^{\frac{1}{2}}/(c_f R^{\frac{1}{2}}) = 1/\sqrt{3}$, and $x_1' = (x_1^3 \sin^2 \alpha)/(3 d^2)$.

In the remainder of this chapter we consider steady two-dimensional motion only.

### 9.3. Von Mises's transformation for steady, two-dimensional motion

Consider steady two-dimensional motion. Introduce $\rho_e U \, dU/dx_1$ for $-\partial p/\partial x_1$ in (199), and $\rho_e v_1 \, dI_e/dx_1$ for $Dp/Dt$ in (200). The equation of continuity is satisfied by the introduction of a stream function $\psi$, such that

(203) $\qquad \rho v_1 = \partial \psi/\partial x_2, \quad \rho v_2 = -\partial \psi/\partial x_1.$

Take $x_1$ and $\psi$ as independent variables in place of $x_1$ and $x_2$, and write $x$ in place of $x_1$. Then

$$v_1(\partial/\partial x_1)_{x_2} + v_2(\partial/\partial x_2)_{x_1} = v_1(\partial/\partial x)_\psi, \quad (\partial/\partial x_2)_{x_1} = \rho v_1(\partial/\partial \psi)_x,$$

## BOUNDARY LAYERS IN GASES

and the equations of momentum and energy are

(204a) $$\rho v_1 \frac{\partial v_1}{\partial x} - \rho_e U \frac{dU}{dx} = \rho v_1 \frac{\partial}{\partial \psi}\left(\mu \rho v_1 \frac{\partial v_1}{\partial \psi}\right)$$

(204b) $$\rho v_1 \frac{\partial I}{\partial x} - \rho_e v_1 \frac{dI_e}{dx} = \frac{\rho v_1}{Pr}\frac{\partial}{\partial \psi}\left(\mu \rho v_1 \frac{\partial I}{\partial \psi}\right) + \mu \left(\rho v_1 \frac{\partial v_1}{\partial \psi}\right)^2.$$

($U$, $I_e$, $\rho_e$, $\mu_e$ are functions of $x_1$ (or $x$) only, and are independent of $\psi$.)

We also find it convenient, later, to introduce non-dimensional variables, and change the $x$-scale. Write

(205a) $r = \rho/\rho_e$, $m = \mu/\mu_e$, $u = v_1/U$, $i = I/I_e$, $t = T/T_e$.

(205b) $$\theta = mr,$$

(205c) $$s = \int_0^x \mu_e \rho_e U \, dx.$$

Note first that through a section of the boundary layer, $p$ is taken as constant and equal to $p_e$, so the equation of state for a perfect gas, $p/\rho T = p_e/\rho_e T_e$, gives

(206a) $$rt = 1, \quad \theta = m/t,$$

and, in the case of constant specific heats,

(206b) $$i = t, \quad ri = 1.$$

With this nomenclature the equations (204) become

(207a) $$\frac{1}{U}\frac{dU}{ds}\left(u^2 - \frac{1}{r}\right) + u\frac{\partial u}{\partial s} = u\frac{\partial}{\partial \psi}\left(\theta u \frac{\partial u}{\partial \psi}\right)$$

(207b) $$\frac{dI_e}{ds}\left(i - \frac{1}{r}\right) + I_e \frac{\partial i}{\partial s} = \frac{I_e}{Pr}\frac{\partial}{\partial \psi}\left(\theta u \frac{\partial i}{\partial \psi}\right) + U^2 \theta u \left(\frac{\partial u}{\partial \psi}\right)^2.$$

For constant specific heats, the first term in (207b) is zero.

If we assume that $\mu$ is proportional to $T$, $\mu/\mu_e = T/T_e$, then $m = t$, and for a perfect gas

(208) $$\theta = 1$$

whether the specific heats are constant or not. For moderate temperatures, say up to 500° K at any rate, the variation of $\mu/T$ for air is neither very large nor very rapid, and such an approximation may give a fair description of the phenomena to be expected, even if the results are not numerically accurate. The variation between 100°K and 500°K is from about $0.7 \times 10^{-6}$ to about $0.55 \times 10^{-6}$ (gm. per cm. per second per degree). The approximation may still serve a useful purpose for somewhat greater temperatures, but becomes unsuitable at high temperatures; moreover dissociation of oxygen and nitrogen molecules in the air must be allowed for, and we cannot apply the thermodynamical relations for a perfect, non-dissociating gas.

D. R. Chapman and M. W. Rubesin (*J. Ae. Sci.* **16** (1949), pp. 547–565) have used a constant value for $\theta$, not necessarily unity, which may be given different values in different calculations, according to the circumstances. In fact, equations (207) retain the same form if $\theta$ is chosen as any function of $x$ only, and $s$ is replaced by $s'$, where $ds' = \theta\, ds$,

$$s' = \int_0^x \theta \mu_e \rho_e U\, dx.$$

What this amounts to is that $\mu\rho$ is taken as constant and equal to some average value across each section, and the method of choosing the average may vary, continuously, from section to section. Then $s'$ is given by (205c) with the selected value of $\mu\rho$ in place of $\mu_e \rho_e$.

### 9.4. *An integral of the energy equation for* $Pr = 1$, *and for a heat-insulated surface or zero main-stream acceleration*

The form of the energy equation in the boundary layer, corresponding with equation (50b), is found by multiplying (204a) by $v_1$ and adding to (204b). The result is (since $U dU/dx_1 + dI_e/dx_1 = 0$)

(209) $$\frac{\partial}{\partial x}(I + \tfrac{1}{2}v_1^2) = \frac{\partial}{\partial \psi}\left[\mu\rho v_1 \frac{\partial}{\partial \psi}(I/Pr + \tfrac{1}{2}v_1^2)\right].$$

If $Pr = 1$, the equation has the particular integral

(210) $$I + \tfrac{1}{2}v_1^2 = \text{constant} = I_e + \tfrac{1}{2}U^2.$$

This makes $I = I_e$ when $v_1 = U$, and the thicknesses of the velocity and temperature boundary layers are the same, which is acceptable for $Pr = 1$; but at the wall (210) makes $\partial I/\partial x_2 = -v_1 \partial v_1/\partial x_2 = 0$, so this particular integral applies only for a *heat-insulated* wall.

On the other hand, if $dU/dx_1 = 0$ (as in the first approximation for flow along a flat plate), a more general integral may be obtained, with $Pr = 1$, by multiplying (204a), divided by $\rho v_1$, by any constant $C$ and adding to (209). This gives

$$(211) \quad \frac{\partial}{\partial x}(I + \tfrac{1}{2}v_1^2 + Cv_1) = \frac{\partial}{\partial \psi}\left[\mu\rho v_1 \frac{\partial}{\partial \psi}(I + \tfrac{1}{2}v_1^2 + Cv_1)\right]$$

with the particular integral

$$(212) \quad I + \tfrac{1}{2}v_1^2 + Cv_1 = \text{constant} = I_e + \tfrac{1}{2}U^2 + CU$$

If we write

$$(213) \quad b = U^2/I_e,$$

and $C' = CU/I_e$, then in terms of the reduced variables of (205), (212) becomes

$$i + \tfrac{1}{2}bu^2 + C'u = 1 + \tfrac{1}{2}b + C',$$

while (210) gives the same result with $C' = 0$. Hence for a *heat-insulated* surface (including a heat-insulated flat plate)

$$(214) \quad i = 1 + \tfrac{1}{2}b(1 - u^2).$$

For a flat plate $(dU/dx = 0)$, with its temperature given, and therefore $i_w$ given, $1 + \tfrac{1}{2}b + C' = i_w$ (since $u_w = 0$), so

$$(215) \quad i = i_w + [1 + \tfrac{1}{2}b - i_w]u - \tfrac{1}{2}bu^2.$$

The subscript $w$ is used to denote conditons at the wall ($x_2 = \psi = 0$).

For a heat-insulated surface, from (210), $I_w = I_e + \tfrac{1}{2}U^2$, so $i_w = 1 + \tfrac{1}{2}b$, and then (215), when it applies, reduces to (214).

For a perfect gas with constant specific heats, in the equation (207a) for $u$, $1/r = i$, and $i$ is a quadratic function of $u$ from either (214) or (215). Also $\theta = m/t$ is a known function of $t$ (since $\mu$ is a known function of $T$); hence $\theta$ is a known function of $i$, and therefore, from (214) or (215), of $u$. Hence in this case, if either of the integrals (214) or (215) applies, (207a) becomes an independent equation for $u$. In the general case, when neither of these energy integrals applies, the two simultaneous equations (207) must be solved for $u$ and $i$.

Note that for a perfect gas with constant specific heats $I = a^2/(\gamma - 1)$, where $a$ is the sound speed, so

(216) $$b = (\gamma - 1)M_e^2.$$

Moreover, $I_e + \tfrac{1}{2}U^2 = $ constant gives

$$U^2/M_e^2 + \tfrac{1}{2}(\gamma - 1)U^2 = \text{constant},$$

and after a little calculation we find that

(217) $$(1 + \tfrac{1}{2}b)U^{-1}\,dU/ds = M_e^{-1}\,dM_e/ds.$$

This result is used to substitute for $U^{-1}dU/ds$ in (207a) in the next section.

Also, for a heat-insulated surface in a perfect gas with constant specific heats it follows from (210) that the wall temperature is the same (with $Pr = 1$, as throughout this paragraph) as the "stagnation" temperature of the main stream when brought adiabatically to rest, namely

$$T_w = T_e + \tfrac{1}{2}U^2/c_p = T_0 + \tfrac{1}{2}U_0^2/c_p,$$

where the subscript zero refers to values in the undisturbed main stream. It follows that

$$T_w = T_0[1 + \tfrac{1}{2}(\gamma - 1)M_0^2].$$

### 9.5. The Illingworth-Stewartson theorem

We now prove that under certain conditions the problem of determining the velocity distribution in a boundary layer in a gas may be transformed into a boundary-layer problem for a fluid of constant properties with a main-stream velocity distribution which is different from, but may be determined in terms of, the main-stream velocity distribution in the gas. The conditions are (i) Perfect gas with constant specific heats. (ii) $Pr = 1$. (iii) $\mu \propto T$. (iv) Heat-insulated boundary. The theorem was discovered independently by Illingworth (*Proc. Roy. Soc. (London)* **A199** (1949), pp. 533–557) and Stewartson (*Proc. Roy. Soc. (London)* **A200** (1949), pp. 84–99). Here we follow Illingworth's method.

Under the conditions stated, $\theta = 1$, $1/r = i$, $i$ is given by (214), $u^2 - 1/r = (1 + \frac{1}{2}b)(u^2 - 1)$. Substitute into (207a), and use (217). Thus the equation for $u$ becomes

$$(218) \qquad \frac{1}{M_e}\frac{dM_e}{ds}(u^2 - 1) + u\frac{\partial u}{\partial s} = u\frac{\partial}{\partial \psi}\left(u\frac{\partial u}{\partial \psi}\right).$$

Now compare with the equation corresponding with (207a) for a fluid of constant properties, for which we use primed quantities. In this case, $\rho' = \rho'_e$, $\mu' = \mu'_e$, i.e., $r' = m' = 1$, so from (205b) $\theta' = 1$. The definition of $s'$ is

$$s' = \mu'\rho' \int_0^{x'} U'\, dx'.$$

(Also the definition of $\psi'$ is

$$\rho' v'_1 = \partial \psi'/\partial x'_2, \quad \rho' v'_2 = -\partial \psi'/\partial x'_1.)$$

Then the equation corresponding with (207a) is

$$(219) \qquad \frac{1}{U'}\frac{dU'}{ds'}(u'^2 - 1) + u'\frac{\partial u'}{\partial s'} = u'\frac{\partial}{\partial \psi'}\left(u'\frac{\partial u'}{\partial \psi'}\right).$$

(218) and (219) are the same if we take $s' = s$, $\psi' = \psi$, $u' = u$, and

$$\frac{1}{U'}\frac{dU'}{ds} = \frac{1}{M_e}\frac{dM_e}{ds},$$

i.e., $U' = AM_e$, where $A$ is a constant, with the dimensions of a velocity. Moreover the boundary conditions are the same, $u = 0$ at $\psi = 0$ and $u \to 1$ exponentially as $\psi \to \infty$.

$A$, $\rho'$, $\mu'$ are still at our disposal. We put them equal to the velocity of sound, the density, and the viscosity at some specified point in the compressible flow. Illingworth took the values at a stagnation point in the main stream. Here we will take the values in the undisturbed main stream in the compressible flow, denoting values there by the subscript zero. Thus $A = a_0$, $\rho' = \rho_0$, $\mu' = \mu_0$, and $U' = a_0 M_e$. Since $U = a_e M_e$, we have also $U'/U = a_0/a_e$.

To calculate the relation between $x$ and $x'$, we have
$$ds = \mu' \rho' U' dx' (= \mu_0 \rho_0 U' dx') = \mu_e \rho_e U dx,$$
so
$$dx'/dx = \mu_e \rho_e a_e / (\mu_0 \rho_0 a_0) = (\rho_e/\rho_0)(T_e/T_0)^{\frac{3}{2}} \quad \text{(since } \mu \propto T, a \propto T^{\frac{1}{2}})$$
$$= (p_e/p_0)^{(3\gamma-1)/2\gamma}.$$
Hence
(220) $$x' = \int_0^x (p_e/p_0)^{(3\gamma-1)/2\gamma} dx.$$

$p_e$ is the pressure in the main stream, just outside the boundary layer, at the section $x$ in the compressible flow, and $p_e/p_0$ is a known function of $x$. Then

(221) $$U'(x') = a_0 M_e(x), \quad U'(x')/U(x) = a_0/a_e(x).$$

Hence we may state the "incompressible" problem corresponding to a given "compressible" one. We suppose we solve the "incompressible" problem, and find $u$ at $x_1'$, $x_2'$ ($x_1' = x'$). $x_1$ is known in terms of $x_1'$; it remains to find $x_2$ in terms of $x_2'$. For fixed and corresponding values of $x_1$ and $x_1'$,

$$d\psi = \rho v_1 dx_2 = \rho' v_1' dx_2'.$$
Hence
$$\frac{dx_2}{dx_2'} = \frac{\rho' v_1'}{\rho v_1} = \frac{\rho_0 U' u'}{\rho_e r U u} = \frac{\rho_0 a_0}{r \rho_e a_e} = \frac{1}{r} \frac{\rho_0}{\rho_e} \left(\frac{T_0}{T_e}\right)^{\frac{1}{2}} = \frac{1}{r}\left(\frac{p_0}{p_e}\right)^{(\gamma+1)/2\gamma},$$

and $r^{-1} = i$, so from (214) and (216) (since $p_0/p_e$ is independent of $x_2$ and $x_2'$),

(222) $$x_2 = (p_0/p_e)^{(\gamma+1)/2\gamma} \int_0^{x_2'} [1 + \tfrac{1}{2}(\gamma - 1) M_e^2 (1 - u^2)] dx_2'$$
Hence
$$x_2/x_2' > (p_0/p_e)^{(\gamma+1)/2\gamma},$$

and it is easy to see that, at any rate if $U \geqq U_0$, $x_2 > x_2'$, and the 'compressible' boundary layer is thicker than the 'incompressible' one.

Let the subscript $w$ denote condition at the wall ($x_2 = 0$), as before. The ratio of the skin-frictions at corresponding values of $x_1$ and $x_1'$ is given by

BOUNDARY LAYERS IN GASES

$$\frac{\tau_w}{\tau'_w} = \frac{\mu_w(\partial v_1/\partial x_2)_{x_2=0}}{\mu'(\partial v'_1/\partial x'_2)_{x'_2=0}} = \frac{\mu_w}{\mu_0}\frac{U}{U'}\frac{(\partial u/\partial x_2)_{x_2=0}}{(\partial u/\partial x'_2)_{x_2=0}}$$

$$= \frac{T_w}{T_0}\frac{a_e}{a_0}\left(\frac{dx'_2}{dx_2}\right)_{x_2=0} = \frac{T_w}{T_0}\left(\frac{T_e}{T_0}\right)^{\frac{1}{2}}r_w\left(\frac{p_e}{p_0}\right)^{(\gamma+1)/2\gamma},$$

and $r_w = 1/i_w = I_e/I_w = T_e/T_w$, so

(223) $\quad \tau_w/\tau'_w = (T_e/T_0)^{\frac{3}{2}}(p_e/p_0)^{(\gamma+1)/2\gamma} = (p_e/p_0)^{(2\gamma-1)/\gamma}.$

In a compressible fluid the displacement thickness $\delta_1$, defined in a similar way as for a fluid of constant properties, is given by

(224) $\quad \rho_e U \delta_1 = \int_0^\infty (\rho_e U - \rho v_1)dx_2,$ i.e., $\delta_1 = \int_0^\infty (1 - ru)dx_2.$

In an incompressible fluid

$$\delta'_1 = \int_0^\infty (1 - u)dx'_2,$$

and a momentum thickness, $\vartheta'$, may also be defined, equal to

$$\int_0^\infty u(1 - u)dx'_2$$

(see *Modern Developments*, Section 52; Schlichting, Section VIII e.). Then

$$\delta'_1 + \vartheta' = \int_0^\infty (1 - u^2)dx'_2,$$

and

(225)
$$\begin{aligned}
\delta_1 &= \int_0^\infty (1 - ru)\left(\frac{p_0}{p_e}\right)^{(\gamma+1)/2\gamma}\frac{1}{r}dx'_2 \\
&= \left(\frac{p_0}{p_e}\right)^{(\gamma+1)/2\gamma}\left[\int_0^\infty (i - u)dx'_2\right] \\
&= \left(\frac{p_0}{p_e}\right)^{(\gamma+1)/2\gamma}\left[\int_0^\infty (1 - u)dx'_2 + \tfrac{1}{2}b\int_0^\infty (1 - u^2)dx'_2\right] \\
&= \left(\frac{p_0}{p_e}\right)^{(\gamma+1)/2\gamma}[\delta'_1 + \tfrac{1}{2}(\gamma - 1)M_e^2(\delta'_1 + \vartheta')].
\end{aligned}$$

The results for a heat-insulated flat plate, $dU/dx = 0$, are easily written down, if the Mach number is not large and all interaction effects between the boundary layer and the main stream, including the presence of a shock-wave, are neglected, and the plate is considered of zero thickness. For we then take $p_e = p_0$, so $x' = x$, $\tau = \tau'$, and from the known numerical values for an incompressible fluid

$$\delta_1/\delta_1' = 1 + 0.693(\gamma - 1)M_e^2,$$

(the last term being $0.28M_e^2$ for $\gamma = 1.4$). $x_2$ is given by the integral in (222).

### 9.6. Remarks on flow past a flat plate ($dU/dx = 0$)

If again we consider a boundary layer for which $dU/ds = dI_e/ds = 0$, the equations (207) become

(226a) $$\frac{\partial u}{\partial s} = \frac{\partial}{\partial \psi}\left(\theta u \frac{\partial u}{\partial \psi}\right)$$

(226b) $$\frac{\partial i}{\partial s} = \frac{1}{Pr}\frac{\partial}{\partial \psi}\left(\theta u \frac{\partial i}{\partial \psi}\right) + b\theta u \left(\frac{\partial u}{\partial \psi}\right)^2.$$

We do not make the assumptions of the preceding paragraph. However, we remark first that, as in that paragraph, $\theta = 1$ both for a fluid of uniform properties and for a perfect gas (not necessarily with constant specific heats) for which $\mu \propto T$. The boundary conditions also correspond exactly in the two cases. When $\theta = 1$, the equation (226a) is an equation for $u$ only; and when $u$ is known, (226b) is an equation for $i$. The solutions for a fluid of constant properties are known, and may be adapted to the case of the perfect gas with $\mu \propto T$. I shall not attempt here to set out the details, which are easily worked out, but shall make some remarks about the general case, when $\theta$ is not necessarily equal to 1. My main purpose is to exhibit the connection between the methods used by Illingworth and by Crocco. In general, in the boundary layer, for a perfect gas $\theta = m/t$ and is a known function of temperature or enthalpy. The boundary conditions are, for the first approximation, $u = i = 1$ at $\psi = \infty$ and at $s = 0$, $u = 0$ at $\psi = 0$, and either $i$ is given or $\partial i/\partial x_2 = 0$, i.e. $u\partial i/\partial \psi = 0$, at $\psi = 0$. If we write

(227a) $$\eta = \psi/(2s)^{\frac{1}{2}},$$

$u$ and $i$ are functions of $\eta$ alone, and

(227b) $$\frac{d}{d\eta}\left(\theta u \frac{du}{d\eta}\right) + \eta \frac{du}{d\eta} = 0$$

(227c) $$\frac{d}{d\eta}\left(\theta u \frac{di}{d\eta}\right) + Pr\,\eta \frac{di}{d\eta} + Pr\,b\theta u \left(\frac{du}{d\eta}\right)^2 = 0.$$

This is a pair of ordinary simultaneous differential equations that can be solved numerically for any given value of $Pr$ and variation of $\theta$ with $i$. When $\theta = 1$, they must be equivalent to known equations, and they may be reduced to these known forms by Illingworth's method, introducing a variable $\zeta$ for which

$$\frac{d\eta}{d\zeta} = u, \quad \zeta = \int_0^\eta \frac{dz}{u(z)}.$$

(The integral converges, since $u(\eta) = O(\eta^{\frac{1}{2}})$ as $\eta \to 0$.) With $\eta$ and $i$ as dependent variables the equations are

$$\frac{d}{d\zeta}\left(\theta \frac{d^2\eta}{d\zeta^2}\right) + \eta \frac{d^2\eta}{d\zeta^2} = 0$$

$$\frac{d}{d\zeta}\left[\theta \frac{di}{d\zeta}\right] + Pr\,\eta \frac{di}{d\zeta} + Pr\,b\theta \left(\frac{d^2\eta}{d\zeta^2}\right)^2 = 0,$$

with the boundary conditions $d\eta/d\zeta = 1$, $i = 1$ at $\zeta = \infty$, and $\eta = 0$, $d\eta/d\zeta = 0$, together with a condition on $i$, at $\zeta = 0$. When $\theta = 1$, the solutions are known. For other $\theta$, the equations may be turned into integral equations and solved by successive approximation, starting from the solution for $\theta = 1$ as a first approximation for any $\theta$. (See Illingworth's paper.)

For our purposes, a different procedure is preferable. Since $i$ and $u$ are functions of $\eta$, $i$ may be expressed as a function of $u$ (*cf.* W. Hantzsche and H. Wendt in *Jahrbuch der deutschen Luftfahrtforschung*, 1940 and 1942). We have seen that, for $Pr = 1$, this function is given by (215) irrespective of the variation of $\theta$. We are concerned mainly with cases in which $Pr$ is fairly near to 1. We first look at the resulting equation when $i$ is considered a function of $u$. From (227c)

$$\frac{d}{d\eta}\left[\theta u \frac{di}{du}\frac{du}{d\eta}\right] + Pr\,\eta \frac{di}{du}\frac{du}{d\eta} + Pr\,b\theta u\left(\frac{du}{d\eta}\right)^2 = 0.$$

Write the first term as

$$\theta u\left(\frac{du}{d\eta}\right)^2 \frac{d^2 i}{du^2} + \frac{di}{du}\frac{d}{d\eta}\left(\theta u \frac{du}{d\eta}\right),$$

substitute for $d(\theta u\,du/d\eta)/d\eta$ from (227b), and divide by $\theta u(du/d\eta)^2$. Hence

(228) $$\frac{d^2 i}{du^2} - \frac{(1-Pr)\eta}{\theta u\,du/d\eta}\frac{di}{du} + Pr\,b = 0.$$

For $Pr = 1$, the integral with the boundary conditions $i = i_w$ at $u = 0$, $i = 1$ at $u = 1$, gives (215). When $Pr \neq 1$, regard for the moment the coefficient of $di/du$ as a known function of $u$. Then if

$$G(u) = \exp\left[-\int_0^u \frac{\eta}{\theta u\,du/d\eta}\,du\right] = \exp\left[-\int_0^\eta \frac{\eta}{\theta u}\,d\eta\right]$$

the solution is

(229) $$i = i_w + C\int_0^u G^{Pr-1}\,du - Pr\cdot b\int_0^u G^{Pr-1}\left[\int_0^u G^{1-Pr}\,du\right]du,$$

where $C$ is a constant, to be found from the condition that $i = 1$ when $u = 1$. For a heat-insulated plate, $di/du = 0$ at $u = 0$, and $C = 0$; the condition $i = 1$ when $u = 1$ then gives the value of $i_w$.

From (227b)

$$\frac{\eta}{\theta u} = -\frac{d}{d\eta}\log\left(\theta u\frac{du}{d\eta}\right),$$

whence

$$G = \frac{\theta u\,du/d\eta}{(\theta u\,du/d\eta)_{\eta=0}}.$$

But if $\tau$ is the shear stress,

$$\tau = \mu \frac{\partial v_1}{\partial x_2} = \mu\rho v_1 \frac{\partial v_1}{\partial \psi} = \mu_e \rho_e U^2 \theta u \frac{\partial u}{\partial \psi} = \left[\frac{\mu_e \rho_e U^2}{(2s)^{\frac{1}{2}}}\right]\theta u\frac{du}{d\eta},$$

so

(230) $$G = \tau/\tau_w.$$

This shows how Illingworth's method is related to that of Crocco, who takes $\tau$ as a dependent variable and $u$ as an independent one.

Unless $\theta = 1$, in which case we can solve for $u$ independently as a function of $\eta$, and so find $G$ as a function of $\eta$ or of $u$, we do not seem to have gained much. We should have to proceed by successive approximation, starting with $\theta = 1$, say. However, $G$ is, in fact, a restricted kind of function; it must be 1 at $u = 0$ and 0 at $u = 1$, where $\tau$ vanishes; and the nature of its variation (i.e., the variation of the shearing stress) will not depend unduly on the variation of $\mu$ with $T$, i.e., on the value of $\theta$. Moreover, $G$ only occurs raised to the power $1 - Pr$, which for gases is not large. So we expect that, for gases, the variation of $i$ with $u$ will be almost independent of the viscosity-temperature law. Since, for a "flat plate," $I_e$ and $U$ are constants, the enthalpy-velocity relation across the boundary layer will be approximately independent of the viscosity-temperature law.

We therefore expect the results that depend on the enthalpy-velocity relation only, and are found with $\theta = 1$, to be good approximations. The most important are as follows. The wall temperature for a heat-insulated wall, or the "final" temperature that would be attained when heat transfer ceases, is approximately

(231) $\qquad T_f = T_e[1 + \tfrac{1}{2}(\gamma - 1)\, Pr^{\frac{1}{2}} M_e^2],$

if $c_p$ is taken as constant. Also, with $c_p$ constant, the ratio of the heat transfer $q$ to the shear stress at the wall is, for a given wall temperature $T_w$,

(232) $\qquad q_w/\tau_w = c_p(T_w - T_f)/[(Pr)^{\frac{2}{3}} U],$

where $T_f$ is given in (231). But to obtain either $q_w$ or $\tau_w$ separately to a sufficient approximation, we must allow for the variation of $\theta$, i.e., of $\mu/T$. For further details see *Modern Developments: High Speed Flow*.

## 9.7. Conclusion

I have made an attempt only to give some of the simplest arguments, that can be put in something like an analytical form.

This does little more than scratch the surface of the subject. Extensions are needed to flows with main-stream accelerations, and wall temperatures that vary along the wall, and to applications for flows past cylindrical and axisymmetrical bodies, with the bow shocks included in the computations. Reference should be made to the literature for other works, and for the valuable results of numerical computations.

It should be pointed out that the results given here will not apply unaltered to high Mach numbers and large temperatures. Some of the formulae stated indicate that very high wall temperatures will be attained at high Mach numbers, and to this extent give a warning of the so-called "thermal barrier." But they are highly unrealistic, and remain so when main-stream accelerations are allowed for. An accurate calculation of the surface temperatures of bodies moving through gases at very high speeds is a very complicated matter. Shock waves have not been allowed for. Knowledge of the "steady-state" temperature is not all that is required; a knowledge of the transient temperature — the rate at which high surface temperatures are built up — is also required. Heat conduction within the solid body must be taken into account. So must the radiation of heat from the surface of the solid. (For a mathematical theory which takes radiation into account, see Lighthill, *Proc. Roy. Soc. (London)*, **A202** (1950), pp. 359–377. With $\mu \propto T$, and a uniform stream along a surface, but with an arbitrary temperature distribution along it, an approximate expression is found for the heat transfer to the surface. The "steady-state" temperature is then deduced by taking this heat transfer to be in local balance at each position on the surface with the heat radiated from it; the calculation requires the solution of an integral equation.) A theory is also required that allows for partial vaporization of the material of the solid, either directly or after partial liquefaction. The gas itself will be at high temperatures in parts of the boundary layer, and "real gas" effects at high temperatures (dissociation, ionization, etc.) will appreciably affect the results. Moreover, in the general case chemical reactions will be proceeding in the boundary layer. Much work still remains to be done.

CHAPTER 10

# Turbulence. Stability

## 10.1. Turbulence

For flow along a circular pipe, under conditions in which the density and viscosity of the fluid are sensibly constant, at sufficiently low Reynolds numbers the calculated laminar motion corresponds with the experimental results at a sufficient distance from the entry for the permanent regime to have become established. If the speed of the flow is increased, a stage is reached at which the rectilinear motion in the permanent regime breaks down, and the motion becomes turbulent. With further increase of velocity the position of the breakdown (or transition to turbulence) may approach the intake, but with a smooth intake it does not reach it. After the transition, the stream is constantly varying, with irregular motions across the tube.

In the turbulent flow, we may consider a mean motion, over a time sufficient for the fluctuations to be smoothed. The distribution of the mean velocity is no longer parabolic; there is a sharp rise near the walls, and over the rest of the tube the distribution is considerably flatter than in laminar flow. If we put the mean pressure gradient $\bar{P}$ proportional to $(\bar{u}_m)^n$, $n$ is no longer unity (cf. Section 6.1); it depends on the Reynolds number and lies between 1 and 2.

Motion in a boundary layer also becomes turbulent if the Reynolds number is too high. There is a laminar portion upstream, followed by a transition to turbulence. Flows in jets and wakes also become turbulent. In fact all fluid motions at large enough Reynolds numbers exhibit turbulence. In the design of wind tunnels, special designs are used and great care is taken in those cases in which it is desired to have as low a degree of turbulence as possible.

Discussion of turbulent motion is necessarily a long business, if it is to be at all worthwhile. Therefore I must be content to give references. Fortunately this has become an easier and more satisfactory procedure because of two publications: G. K. Batchelor, *The Theory of Homogeneous Turbulence*, Cambridge Univ. Press, 1953 and A. A. Townsend, *The Structure of Turbulent Shear Flow*, Cambridge Univ. Press, 1956.

First, I would remark that for qualitative descriptions and the older, semi-empirical, theories of technical usefulness, reference may be made to *Modern Developments*, and to Schlichting's book. In this connection, I should remark that "mixing-length" theories are now held to be physically unsound and are out of favor; and every effort is made to present the "theory" on a basis of dimensional and similarity arguments. Of course, "mixing-length" theories always were physically unsound, and this was known: they were used merely as a convenient device for building semi-empirical theories, which involved dimensional and similarity arguments. Probably it is best now to dispense with them as much as possible. In certain cases, however, the ideas may still be useful, e.g. when a more general relation is required for the turbulent stress-tensor than in a simple shearing flow, and the concept must be generalized to a tensor concept; such calculations, even if mathematically difficult, must still be regarded as semi-empirical. Among recent additional papers on turbulent boundary layers for incompressible fluids I may mention G. B. Schubauer, *J. Appl. Phys.* **25** (1954); D. Coles, *J. Fluid Mechs.* **1** (1956), pp. 191–226; E. R. van Driest, *J. Ae. Sci.* **23** (1956), pp. 1007–11; A. A. Townsend, *J. Fluid Mechs.* **1** (1956), pp. 561–73, in which further references may be found. For turbulent boundary layers in compressible fluids, and for shock-wave boundary-layer interactions, reference may be made to *Modern Developments*: *High Speed Flow*, A. D. Young, Chap. X, Sections 16–18, and the references there given. (When Vols. 4 and 5 of the Princeton series on High Speed Aerodynamics and Jet Propulsion appear, these should be added.)

There is some brief discussion of the theory of isotropic turbulence in *Modern Developments*; at the time of publication, the theory, begun by G. I. Taylor, was in the early stages of development. Reference may now be made to Batchelor's book, mentioned above. A beginning of a fundamental physical consideration of turbulent shear flow is made by Townsend in his book.

## *10.2. Stability*

Here again, I am fortunate in being able to content myself with a reference to a recently published book: C. C. Lin, *The Theory of Hydrodynamic Stability*, Cambridge Univ. Press, 1955. The mathematical theory investigates the stability of a flow for infinitesimal disturbances, and involves the solution of a characteristic-value problem, usually for a linear fourth-order differential equation. The results obtained are well substantiated by experiment. Theoretically, much has been accomplished; but much remains to be done, even for infinitesimal disturbances; for finite disturbances, there has been only some initial "breaking of the ground." Even for infinitesimal disturbances, with a linear differential equation, the mathematical questions involved are far from easy in many cases, for which the instability that arises is intimately connected with the existence of a singularity in the equation for inviscid fluids at the point where the wave velocity of the disturbance is equal to the undisturbed fluid velocity; viscous forces then act to produce instability. For details and references, see Lin's book.

CHAPTER 11

# Dynamics of Inviscid Gases

## 11.1. Introduction

In the short space that is available, I find it impossible to give anything approaching a satisfactory and intelligible résumé of the development of the theory of the dynamics of inviscid gases. Moreover, almost every topic that has been considered is well discussed in one or more of the published works. It therefore seems best simply to select parts of a particular topic to discuss here; I have selected slender-body theory. I hope that such details as are discussed will be fairly intelligible on their own. However, what must of necessity be unintelligible without considerable further reading and study is the connection of this topic with others, and its place in the general development of the subject. Therefore I should have liked to say sufficient here to enable the next two sections to be placed in some kind of pattern, but even that seems impossible without a disquisition. The following remarks are therefore intended to be considered in conjunction with, and to be amplified by, reading in the standard books, of which several (Liepmann and Roshko, Howarth, Sears, Courant and Friedrichs, Ward, Frankl and Karpovich) are listed in the bibliography. In addition, reference may be made to three general lectures — Karman, *J. Ae. Sci.* **8** (1941), pp. 337-356 or *Collected Works*, Vol. 4, pp. 127 164, and *J. Ae. Sci.* **14** (1947), pp. 373–409 or *Collected Works*, Vol. 4, pp. 271-326; Lighthill, *J. Ae. Sci.* **16** (1949), pp. 69-83 — which elucidate the standing of the theory and the main points of general interest at the times of delivery of the lectures. A more recent lecture by Karman, *Proc. Conference on High-Speed Aeronautics*, Polytechnic Institute of Brooklyn, 1955, pp. 11-39, which deals briefly with this and other subjects, may also be consulted.

The greater difficulty encountered in attempting a unified

résumé of the theoretical development of the dynamics of inviscid compressible fluids, as compared with that for inviscid incompressible fluids, appears to arise from four main features.

For incompressible fluids we may say, mathematically, that we are concerned with harmonic functions satisfying certain boundary conditions, with possible lines and, more importantly, possible surfaces of discontinuity, i.e., we are concerned with potential flows with possible line vortices and vortex sheets. For unsteady motions — motions for which there is no frame of reference relative to which the motion is steady — we may have to consider the development of vortex sheets (and sometimes the deformation of vortex sheets and motions of line vortices); apart from this, and differences in the calculation of the pressure, we still have to consider solutions of the same differential equation — Laplace's equation. (The motion of a uniformly rotating inviscid fluid of constant density, or any motion of such a fluid with diffused vorticity, may be treated as a separate topic: see the remarks at the beginning of Appendix III.)

For the motion of a gas, we see from equation (98) that even for a continuous, irrotational motion with constant entropy, because of the finiteness of the velocity of propagation of a disturbance, a problem of unsteady motion is different from, and harder than, any comparable problem of steady motion. Secondly, even for steady motions, the governing equation is non-linear. Thirdly, the equation is of "mixed type," elliptic where the motion is subsonic and hyperbolic where it is supersonic, and the position of the boundary between the two — the sonic line or lines — is *a priori* unknown. Fourthly, we have seen in Section 4.4(b) that a continuous solution of rather a simple problem may not, in fact, exist. The problem of Section 4.4(b) was an unsteady problem, but the same remark is true for many problems of steady motion involving supersonic flow. Thus discontinuities must now be introduced not only, as for incompressible fluids, to obtain better agreement with experiment, but in order to obtain a solution at all; and the discontinuities must now certainly include shock discontinuities. Moreover, a motion which is isentropic and irrotational upstream of a shock wave is, in general, neither isentropic nor irrotational downstream, and is approximately so only if the shock wave is sufficiently weak.

It is therefore not surprising that when we look over the accounts of work on various topics, the whole subject appears to have lost any simple unifying theme, and it is only with effort that we recall the physical unity, and are encouraged to search for greater mathematical unity, simplicity, and generality. Much of this, of course, stems from our use of a rather crude model for any real gas; if we could only achieve a mastery, for example, of the equations for a viscous, heat-conducting gas, learn to understand them and to forecast from them for particular situations, the unity of the whole subject would certainly be seen.

In fact most of the calculations for inviscid gases of which accounts have been published are for perfect gases with constant specific heats, and from the physical point of view it should be remembered that (i) the gas has been taken as a continuum (ii) viscosity and heat conduction, boundary-layer effects, etc. have been neglected (iii) variations of the specific heats and departures from the perfect-gas laws (of particular importance at high temperatures) have not been allowed for. However, the calcuations have both mathematical interest and physical and technical importance; but care must be taken not to apply the results where neglected effects are important.

Topics from the dynamics of inviscid gases may, as we have seen, be classified as topics involving steady or unsteady motion. They may also be classified by the number of independent space variables involved — one, two, or three. Also, according to the disposition of the boundaries of the gas, they may be classified as topics arising from external or internal aerodynamics (or ballistics). A typical "simple" problem from external aerodynamics is the determination of the flow arising from motion, relative to the gas, of a solid obstacle in a gas which is otherwise unbounded; a typical problem from internal aerodynamics is the motion of a gas through a tube or nozzle. Of course there is nothing absolute about these classifications; but for the sake of brevity we do not enter here into questions of transformations and approximations which cross from one classification to another.

Problems involving only one space coordinate do not arise from external aerodynamics. For unsteady motion, they are largely problems of the propagation of finite pressure waves and shock and blast waves in one direction, or with circular-cylindrical or spherical symmetry; contact discontinuities, and the theory of the shock tube, are also to be included. For steady motion there is an approximate theory of motion along a straight tube of varying cross-sectional area, in which the variation of the cross-section is sufficiently gradual to make it a good approximation to neglect the velocity components of the gas parallel to a cross-section, and to take conditions across each section as uniform. The equation of continuity is replaced by $\rho u A =$ constant, where $A$ is the cross-sectional area, $u$ the velocity along the tube, and $\rho$ the density. The beginning of the theory is the same as that for flow along a narrow stream-tube (Lamb, Section 24a): the result is very well known, but is so fundamental to the physical understanding of the difference between subsonic and supersonic flow that I repeat it here. From the equation of continuity, with $A$ as the (small) cross section of the narrow stream-tube

(233) $$d\rho/\rho + du/u + dA/A = 0.$$

From Bernoulli's equation

(234) $$u du = -dp/\rho = -a^2 d\rho/\rho, \text{ i.e., } d\rho/\rho = -M^2 du/u,$$

where $a$ is the local speed of sound and $M$ the local Mach number. Hence, if $s$ is distance along the axis of the stream-tube

(235) $$\frac{du}{ds} = -\frac{u}{1-M^2}\frac{1}{A}\frac{dA}{ds}.$$

Thus when the flow is subsonic, in a converging tube $u$ increases, and the density, pressure, temperature, and local speed of sound all decrease. But when the flow is supersonic, the opposite is true — in a converging tube, $u$ decreases, and the density, pressure, temperature, and local sound-speed increase. In a diverging tube, of course, $u$ decreases in subsonic and increases in supersonic motion, etc. Also, at the narrowest section either $u$ is a maximum or minimum, or $M = 1$. If the velocity is increasing continually from a subsonic value, the tube first converges, and $M = 1$ is reached at the narrowest section; thereafter the tube diverges. When $M$ is near 1, a small fractional change in $A$ is accompanied by a large fractional change in $u$.

For further developments — the necessity under certain conditions of the appearance of shock waves, and the application to the design of supersonic wind-tunnels — reference may be made to the literature. (For example, Liepmann and Roshko, Chap. 5.)

We shall next consider briefly some aspects of two-dimensional steady flow past an obstacle. For a general description of the pattern of the flow past a symmetrical airfoil at zero incidence at various Mach numbers, reference may be made to Howarth, Chap. 1, Section 8. Before we start, however, to consider the mathematical questions that arise in connection with these patterns, it is convenient to make three separate remarks.

(a) If a uniform supersonic stream is incident on a (semi-infinite) wedge, there is no possible continuous flow pattern. If the wedge is placed symmetrically in the stream, and the half-angle of the wedge is less than the maximum possible deflection through a plane oblique shock at the incident Mach number (Appendix IV), then the solution is obtained with two symmetrically placed plane shocks through the leading edge of the wedge, with uniform flow upstream of the shock, and a uniform flow on each side of the wedge downstream of the shock. The inclinations of the shocks are such as just to give the desired deflections, making the uniform flows downstream parallel to the faces of the wedge. There are, mathematically, two possible solutions for each inclination. In practice, we cannot, of course, use a semi-infinite wedge, but the indications from finite wedges are that in ordinary circumstances it is the weaker of the two possible shocks that is found.

If the wedge is not placed symmetrically, then on each side the flow has to turn through an angle, and the two sides may be considered independently. If the corner to be turned through is concave to the relevant part of the flow field, the turning is still accomplished by a shock wave attached to the leading edge if the required stream deflection is not too large. If the wedge has been turned so far from the symmetrical position that one corner to be turned through is convex to the relevant part of the flow field, the turning is accomplished by a centered simple wave, or Prandtl-Meyer expansion (Appendix V). In any case, there is a uniform flow on each side of the wedge downstream of the shock or Prandtl-Meyer expansion. (In experiment, the gas usually overexpands through such a Prandtl-Meyer expansion, and returns by a shock to the required direction, but that is a nicety in our present consideration: it illustrates, however, the lack of uniqueness when discontinuities are allowed but not prescribed.)

If the required angle of deflection of the stream (half the wedge angle for a symmetrical wedge) is too large, no solution with an attached shock is possible. In that case, there must be a continuous (not plane) detached shock upstream of the leading edge of the wedge. For a symmetrical wedge, the shock is symmetrical; on the line of symmetry it is normal to the incident flow, and the flow behind it is subsonic. There is therefore some region of subsonic flow between the detached shock and the surface of the wedge. This must be the state of affairs for steady flow past a finite wedge, or past any body whose front end consists of a finite wedge, and this is confirmed by observation. It may also be the correct solution for a semi-infinite wedge, if the leading edge is at all rounded, and is not truly sharp. But Dr. Ludford has pointed out to me that for a semi-infinite wedge with a sharp leading edge, there is probably no unique solution of the steady flow problem, since no length enters into the problem by which the "stand-off" distance of a detached shock from the leading edge could be defined.

(b) The equations of motion for a steady, two-dimensional, isentropic, irrotational flow become linear if the velocity components $u$, $v$ are taken as independent variables, and the velocity potential $\phi$ and stream function $\psi$ as dependent variables. This is the hodograph transformation. (Polar coordinates in the hodograph or velocity plane are, in fact, used; a single equation may be obtained for $\psi$; alternatively $ux + vy - \phi$ may be taken as dependent variable and the transformation (with $u$ and $v$ as independent variables) is then the ordinary Legendre contact transformation of the equation for $\phi$.)

The resulting linear equations for $\phi$ and $\psi$ may be transformed into the simple Cauchy-Riemann relations if the equation of state connecting $p$ and $\rho^{-1}$ is linearized (the Karman-Tsien approximation). This approximation cannot be applied to flows which are partly subsonic, partly supersonic. Even without this approximation, if the usual relation, $p/\rho^\gamma =$ constant, for a perfect gas with constant specific heats is taken, mathematical discussion of the equation for $\psi$ has proceeded rather far. (Also, other approximations have been suggested to simplify the mathematics and to keep rather near the required physical relation over as large a range as possible.)

Essentially, it is the hodograph method that is used to solve "free stream-line" problems in incompressible fluids, such as the problem of a two-dimensional jet (Lamb, Section 75), and the application to a subsonic compressible jet was given by Chaplygin in his original paper (1904) on the use of the hodograph transformation in compressible flow. (See Lighthill, Chap. 7 of *Modern Developments: High Speed Flow*.) Its most important use in recent times has been to attempt to solve problems of flow past cylindrical obstacles when the undisturbed stream 'is subsonic, but at a sufficiently high Mach number for part of the field of flow next to the obstacle to be supersonic.

Two difficulties immediately appear. (i) For flows past obstacles, boundary conditions cannot be assigned in the hodograph plane: a boundary-value problem cannot be specified and then solved; solutions which are modifications of specified problems (such as flow past a circular cylinder) at Mach number zero (incompressible fluid) are found, and at each Mach number of the undisturbed stream the shape of the obstacle to which the solution applies (if there is one) is found *a posteriori*. We cannot study flows past a given and fixed contour for varying Mach numbers of the undisturbed flow. (ii) Singularities in the transformation from the hodograph plane to the physical plane, and *vice versa*, must be fully studied; they turn out to be most important. Certainly a solution of the hodograph equations ceases to be physically applicable when the physical plane becomes multiply covered.

(c) A knowledge of the applications of the theory of characteristics to unsteady flows depending on one space coordinate, and to the supersonic parts of steady two-dimensional flows, is of great value both for analytical and numerical purposes. A numerical (or graphical) solution may always be found if we have sufficient information about the appearance of shock waves and other discontinuities. Fruitful engineering approximations have been used to shorten and simplify the work. Applications with more than two independent variables are necessarily much more complicated.

Consider now the two-dimensional, steady flow of an otherwise unbounded gas past a cylindrical obstacle. The undisturbed flow (at infinity upstream) is supposed uniform and isentropic.

When the Mach number, $M$, of the undisturbed flow is sufficiently low, a continuous solution is possible in which the flow is everywhere subsonic, isentropic and irrotational. The flow pattern is, qualitatively, of the same character as that for $M = 0$ (incompressible fluid). The governing equation is everywhere elliptic. A uniqueness theorem may be proved, showing that, as for incompressible flow, the solution for a given velocity of the undisturbed stream is determined if the circulation is given, or if, for an airfoil section, a Kutta-Joukowski condition is imposed at the trailing edge; and the form of the asymptotic expansion of $\phi$ at a large distance from the obstacle may be found. (For the latest published paper on the subject, and other references, see Finn and Gilbarg, *Comm. Pure Appl. Math.* **10** (1957), pp. 23–63). But because the governing equation is non-linear, analytical methods of calculating exactly the details of flows past even the simplest cylindrical shapes, for perfect gases with constant specific heats, at non-zero $M$ are not available. A numerical solution may be obtained for any particular case, or electrical analogies may be used to find the solution approximately by the use of an "analogue apparatus." Analytical approximate methods have been developed. One of these proceeds by successive approximation starting from the solution for $M = 0$, and is essentially an expansion of the solution in powers of $M^2$. Another, useful for thin cylinders (e.g. airfoils) at low incidences, proceeds by successive approximation starting from a solution of the linearized equation ((99) with $\partial/\partial t = 0$, $\partial/\partial x_1 = 0$). The hodograph equations may also be used; this is usually done with the Karman-Tsien approximation, and engineering approximations are made, to balance the error so introduced and the change of shape of the cylinder (airfoil) from that at $M = 0$. For details and references see the works previously cited and a recent report on "Approximation Methods in Compressible Fluid Dynamics," by Prof. Isao Imai, from the Institute for Fluid Dynamics and Applied Mathematics of the University of Maryland, March, 1957.

It is sufficient for our purposes to continue the discussion for a perfect gas with constant specific heats and for the case when the section of the cylinder is symmetrical, and symmetrically placed, about a line in the direction of the undisturbed flow, and the circulation is zero. The flow pattern is then also symmetrical.

The regions in which the local Mach numbers are largest are adjacent to the contour of the cylinder, one on each side. For a sufficiently large value of the Mach number, $M$, of the undisturbed flow (still, of course, less than unity), the local Mach number will be just equal to unity at each of two symmetrically placed points on the contour. If continuous, isentropic, irrotational flows exist for still larger values of $M$, then it seems that there will be two symmetrically situated regions of supersonic flow next to the contour. Do such solutions exist? For certain contours, the answer is certainly yes, for such solutions have been calculated by the hodograph method. But this method will not enable us to vary $M$ and the contour at will, keeping either fixed and changing the other in a prescribed manner, so this does not show that such solutions exist for all contours, even for a restricted range of values of $M$. It appears, in fact, that if certain assumptions are satisfied, this is not the case, and that if there is such a solution past a certain contour at a given $M$, then, in a certain sense, there is no solution for the same $M$ and a "neighboring" contour. (For the latest published paper on the subject, and other references, see Morawetz, *Comm. Pure Appl. Math.* **10** (1957), pp. 107–131.) The physical significance of such results needs further elucidation. Note that experiments are not carried out on inviscid gases which slip freely along a solid surface without boundary layers (or in which concentrated vortex sheets along a solid surface can continue to exist). Nor do experiments even answer conclusively the question of the possible existence in real gases of flows without shock waves in which the streamlines pass from a region of subsonic flow to one of supersonic flow and back again to subsonic flow, as in the patterns here contemplated. Such regions of supersonic flow have, in fact, been observed, and no shocks could be found; but the objection that that showed only that the shocks were too weak to be found with the instruments used, not that they did not exist at all, is, of course, unanswerable (except by theory). In any case the ranges of Mach numbers for which such regions of supersonic flow may exist are certainly small, and either when the local Mach number is unity, or at slightly higher speeds, although the flow becomes supersonic in a continuous manner in some region or regions, the return to subsonic flow is through a shock wave. For all larger values of $M$, shock waves appear in the flow. (See Howarth, *op. cit.*)

The theory of such flows past cylinders, when the flow in the undisturbed stream is subsonic, and there are regions of supersonic flow, with shock waves present as the flows return to subsonic, is in a very imperfect state. It even appears likely that no satisfactory theory can be found without consideration of the interaction of the shock waves with boundary layers. Changes of the flow pattern, as the Mach number of the undisturbed flow increases, are known from observation. Here we point out only that when the incident flow becomes supersonic, a detached shock wave appears upstream of the cylinder, which, for the symmetrical system we are now considering, is normal to the incident stream on the plane of symmetry; there is a region of subsonic flow between the detached shock and the surface of the cylinder. As the Mach number of the undisturbed stream increases, the detached shock approaches the cylinder. If the cylinder is blunt-nosed, the shock is always detached. If the cylinder is an airfoil with a sharp leading edge, at some Mach number the shock becomes attached at the leading edge. For a very small range of Mach numbers, the flow behind the shock is still in part subsonic. With a further increase in Mach number, the flow past a sharp-nosed airfoil becomes everywhere supersonic, with shock waves at the leading and trailing edges; in spite of the shock waves, computations are now easier than for "mixed" flows; the theory of characteristics (and the approximations which have been devised to shorten the work) may be used.

All this must be considered for cylinders not symmetrically placed, and for three-dimensional flow past bodies of revolution, and past general three-dimensional bodies. The flow past a circular cone, with an attached shock, is fundamental. (In addition to the works cited, see also Z. Kopal, *Tables of Supersonic Flow about Cones*, M. I. T., Cambridge, Mass., 1947.) For applications we are concerned with airfoils of finite span (wings), and with bodies with wings attached.

Much of the theory used in applications starts from the linearized equations; sometimes closer approximations are found. In the linearized equations, we consider only first-order terms for a small perturbation about an undisturbed flow (*cf.* equation (99)). We cannot pause here to consider definite conditions for the validity of such approximations — it must suffice to say that even when the elementary linearized theory of steady flow (which we are now discussing) may be valid because of the geometry of the fluid boundaries, it will still not hold if the Mach number is too near to unity (transonic flow) or is too great (hypersonic flow). Also, as regards the geometry, two cases must be distinguished in flow past a body when it comes to applying the boundary condition at the surface of the body. If the boundary of the body lies near to a surface — mathematically, the limit of the body boundary as its "thickness" tends to zero is a surface — then the boundary condition may be applied on this limiting surface. But if the boundary of the body lies near to a line, i.e., the limit of the body boundary as its thickness tends to zero is a line — then the approximate velocity potential has a logarithmic singularity on the line if continued into the body, and the boundary condition certainly cannot be applied on the line. There is a vast literature on "linearized" flow, concerned in the first instance with similarity rules, and then with the forces and moments on bodies in flows, and with methods of solving the linearized equation with given boundary conditions — methods of singularities, of "conical fields" and their superposition, of transform theory, etc. (There are similarity rules also for approximate theories of transonic and hypersonic flow.) For all of these, together with methods of making the approximation uniformly valid, and of proceeding to higher approximations, we must now refer to the literature.

Two references may be given here for those who wish to pursue the mathematical study of hypersonic flow: J. D. Cole, Newtonian flow theory for slender bodies, *J. Ae. Sci.* **24** (1957), pp. 448–455, and W. Chester, Supersonic flow past a bluff body with a detached shock: Part I, Two dimensional body, *J. Fluid Mechs.*, **1** (1956), pp. 353–365; Part II, Axisymmetrical body, *ibid.*, 490–496. The former is concerned with flow past slender bodies, and the latter with flow past bluff bodies. Some other references will be found in the cited papers, and other works are expected to appear shortly. (Added September 1959: see also W. D. Hayes and R. F. Probstein, *Hypersonic Flow Theory*, Academic Press, New York and London, 1959.) There are, however, still interesting and unanswered mathematical questions, particularly in connection with hypersonic flow past blunt bodies. (This is apart from such complications as arise in practical computations from the fact that the gas is so heated in passing through the shock wave that it cannot be treated as a perfect gas with constant specific heats.) There is also much we have not yet mentioned, including the theory of unsteady external aerodynamics, which is concerned with the relative motion of air and an oscillating airfoil, with the theory of gusts, with accelerated motions of airfoils, and so on.

After this rapid attempt to list some of the topics usually discussed under the heading of the dynamics of inviscid gases, we turn to a rather more detailed treatment of one particular topic — slender-body theory. Essentially, this is the theory of an approximation to an approximation. We consider flow past a slender body, and start from the linearized equations. The body is considered to be very elongated in the direction of the relative motion of the body and the air( or other gas), and it will appear that the equation for the velocity potential may, near the body, be approximated by Laplace's equation in two dimensions. But such an approximation would still leave undetermined those quantities which cannot be found without applying boundary conditions at infinity; this is accomplished by the use of transform theory. Reference may be made to Chap. 9 of G. N. Ward, *Linearized Theory of Steady High-Speed Flow*, to Mac C. Adams and W. R. Sears, Slender-Body Theory — Review and Extension, *J. Ae. Sci.*, **20** (1953), pp. 85-98, and to a forthcoming article by A. E. Bryson in the *Handbook of Engineering Mechanics*, ed. by W. Flugge (McGraw Hill). Conditions for the validity of the theory, other references, formulae for the forces and moments, and references to extensions of the theory may be found in these works.

## 11.2. Slender-body theory. Subsonic flow

(a) *Flow past a body of revolution at zero yaw by the method of sources*

It is of some mathematical interest to obtain the approximate solution for flow past a body of revolution by the method of sources (*cf.* Ward), before proceeding to the use of Fourier transforms (*cf.* Adams and Sears). We consider irrotational flow with constant entropy. With an undisturbed flow $W$ in the direction of the axis of $z$, if the velocity potential is $Wz + \phi$, the linearized equation for $\phi$ is

(236) $$\partial^2\phi/\partial x^2 + \partial^2\phi/\partial y^2 + B^2\, \partial^2\phi/\partial z^2 = 0,$$

where

(237) $$B^2 = 1 - M^2, \qquad M^2 = W^2/a_0^2,$$

and $a_0$ is the velocity of sound in the undisturbed stream.

Corresponding with a source at the origin in incompressible flow $(B = 1)$, we may consider

(238) $$\phi = A/R, \qquad R = (B^2 x^2 + B^2 y^2 + z^2)^{\frac{1}{2}}.$$

Then
$$u = \partial\phi/\partial x = -AB^2 x/R^3, \quad v = \partial\phi/\partial y = -AB^2 y/R^3,$$
$$w = W + \partial\phi/\partial z = W - Az/R^3.$$

The mass flux out of any closed surface $S$ for which the origin is an interior point is

$$F = \int \rho(lu + mv + nw) dS$$

over $S$, where $\rho$ is the density. If $\rho_0$ is the density in the undisturbed flow, since $dp/d\rho = a^2$, Bernoulli's equation may be written in the form

$$\int_{\rho_0}^{\rho} a^2 \frac{d\rho}{\rho} = \tfrac{1}{2} \{W^2 - u^2 - v^2 - (W + w_1)^2\},$$

where $w_1 = \partial\phi/\partial z$. From this it is easily seen that on a linear approximation

$$\rho - \rho_0 = -\rho_0 M^2 w_1/W.$$

Hence, to a linear approximation,

$$F/\rho_0 = \int \left( l \frac{\partial\phi}{\partial x} + m \frac{\partial\phi}{\partial y} + nB^2 \frac{\partial\phi}{\partial z} \right) dS$$

(since $\int nW dS = 0$). From (236) the integral is the same for all surfaces enclosing the origin. Take $S$ to be a sphere of radius $K$. Then

$$\frac{F}{\rho_0} = -\frac{AB^2}{K} \int \frac{x^2 + y^2 + z^2}{R^3} dS = -2\pi AB^2 \int_0^{\pi} \frac{\sin\theta \, d\theta}{(B^2 \sin^2\theta + \cos^2\theta)^{\frac{3}{2}}}$$
$$= -2\pi AB^2 \int_{-1}^{1} \frac{dc}{(B^2 + M^2 c^2)^{\frac{3}{2}}} = -2\pi A \left[ \frac{c}{\{B^2 + M^2 c^2\}^{\frac{1}{2}}} \right]_{-1}^{+1}$$
$$= -4\pi A,$$

and is independent of $M$. (We have put $(x^2 + y^2)^{\frac{1}{2}} = K \sin \theta$, $z = K \cos \theta$ on the sphere, and $c = \cos \theta$.) For a source of strength $m$ in a fluid of constant density $\rho_0$ the mass flux out of a surface enclosing the source is $4\pi\rho_0 m$, and the velocity potential is $-m/(x^2 + y^2 + z^2)^{\frac{1}{2}}$. We may therefore say that in linearized subsonic flow, the perturbation potential for a source of strength $m$ at the origin is $-m/R$, where $R$ is given by (238).

Now consider the flow, with an undisturbed velocity $W$ along the $z$-axis, past an elongated body of revolution, of length $l$, at zero yaw—i.e., with its axis along the undisturbed flow. Let us try to find a distribution of sources, of strength $f(z)$ per unit length, which gives, to a sufficient approximation, the same perturbation as the given body. (*cf.* Appendix III for an incompressible fluid.) Then if the origin is at the upstream end of the body, we should take

$$\phi = -\int_0^l \frac{f(\lambda)d\lambda}{\{(z-\lambda)^2 + B^2 r^2\}^{\frac{1}{2}}},$$

where we have written $r^2 = x^2 + y^2$. Now, as pointed out in Appendix III, it is only in exceptional cases that there is a solution of this form of equation (236) with the necessary boundary conditions. But it appears that to the order of approximation to which we shall proceed, this is sufficiently accurate, and the value of $f$ may be found by an elementary argument. Take two sections normal to the axis at $z$ and $z + dz$, and consider the flux out of the surface formed by these sections and the part of the surface of the body connecting them. This flux must be $4\pi\rho_0 f(z)dz$. The flow is to be tangential to the body surface, so there is no flow across it. Let $S(z)$ be the area of the cross-section at $z$. To the lowest approximation the flux out through the two cross-sections is

$$\rho_0 W S(z + dz) - \rho_0 W S(z) = \rho_0 W S'(z)dz,$$

so we must take

$$4\pi f(z) = W S'(z),$$

and

(239) $$\frac{\phi}{W} = -\frac{1}{4\pi}\int_0^l \frac{S'(\lambda)d\lambda}{\{(z-\lambda)^2 + B^2 r^2\}^{\frac{1}{2}}}.$$

To find $\phi$ on and near the surface of the body, we require only an approximation when $r$ is small. But if we put $r = 0$ in the integrand, the integral becomes an improper integral, and care is needed in finding the correct form for small $r$. In what follows, we shall assume that $S(z)$ has a bounded second derivative. With the above value of $f$, we write

$$-\phi = I_1 + I_2, \quad I_1 = \int_0^z \frac{f(\lambda)d\lambda}{\{(z-\lambda)^2 + B^2r^2\}^{\frac{1}{2}}},$$

$$I_2 = \int_z^l \frac{f(\lambda)d\lambda}{\{(z-\lambda)^2 + B^2r^2\}^{\frac{1}{2}}}$$

and

$$f(\lambda) - f(z) = (\lambda - z)F(\lambda, z),$$

where $F$ is bounded. Then

$$I_1 = f(z)\int_0^z \frac{d\lambda}{\{(z-\lambda)^2 + B^2r^2\}^{\frac{1}{2}}} + \int_0^z \frac{(\lambda-z)F}{\{(z-\lambda)^2 + B^2r^2\}^{\frac{1}{2}}} d\lambda$$

$$= f(z)\sinh^{-1}\frac{z}{Br} + \int_0^z \frac{(\lambda-z)F}{\{(z-\lambda)^2 + B^2r^2\}^{\frac{1}{2}}} d\lambda,$$

and similarly

$$I_2 = f(z)\sinh^{-1}\frac{l-z}{Br} + \int_z^l \frac{(\lambda-z)F}{\{(\lambda-z)^2 + B^2r^2\}^{\frac{1}{2}}} d\lambda.$$

When $z/(Br)$ is large, $\sinh^{-1} z/Br \sim \log 2z/Br$. Similarly $\sinh^{-1}(l-z)/Br \sim \log 2(l-z)/Br$. As $r$ tends to zero, the square root in the second term for $I_1$ tends to $z - \lambda$, and that in $I_2$ tends to $\lambda - z$. Hence as $r \to 0$,

$$I_1 + I_2 + 2f(z)\log\frac{Br}{2} \to f(z)\log z(l-z) - \int_0^z F d\lambda + \int_z^l F d\lambda.$$

Now consider

$$J = \int_0^{z-\frac{1}{2}Br} \frac{f(\lambda)d\lambda}{z-\lambda} + \int_{z+\frac{1}{2}Br}^l \frac{f(\lambda)d\lambda}{\lambda-z}$$

$$= f(z)\log z(l-z) - 2f(z)\log\frac{Br}{2} - \int_0^{z-\frac{1}{2}Br} F d\lambda + \int_{z+\frac{1}{2}Br}^l F d\lambda.$$

Hence, as $r \to 0$

(240)
$$I_1 + I_2 - J \to 0.$$

By integration by parts

$$\int_0^{z-\frac{1}{2}Br} \frac{f(\lambda)d\lambda}{z-\lambda} = f(0) \log z - f(z - \tfrac{1}{2}Br) \log \tfrac{1}{2}Br$$
$$+ \int_0^{z-\frac{1}{2}Br} f'(\lambda) \log (z-\lambda)d\lambda,$$

$$\int_{z+\frac{1}{2}Br}^{l} \frac{f(\lambda)d\lambda}{\lambda - z} = f(l) \log (l - z) - f(z + \tfrac{1}{2}Br) \log \tfrac{1}{2}Br$$
$$+ \int_{z+\frac{1}{2}Br}^{l} f'(\lambda) \log (\lambda - z)d\lambda.$$

Hence finally, for small $r$,

(241)
$$\frac{\phi}{W} \sim \frac{1}{2\pi} S'(z) \log \tfrac{1}{2}Br + g(z),$$

where

(242)
$$g(z) = -\frac{1}{4\pi}\Big\{S'(0) \log z + S'(l) \log (l - z)$$
$$+ \int_0^z S''(\lambda)\log(z-\lambda)d\lambda - \int_z^l S''(\lambda) \log (\lambda - z)\, d\lambda\Big\}.$$

This approximation to (239) may be obtained otherwise. The interest of the above method is that it relates the limiting value, as $r \to 0$, of $I_1 + I_2$ to a particular "principal value" of the integral of $f(\lambda)/|z - \lambda|$ from 0 to $l$, namely $\lim J$.

If the body is pointed at the forward end, $S'(0) = 0$, and if it is pointed at the rear end $S'(l) = 0$. Note that $\phi$ is logarithmically infinite at the forward end unless $S'(0) = 0$, and at the rear end unless $S'(l) = 0$. We have, in fact, assumed that the perturbation is small; if the forward end is blunt, there will be a stagnation point there, and the perturbation will not be small. But even if the forward end is pointed, there will be a stagnation point there, unless the end is a cusp of each meridian section. Thus unless the forward end is a cusp, linearized theory (and not only the slender-body approximation) will not be valid there.

Exactly similar remarks apply to the rear end (though in a real fluid viscous boundary-layer effects would in any case usually invalidate any practical applications of inviscid theory at the rear end.) It appears that this lack of validity at the ends does not vitiate the applicability of the results away from the ends.

On the other hand, if the linearized equation is accepted as valid (as for an incompressible fluid) then the process of approximation for small $r$ is itself not uniformly valid, and does not hold near the ends of the body, unless the ends are cusped. This is also clearly seen when we consider the radial and longitudinal velocity components, $\partial\phi/\partial r$ and $\partial\phi/\partial z$.

We have

$$\frac{1}{W}\frac{\partial \phi}{\partial r} = \frac{B^2 r}{4\pi} \int_0^l \frac{S'(\lambda)\,d\lambda}{\{(z-\lambda)^2 + B^2 r^2\}^{\frac{3}{2}}}$$

$$= -\frac{1}{4\pi r} \int_0^l S'(\lambda) \frac{\partial}{\partial \lambda} \left\{ \frac{z-\lambda}{[(z-\lambda)^2 + B^2 r^2]^{\frac{1}{2}}} \right\} d\lambda$$

$$= \frac{1}{4\pi r} \left[ \frac{zS'(0)}{(z^2+B^2r^2)^{\frac{1}{2}}} - \frac{(z-l)S'(l)}{[(z-l)^2 + B^2 r^2]^{\frac{1}{2}}} \right]$$

$$+ \frac{1}{4\pi r} \int_0^l \frac{S''(\lambda)(z-\lambda)}{\{(z-\lambda)^2 + B^2 r^2\}^{\frac{1}{2}}}\,d\lambda.$$

So long as $Br$ is small compared with $z$ and with $l - z$, it may be shown that $\partial\phi/\partial r$ is given approximately by

$$\frac{1}{W}\frac{\partial \phi}{\partial r} = \frac{1}{4\pi r}[S'(0) + S'(l)] + \frac{1}{4\pi r}\left[\int_0^z S''(\lambda)\,d\lambda - \int_z^l S''(\lambda)\,d\lambda\right]$$

(243a)
$$= \frac{S'(z)}{2\pi r},$$

which is the result obtained by differentiating (241).

The longitudinal perturbation velocity $\partial\phi/\partial z$ may be treated similarly; assuming the existence and boundedness of $S'''(z)$, it is found that

(243b)
$$\frac{1}{W}\frac{\partial \phi}{\partial z} = \frac{1}{2\pi}S''(z) \log \tfrac{1}{2}Br - \frac{1}{4\pi}\left[\frac{S'(0)}{z} - \frac{S'(l)}{l-z}\right.$$
$$+ S''(0) \log z + S''(l) \log (l-z)$$
$$\left. + \int_0^z S'''(\lambda) \log (z-\lambda) d\lambda - \int_z^l S'''(\lambda) \log (\lambda - z) d\lambda\right],$$

and it may be shown that this is the result obtained by differentiating (241).

On the surface of the body, the velocity must be tangential. If we write $R_0(z)$ for the radius of the cross-section at $z$, the boundary condition is

$$\frac{\partial \phi/\partial r}{W + \partial \phi/\partial z} = \frac{dR_0}{dz} \text{ at } r = R_0,$$

When we linearize by dropping $\partial \phi/\partial z$ compared with $W$ in the denominator, the condition becomes

(244)
$$\frac{1}{W}\frac{\partial \phi}{\partial r} = \frac{dR_0}{dz} \text{ at } r = R_0.$$

Since $S(z) = \pi R_0^2$, $S'(z) = 2\pi R_0 R_0'$, we see from (243) that this condition is satisfied.

The approximate solution in (241) is, for each section ($z$ fixed) a solution of Laplace's equation in two dimensions

(245)
$$\partial^2 \phi/\partial x^2 + \partial^2 \phi/\partial y^2 = 0.$$

That this would be so could have been guessed, intuitively. For if variations of the geometry with $z$ are very slow, we might have expected that the $z$-derivatives of $\phi$ would be much smaller than the $x$- or $y$-derivatives, and that in (236) the term $B^2 \partial^2 \phi/\partial z^2$ could be dropped. The matter is, in fact, not quite so simple, because of the logarithmic infinity in $r$ in (241). Let $F(x, y, z) = 0$ be the equation of the surface of the body. Let $t$ be the maximum diameter of a cross-section, $\varepsilon = t/l$, and $\xi = x/\varepsilon$, $\eta = y/\varepsilon$. If the equation of the surface of the body becomes $f(\xi, \eta, z) = 0$, then on and near the body, $f_\xi$, $f_\eta$, $f_z$ are all of the same magnitude.

(A spheroid may be considered as an example.) But $F_x = f_\xi/\varepsilon$, $F_y = f_\eta/\varepsilon$, $F_z = f_z$. So $F_z/F_y$ and $F_z/F_x$ will be $O(\varepsilon)$.

If this were true also of $\phi$, $\phi_z/\phi_r$ would be $O(\varepsilon)$ near the body. Because of the logarithmic singularity, this is not correct. $S(z)$ and $S'(z)$ are $O(\varepsilon^2)$, and $\phi = O(\varepsilon^2 \log \varepsilon)$, $\phi_r = O(\varepsilon)$, $\phi_z = O(\varepsilon^2 \log \varepsilon)$. So $\phi_z/\phi_r = O(\varepsilon \log \varepsilon)$. With these orders of magnitude, the approximations used in obtaining first (236) and then (241) may be re-examined; it turns out that the errors in $\phi$ and $\phi_z$ are $O(\varepsilon^4 \log^2 \varepsilon)$ and in $\phi_r$ are $O(\varepsilon^3 \log \varepsilon)$.

The approximation, while not quite as good as we might have expected, is still good enough to justify further work along the same lines.

We should expect the arguments set out immediately above to be capable of generalization from a body of revolution at zero yaw to the case of any slender body at a small yaw. We should expect for this more general case that a good approximation to $\phi$ would be the form $\phi'(x, y; z) + g(z)$, where $\phi'$ is a solution of (245) and $z$ enters as a parameter.

Before we proceed to the use of Fourier transforms we consider the approximate boundary condition to be satisfied by $\phi$ at the surface of the general slender body.

(b) *Boundary condition at the surface*

Let $F(x, y, z) = 0$ be the equation of the surface, and $l$, $m$, $n$ the direction cosines of the normal to the surface. If $\varepsilon$ is the ratio of the maximum chord of a cross-section to the length of the body, $F_z/F_x = O(\varepsilon)$, $F_z/F_y = O(\varepsilon)$, and

$$l = F_x/[F_x^2 + F_y^2 + F_z^2]^{\frac{1}{2}}$$

is approximately equal to $F_x/[F_x^2 + F_y^2]^{\frac{1}{2}}$. Similarly $m$ and $n$ are approximately $F_y/[F_x^2 + F_y^2]^{\frac{1}{2}}$ and $F_z/[F_x^2 + F_y^2]^{\frac{1}{2}}$. Hence $l$ and $m$ may be replaced by $l'$ and $m'$, respectively, where $l'$ and $m'$ are the direction cosines of the normals to sections of the surface in the planes $z = $ constant. Also $n/l$ and $n/m$ are $O(\varepsilon)$. Hence the condition of zero normal velocity

$$l\phi_x + m\phi_y + n(W + \phi_z) = 0$$

may be replaced by
$$l'\phi_x + m'\phi_y + nW = 0.$$

If we accept the form $\phi'(x, y; z) + g(z)$ for $\phi$, where $\phi'$ is a solution of (245), the boundary condition for $\phi'$ is

(246) $$W^{-1}\,\partial\phi'/\partial\nu = -F_z/[F_x^2 + F_y^2]^{\frac{1}{2}},$$

where the differentiation is along the outwardly directed normal to the section of the body surface. Since the gradient of $\phi'$ must vanish at infinity, we have a definite two-dimensional potential problem to solve for $\phi'$, with a unique solution apart from an additive function of $z$. (The determination of $g(z)$, however, requires other considerations.)

With $u' = \phi'_x$, $v' = \phi'_y$, the boundary condition may also be written in the form

(247) $$u'F_x + v'F_y + WF_z = 0.$$

If we make the substitution $z/W = t$, the two-dimensional boundary condition may therefore be written $DF/Dt = 0$, which is the boundary condition at the surface of a deforming cylinder for two-dimensional flow (see equation (13), Section 1.4); if at any time $t$ the section of the cylinder is the section of the body at $z$, then after a time $dt = dz/W$ the section is deformed into the section of the body at $z + dz$. The simple-minded view of the matter is this: $\phi$ is the disturbance potential for the flow caused by the motion of the body with velocity $W$ along the negative direction of the $z$-axis through otherwise still air. If we neglect the motion of the air out of any section, then if the air in any plane perpendicular to the $z$-axis "sees" at time $t$ the section by $z = z_1$, after a time $dt$ it "sees" the section which was at $z_1 + dz = z_1 + Wdt$ at time $t$.

Note also that the normal distance between the boundaries at time $t$ and time $t + dt$ would be $\phi'_\nu dt$, so the change in the area of the section would be $dt \int \phi'_\nu ds$, where the integral is taken round the contour of the section at time $t$. With $dt = dz/W$, this is the change in the area of the section between $z$ and $z + dz$, and is $S'(z)dz$, where $S(z)$ is the cross-sectional area at $z$. So

(248)
$$\oint_C \frac{\partial \phi'}{\partial \nu} ds = WS'(z),$$

where $C$ is the contour of the section of the body at $z$. The two-dimensional potential motion represented by $\phi'$ therefore includes a two-dimensional source of strength $WS'(z)/2\pi$.

It is convenient to separate, in the boundary condition, the effect of incidence or yaw. Of course, the incidence must be small. (We recall that we are concerned only with cases in which every point of the body surface is sufficiently near the $z$-axis, and slopes are gentle, and the velocity disturbances are small — except, perhaps, at the ends, where this approximate theory will not be valid.) We have not so far defined, for the general slender body, the position of zero incidence, nor have we fixed the directions of the axes of $x$ and $y$. We must now agree to adopt one position of the body as the standard position of zero incidence, and, the axis of $x$ having been suitably chosen, we define an incidence $\alpha$ to denote a (small) positive rotation of the body about the $x$-axis. Then if $F(x, y, z) = 0$ is the equation of the body surface at zero incidence, to the first order in $\alpha$ the equation at an incidence $\alpha$ is $F(x, y + \alpha z, z - \alpha y) = 0$. This form must be used in (247), which becomes

$$u'F_x + v'(F_y - \alpha F_z) + W(F_z + \alpha F_y) = 0,$$

all values being taken on the new position of the surface. But $u'$, $v'$ will be $O(\varepsilon)$ and $F_z/F_x$, $F_z/F_y$ are $O(\varepsilon)$. So the condition may be replaced by

$$u'F_x + v'F_y + WF_z + \alpha W F_y = 0,$$

applied at the position of the body at zero incidence. We may write the condition as

(249)
$$W^{-1} \partial\phi'/\partial\nu = -F_z/(F_x^2 + F_y^2)^{\frac{1}{2}} - m'\alpha,$$

and we see that we may take $\phi'$ as the sum of two parts. The first allows for the change of shape and size in the cross-sections, and is exactly the same as for zero incidence. The second is the solution for two-dimensional potential flow for a cylinder with the given cross-section (the cross-section of the body at $z$) moving with velocity $\alpha W$ along the negative direction of the $y$-axis through otherwise undisturbed, incompressible, fluid.

The two-dimensional theory gives $\phi'$ only near the body, and will not allow the determination of $g$, which requires application of the boundary conditions at infinity. The perturbation potential, $\phi$, may be taken to vanish, and its derivatives must vanish, at infinity. To apply these conditions, we turn, following Adams and Sears, to Fourier transforms. (The first use of transforms in this way was by Ward, who used Laplace transforms for supersonic flow. See Section 11.3(b).)

(c) *Fourier transforms*

Write
$$\overline{\phi}(x, y, k) = \int_{-\infty}^{\infty} \exp(ikz)\, \phi(x, y, z)\, dz,$$
$$\phi(x, y, z) = \frac{1}{2\pi} \int_{-\infty}^{\infty} \exp(-ikz)\, \overline{\phi}(x, y, k)\, dk.$$

Since
$$\int \frac{\partial^2 \phi}{\partial z^2} \exp(ikz)\, dz = \frac{\partial \phi}{\partial z} \exp(ikz) - ik\phi \exp(ikz)$$
$$- k^2 \int \phi \exp(ikz)\, dz,$$

and $\phi$, $\partial \phi / \partial z$ must vanish at $z = \pm \infty$, the transform of equation (236) is
$$\partial^2 \overline{\phi}/\partial x^2 + \partial^2 \overline{\phi}/\partial y^2 - B^2 k^2 \overline{\phi} = 0.$$

Use polar coordinates $r$, $\theta$ in the $(x, y)$ plane and expand $\overline{\phi}$ in a Fourier series in $\theta$.

$$\overline{\phi} = A_0(k) K_0(B\,|k|\,r) + \sum_{n=1}^{\infty} \{A_n(k) \cos n\theta + C_n(k) \sin n\theta\} K_n(B|k|r),$$

where $K_n$ is the "Bessel function of imaginary argument of the second kind," as defined by Watson, *Theory of Bessel Functions*,

Section 3.7. Only the $K_n$ enter and not the $I_n$, since $\bar{\phi}$ must vanish when $r$ is infinite.

Now let $r$ be small, and keep only the leading terms in the expansion of each $K_n$:
$$K_0(z) = -[\log \tfrac{1}{2}z + \gamma], \quad K_n(z) = 2^{n-1}(n-1)!/z^n \quad (n \geqq 1)$$
where $\gamma$ is Euler's constant, $0.5772\cdots$.
$$\bar{\phi} = -A_0(k)\,[\log \tfrac{1}{2}B\,|k|r + \gamma]$$
$$+ \sum_{n=1}^{\infty} \frac{A_n(k)\cos n\theta + C_n(k)\sin n\theta}{B^n|k|^n r^n} 2^{n-1}(n-1)!$$

Let $a_0(z)$ be the inverse transform of $-A_0(k)$, $g_1(z)$ the inverse transform of
$$-A_0(k)\,[\log \tfrac{1}{2}B\,|k| + \gamma],$$
and $a_n(z)$, $b_n(z)$ the inverse transforms of
$$2^{n-1}(n-1)!A_n(k)/B^n|k|^n, \quad 2^{n-1}(n-1)!C_n(k)/B^n|k|^n.$$
Then near the body
$$\phi = a_0(z)\log r + \sum_{n=1}^{\infty} [a_n(z)\cos n\theta + b_n(z)\sin n\theta]/r^n + g_1(z).$$

The first two terms represent the two-dimensional potential function, $\phi'$. Also $a_0(z)$ is easily found from (248): we have

(250) $$a_0(z) = \frac{W}{2\pi} S'(z),$$

$g_1(z)$ depends on the shape of the body only through $A_0$, and therefore only through $a_0$, i.e., through $S'(z)$. It is therefore the same for any body as for a body of revolution, if the cross-sectional areas at any $z$ are the same. The resulting transform is

(251) $$g_1(z) = [WS'(z)\log \tfrac{1}{2}B]/2\pi + Wg(z),$$

where $g(z)$ is given by (242).

For a body of revolution at zero yaw, $a_n = b_n = 0$. More generally, for any slender body at any small incidence, $\phi = \phi' + g_1(z)$, where $\phi'$ is a two-dimensional potential function satisfying (249) on the boundary of the section (taken at zero incidence) at $z$, and such that $\phi' - [WS'(z)\log r]/2\pi$ tends to zero at an infinite distance in the plane. (With this last condition, $g_1$ is given correctly by (251) and (242).)

We may now return to the consideration of orders of magnitude, as mentioned near the end of 11.2(a) for a body of revolution at zero yaw. We see that the statements may be easily generalized.

Since $\phi_z$ is not of the same order as $\phi_x$ and $\phi_y$, we note that in calculating the pressure we must include $\phi_z$, $\phi_x^2$ and $\phi_y^2$, and that if $p_0$ and $\rho_0$ are the pressure and density in the undisturbed stream, it may be shown that the required formula is

$$(252) \qquad (p - p_0)/\rho_0 = -[W\phi_z + \tfrac{1}{2}(\phi_x^2 + \phi_y^2)].$$

(d) *Example*

When the body shape is given, the determination of $g_1(z)$ or $g(z)$ is a straightforward matter of integration. The determination of the part of $\phi'$ which depends on the incidence is a potential problem of the usual type that is considered in elementary classical hydrodynamics. The determination of the part of $\phi'$ which depends on changes in the size and shape of the sections is, however, a two-dimensional potential problem with a boundary condition of a type which has not been extensively studied. I know of no classification of geometrical shapes which give "simple" solutions. Of course, if the sections are circles (as for a body of revolution) the solution is already known. There is also a simple solution if the sections are ellipses. Let the surface of the body have the equation

$$F(x, y, z) \equiv x^2/a^2 + y^2/b^2 - f(z) = 0.$$

The section at $z$ is

$$x^2/A_1^2 + y^2/B_1^2 = 1,$$

where

$$A_1^2 = a^2 f(z), \ B_1^2 = b^2 f(z),$$

and

$$S(z) = \pi A_1 B_1 = \pi a b f(z).$$

The potential problem is to determine a potential function outside the elliptic section such that, on the contour of the ellipse

$$\frac{1}{W}\frac{\partial \phi'}{\partial \nu} = \frac{f'(z)}{2(x^2/a^4 + y^2/b^4)^{\frac{1}{2}}}.$$

At an infinite distance $\phi' - \frac{1}{2}W\, abf'(z) \log r \to 0$.

We may either transform the ellipse conformally into a circle, or introduce elliptic coordinates. We introduce elliptic coordinates, $\xi$, $\eta$, for which

$$x + iy = c \cosh(\xi + i\eta), \quad x = c \cosh \xi \cos \eta, \quad y = c \sinh \xi \sin \eta.$$

If $\xi = \xi_0$ is the ellipse,

$$c \cosh \xi_0 = A_1, \quad c \sinh \xi_0 = B_1, \quad c^2 = A_1^2 - B_1^2 = (a^2 - b^2)f(z).$$

On the ellipse,

$$\frac{x^2}{a^4} + \frac{y^2}{b^4} = \frac{A_1^2}{a^4}\cos^2\eta + \frac{B_1^2}{b^4}\sin^2\eta = \frac{f(z)}{a^2 b^2}[a^2 \sin^2\eta + b^2 \cos^2\eta],$$

$$\left|\frac{d(x+iy)}{d(\xi+i\eta)}\right| = |c \sinh(\xi_0 + i\eta)| = (B_1^2 \cos^2\eta + A_1^2 \sin^2\eta)^{\frac{1}{2}}$$

$$= [f(z)]^{\frac{1}{2}}[a^2 \sin^2\eta + b^2 \cos^2\eta]^{\frac{1}{2}},$$

and

$$\frac{\partial \phi'}{\partial \nu} = \frac{\partial \phi'}{\partial \xi} \bigg/ \left|\frac{d(x+iy)}{d(\xi+i\eta)}\right|.$$

Hence the boundary condition becomes

$$W^{-1}\, \partial\phi'/\partial\xi = \tfrac{1}{2}abf'(z),$$

and the solution is

$$\phi'/W = \tfrac{1}{2}abf'(z)[\xi + \text{constant}].$$

At infinity, $\xi = \log(2r/c) + 0(r^{-2})$, where $r^2 = x^2 + y^2$. To conform with the boundary condition laid down at infinity to keep $g_1(z)$ as in (251), we therefore take

$$\phi'/W = \tfrac{1}{2}abf'(z)[\xi + \log \tfrac{1}{2}c].$$

Note that $\tfrac{1}{2}abf'(z)$ is $S'(z)/2\pi$, as it should be.

When $f(z)$ is given explicitly (as for an ellipsoid, for example)

$g(z)$ is easily computed and the solution for zero incidence completed. The computation of the effect of incidence is standard (motion of an elliptic cylinder parallel to an axis). It is of some interest to complete the solution for an ellipsoid and, for incompressible flow ($B = 1$), to compare it with the full solution (Lamb's *Hydrodynamics*, Section 114.)

We now turn to supersonic flow, and we shall see that a very similar theory applies in that case. In fact, to a certain extent slender-body theory applies approximately for some classes of bodies over the range of Mach numbers from zero to moderately high supersonic speeds (but not at hypersonic speeds). In particular, the expression of the perturbation potential $\phi$ as $\phi'(x, y; z) + g_1(z)$, where $\phi'$ is a two-dimensional potential function, appears to be a good approximation in certain cases even at transonic speeds, where the Mach number is near unity. However, the occurrence of $\log B$ in $g_1$ in (251) serves as a warning that anything like the above determination of $g_1$ will fail in the transonic range. We show how to determine $g_1$ in supersonic, as we have done for subsonic, flow, but make no attempt to consider $g_1$ for transonic flow.

### 11.3. Slender-body theory. Supersonic flow

For supersonic flow, we consider only flow past slender pointed bodies—i.e., the forward end is taken to be pointed, so that, with a previous notation, $S'(0) = 0$. We are, in fact, approximating to flows past slender pointed bodies with attached shock waves. In order that a body may be considered slender in supersonic flow it must be well within the downstream half of the Mach cone from the upstream vertex. The body may have a pointed rear end or a flat base.

On the body the boundary condition is the same as for subsonic flow; for a body of revolution it may be expressed in the form (244), and for more general cases in the forms (246) or (247) for zero incidence, and (249) at an incidence $\alpha$. However, the boundary conditions at infinity take an entirely different form. Ahead of the body ($z < 0$), there is no disturbance; $\phi$ and its derivatives are zero. This is, in fact, the case ahead of the downstream half of the Mach cone of the vertex. As we go from any point of the body to infinity along a characteristic, the disturbance must have the character of waves propagating outwards from the body (Helmholtz's radiation condition).

The linearized equation for the perturbation potential $\phi$ is now

(253) $\qquad \partial^2\phi/\partial x^2 + \partial^2\phi/\partial y^2 = \beta^2\,\partial^2\phi/\partial z^2,$

where

(254) $\qquad\qquad\qquad \beta^2 = M^2 - 1.$

(a) *Flow past a body of revolution at zero yaw*

Note that if we substitute $ct$ for $z/\beta$ in (253), this equation becomes the standard equation for two-dimensional wave motion. This is accomplished, for example, if we put $t = z/W$ and

(255) $\qquad\qquad c^{-2} = a^{-2} - W^{-2} = \beta^2 W^{-2}.$

Thus we are concerned with the same kind of problem as is treated, for example, in the theory of sound waves of infinitesimal amplitude. In particular, over the length of the body ($0 \leqq z \leqq l$), the body acts as a source of outwardly propagated waves, in such a way that, at $r = R_0$,

(256) $\qquad\qquad r\dfrac{\partial \phi}{\partial r} = \dfrac{W}{2\pi} S'(z)$

from (244). Now according to the essence of slender-body theory, $R_0$ is small, and we anticipate that although $\phi_r$ will not have a limit as $r \to 0$, $r\phi_r$ will have such a limit, and that our "slender-body" approximation will be given by taking this limit as the right-hand side of (256). This we shall later check, *a posteriori*. But when we make the substitutions involving $c$ and $t$, the problem of finding a solution of

$$\partial^2\phi/\partial x^2 + \partial^2\phi/\partial y^2 = c^{-2}\,\partial^2\phi/\partial t^2$$

which has circular symmetry, represents outgoing waves, and for which

$$\lim_{r \to 0} (2\pi r\,\partial\phi/\partial r) = f(t),$$

## DYNAMICS OF INVISCID GASES

is exactly the problem of a "line source of sound." The solution may be found by any one of a number of methods. (i) $f(t)$ may be taken first as a simply-harmonic periodic function, and the solution found in terms of Bessel functions (Hankel functions). This solution may then be generalized for any function $f(t)$ by Fourier's integral. (Lamb's *Hydrodynamics*, Sections 302, 195, 196). (ii) The solution may be found by using Laplace transforms (Lighthill, *Reports and Memoranda of the Aeronautical Research Council*, London, No. 2003, 1945). Note that the case in which we are interested corresponds with no disturbance for $t < 0$. (iii) We may find the solution for a line source by integrating the solution for a point source of sound. (Lamb, Sections 302, 285). If $R$ is distance from the point source, the equation to be solved is

$$c^2 \partial^2(R\phi)/\partial R^2 = \partial^2(R\phi)/\partial t^2,$$

with

$$\lim_{R \to 0} (4\pi R^2 \, \partial \phi / \partial R) = f(t),$$

and the solution for outgoing waves is

$$\phi = -f(t - R/c)/(4\pi R)$$

Hence for a point source at $(0, 0, \zeta)$ of strength $f(t)d\zeta$ the contribution to $\phi$ at the point $(x, y, 0)$ is

$$-\frac{1}{4\pi (r^2 + \zeta^2)^{\frac{1}{2}}} f\!\left(t - \frac{(r^2 + \zeta^2)^{\frac{1}{2}}}{c}\right) d\zeta,$$

so the potential for a line source is

$$\phi = -\frac{1}{4\pi} \int_{-\infty}^{\infty} f\!\left(t - \frac{(r^2 + \zeta^2)^{\frac{1}{2}}}{c}\right) \frac{d\zeta}{(r^2 + \zeta^2)^{\frac{1}{2}}}$$

$$= -\frac{1}{2\pi} \int_{0}^{\infty} f\!\left(t - \frac{r}{c} \cosh u\right) du$$

$$= -\frac{1}{2\pi} \int_{-\infty}^{t-r/c} \frac{f(\tau)d\tau}{\{(t-\tau)^2 - r^2/c^2\}^{\frac{1}{2}}}.$$

(For the first result substitute $\zeta = r \sinh u$, and note that the integrand is even; for the second, substitute $\tau = t - (r/c) \cosh u$.)

When there is no disturbance for $t < 0$, the lower limit in the last integral may be replaced by zero.

If we now return to (253), with the condition (256), the solution we require is

$$(257) \qquad \frac{\phi}{W} = -\frac{1}{2\pi} \int_0^{z-\beta r} \frac{S'(\lambda) d\lambda}{[(z - \lambda)^2 - \beta^2 r^2]^{\frac{1}{2}}}.$$

The solution is thus expressed in terms of what may be called "supersonic sources." But the value of $\phi$ at $(r, z)$ depends only on values of $S'(\lambda)$ for $\lambda < z - \beta r$. Now if $\mu$ is the Mach angle ($\sin^{-1} 1/M$), $\beta = \cot \mu$, and the point $(r, z)$ lies on the Mach cone (downstream half) which has its vertex on the axis at $(0, z - \beta r)$. The influence of the "source strength," and therefore of $S'(\lambda)$, for values of $\lambda > z - \beta r$ is restricted to points in the interior of this Mach cone.

We may now approximate to the right-hand side of (257) by the method used in Section 11.2(a). It will suffice to point out that

$$\int_0^{z-\beta r} \frac{d\lambda}{[(z - \lambda)^2 - \beta^2 r^2]^{\frac{1}{2}}} = \cosh^{-1} \frac{z}{\beta r}$$

(exactly,) that when $\beta r/z$ is small, $\cosh^{-1} z/\beta r$ is approximately $\log 2z/\beta r$, and that it may be proved that

$$\lim_{r \to 0} \left\{ \int_0^{z-\beta r} \frac{S'(\lambda) d\lambda}{[(z - \lambda)^2 - \beta^2 r^2]^{\frac{1}{2}}} - \int_0^{z-\frac{1}{2}\beta r} \frac{S'(\lambda)}{z - \lambda} d\lambda \right\} = 0.$$

Hence the required approximation for $\phi$ (with $S'(0) = 0$) is

$$(258) \qquad \frac{\phi}{W} = \frac{1}{2\pi} \left[ S'(z) \log \tfrac{1}{2}\beta r - \int_0^z S''(\lambda) \log(z - \lambda) d\lambda \right].$$

The boundary condition (256) is clearly satisfied. The discussion may be continued as in Section 11.2(a), the results being directly similar.

(b) *Laplace transforms*

We use $p$-multiplied Laplace transforms (or Heaviside operators) in $z$, and since $\phi$ and $\partial\phi/\partial z$ are zero for $z < 0$, the equation for the transform $\bar{\phi}$ is

$$\partial^2 \bar{\phi}/\partial x^2 + \partial^2 \bar{\phi}/\partial y^2 - \beta^2 p^2 \bar{\phi} = 0.$$

With polar coordinates $r$, $\theta$ in the $(x, y)$ plane,

$$\bar{\phi} = A_0(p) K_0(\beta p r) + \sum_{n=1}^{\infty} \{A_n(p) \cos n\theta + B_n(p) \sin n\theta\} K_n(\beta p r).$$

Again the functions $I_n$ do not enter. For large $r$

$$K_n(\beta p r) \sim (\pi/2\beta p r)^{\frac{1}{2}} \exp(-\beta p r),$$

whereas $I_n$ involves positive exponentials. Only the negative exponentials arising from $K_n$ are permitted, in order that the solution should represent outgoing waves.

Again we let $r$ be small, and keep only the leading terms in the expansion of each $K_n$. Then

$$\bar{\phi} = -A_0(p) [\log \tfrac{1}{2}\beta p r + \gamma]$$
$$+ \sum_{n=1}^{\infty} 2^{n-1}(n-1)! \, \frac{A_n(p) \cos n\theta + B_n(p) \sin n\theta}{(\beta p r)^n}.$$

Let $a_0(z)$, $g_1(z)$, $a_n(z)$, $b_n(z)$ be the inverse transforms of $-A_0(p)$, $-A_0(p)[\log \tfrac{1}{2}\beta p + \gamma]$, $2^{n-1}(n-1)! A_n(p)\beta^{-n} p^{-n}$ and $2^{n-1}(n-1)! B_n(p)\beta^{-n} p^{-n}$, respectively. Then

$$\phi = a_0(z) \log r + \sum_{n=1}^{\infty} \frac{a_n(z) \cos n\theta + b_n(z) \sin n\theta}{r^n} + g_1(z).$$

Again the first two terms represent the two-dimensional potential function $\phi'(x, y; z)$, and from (248) $a_0(z) = (W/2\pi) S'(z)$ as before. Then it may be shown that

$$g_1(z) = \frac{W}{2\pi} \left[ S'(z) \log \tfrac{1}{2}\beta - \int_0^z S''(\lambda) \log (z - \lambda) d\lambda \right],$$

in agreement with (258).

The boundary conditions on $\phi'$ are the same as in the subsonic case. Only $g_1(z)$ is different.

(c) *Plane airfoils of small aspect ratio*
*(subsonic or supersonic flow)*

As Ward has pointed out, the conditions for the validity of

the approximations do not obtain when at some point or points the boundary of a section of the body is concave and has a large curvature, since then the perturbation velocity is no longer small. Nevertheless, the lack of validity may be expected in most cases to be local only, and the application to plane airfoils of small aspect ratio is both simple and interesting.

For the general "lifting-surface" theory, reference may be made to the remarks in Appendix III(c) and some of the works cited there. Slender-body theory provides an approximation to linearized lifting-surface theory for airfoils of small aspect ratio. In particular, note that for a lifting airfoil there will be a trailing vortex sheet behind the trailing edge of the airfoil. In a linearized theory, the vortex lines are taken to be in the direction of the undisturbed flow. We are concerned here only with the effect of incidence of an infinitely thin plane airfoil—i.e., with the part of the perturbation potential which is called $\phi_2$ in Appendix III. At zero incidence we suppose the airfoil lies in the $(z, x)$ plane; the undisturbed velocity is $W$ along the $z$-axis, and the incidence $\alpha$ is obtained by a (small) positive rotation about a line parallel to the $x$-axis. If the velocity components are denoted by $u$, $v$, $W + w$, and if $\Sigma$ is the projection of the airfoil surface on the plane $y = 0$, the boundary condition on $\Sigma$ is $v = -W\alpha$ (on both sides). $v$ is an even function of $y$, and $\phi$, $u$, $w$ are odd functions of $y$. Let $T$ be the trailing vortex sheet, and $R$ the part of the plane $y = 0$ other than $T$ and $\Sigma$. We know that for general lifting-surface theory the Kutta-Joukowski condition must be satisfied at the trailing edge; $v$ is given on $\Sigma$ (as above); $\phi$ and $v$ must be continuous on $R$, and, since $\phi$ is odd in $y$, $\phi = 0$ on $R$; the pressure must be continuous on $T$, and since the squares of the velocity components are even in $y$, while $w$ is odd, it follows that, even with second-order terms retained in the expression for the pressure, we must have $w = 0$ on $T$. Hence on $T$, $\phi$ is independent of $z$, and, for a given $x$, $\phi$ must have the same value at any point of $T$ as it has at the trailing edge.

Since $\alpha$ is small, to a first approximation $u$ and $w$ as calculated on $\Sigma$ are the components of the tangential velocity on the airfoil.

Since $u$ and $w$ are odd functions of $y$, if we suppose the airfoil removed, there would be discontinuities $2u$, $2w$ in these components, where $u$, $w$ are the values calculated on one side. This would mean a vortex sheet at the position of the airfoil, and such a vortex sheet is called a "bound" vortex sheet if forces are supposed to act to constrain the vortex lines of the sheet to keep the positions dictated in this way by the position of the airfoil. The bound vortex lines and the trailing vortex lines then join to produce unbroken vortex lines.

With this introduction we may now write down the results from slender-body theory. We begin with the case when the airfoil has a straight trailing edge, and the span at the trailing edge is not exceeded anywhere on the airfoil (Fig. 12a).

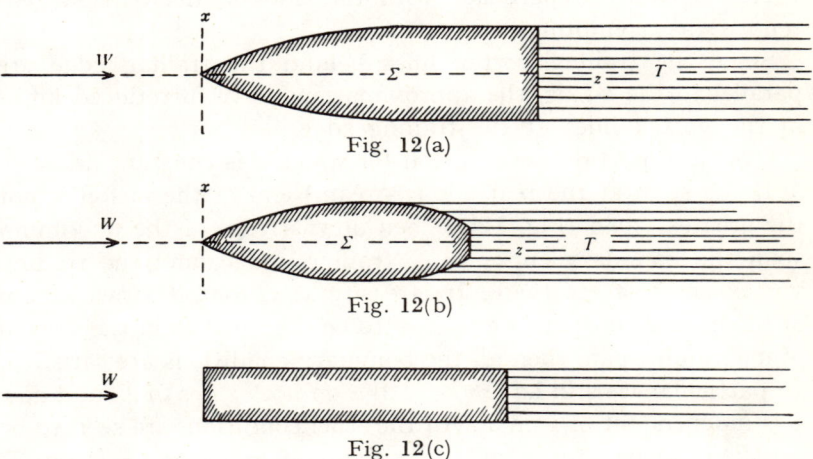

Fig. 12(a)

Fig. 12(b)

Fig. 12(c)

Since the cross-sectional area $S(z) = 0$, $g_1(z) = 0$. Let $2s$ be the span at any section $z$. Then $\phi'$ is the two-dimensional velocity potential for a flat plate moving with velocity $W\alpha$ in the negative direction of the axis of $y$. The determination of $\phi'$ is a simple and familiar problem. On $y = 0+$, we find

$$\phi' = \phi'_0 = \alpha W(s^2 - x^2)^{\frac{1}{2}}, \quad u = u_0 = \partial \phi'_0/\partial x = -\alpha W x(s^2 - x^2)^{-\frac{1}{2}},$$
$$w = w_0 = \partial \phi'_0/\partial z = \alpha W(s^2 - x^2)^{-\frac{1}{2}} s \, ds/dz.$$

From the formula $(\mathbf{v}_1 - \mathbf{v}_2) \times \boldsymbol{\nu} = \boldsymbol{\omega}'$ in Section 1.8 (p. 20) for the strength of a vortex sheet, it follows that the strength of the bound vorticity discussed above has components

$$\omega'_x = 2\partial\phi'_0/\partial z, \quad \omega'_z = -2\partial\phi'_0/\partial x.$$

The equation of the bound vortex lines

$$dx/\omega'_x = dz/\omega'_z$$

is therefore simply $\phi'_0 =$ constant, i.e.,

$$s^2 - x^2 = \text{constant}.$$

For example, if the plate is a symmetrically placed isosceles triangle with vertex angle $2\chi$, for which $s = z \tan \chi$, the bound vortex lines are hyperbolae, with the sides of the triangle (extended) as asymptotes.

Since the trailing vortex lines behind the trailing edge are parallel to the $z$-axis, the approximations have introduced kinks in the vortex lines at the trailing edge.

Over any portion of an airfoil for which $s$ is constant, $w_0 = 0$.

If the span at the rear (downstream) end of the airfoil is not the greatest span (Fig. 12b), then downstream of the maximum span the boundary curve is a "trailing" edge, and the trailing vortex sheet starts there. It is easily seen that if, downstream of the maximum span, we take $\phi'$ to be the same as in the section of maximum span, then all the boundary conditions are satisfied. In particular we still have $v = -W\alpha$ on both sides of $\Sigma$, and now $\phi$ is independent of $z$ on $T$. All the other conditions are seen to be satisfied. Downstream of the maximum span, $w_0 = 0$ on $\Sigma$. The bound vortex lines are parallel to the $z$-axis there.

The difference in pressure on the two sides of the plate is $2\rho_0 W w_0$, since $u^2$, $v^2$ are the same on the two sides. The resultant normal force is therefore

$$N = 2\rho_0 W \iint (\partial\phi'_0/\partial z) dz\, dx.$$

The integration for $z$ proceeds from the leading edge, where $\phi'_0 = 0$, to the section of maximum span, where $\phi'_0 = \alpha W(s_m^2 - x^2)^{\frac{1}{2}}$, if $2s_m$ is the maximum span. Hence

$$N = 2\rho_0 \alpha W^2 \int_{-s_m}^{s_m} (s_m^2 - x^2)^{\frac{1}{2}} dx = \pi \rho_0 W^2 \alpha s_m^2.$$

This formula is, in fact, general, but in the derivation above we have tacitly assumed that $s$ does not begin with a finite value when $z = 0$. This was, in fact, part of the basis of our approximation, but if we now ignore it, then, if $z$ is the section of maximum span, and we take, for any given $x$, $\phi_0'(-0) = 0$, we should write the Stieltjes integral

$$N = 2\rho_0 W \int_{-s_m}^{s_m} \int_{-0}^{z} d\phi_0'(z) dx,$$

which allows for the effect of an infinite pressure difference at a leading edge. For example, in the case of a rectangular wing (Fig. 12c), the span is constant, and on slender-body theory, $w_0 = 0$ everywhere on the wing (apart from the leading edge), and the normal force arises entirely from the infinite pressure difference at the leading edge. This is, of course, an extreme case. It may be shown that the expression for the total normal force is still a good approximation, even though the above result for its distribution is not.

More complicated calculations are needed if there is a re-entrant trailing edge. For this, for other problems such as flow past winged bodies, and for further developments of the theory, reference may be made to the works cited.

CHAPTER 12

# Mixtures, and High Temperature Effects in Gases

The mathematics of gasdynamics, as we have described it — and as it appeared until lately in the bulk of the literature — is based on six equations for three velocity components and three thermodynamic variables, namely, the equation of continuity, three Navier-Stokes equations of momentum, an energy equation, and a thermodynamic equation of state. (For electrically conducting fluids in the presence of electromagnetic forces, we add the electrodynamic variables on the one hand, and Maxwell's equations and the constitutive relations on the other.) Some mention should be made of circumstances in which these equations do not sufficiently govern the processes involved.

On the question of very rarefied gases, and the conditions under which macroscopic equations may be usefully obtained and applied, we remark that the possibility of using macroscopic equations depends, in general, both on the ratio $\lambda/L$ (called the Knudsen number) of the mean free path to a characteristic length of the macroscopic system, and on the ratio of a characteristic frequency of the macroscopic system to the collision frequency. As mentioned in Section 5.4, the physical order of magnitude of $\lambda/L$ may vary from one part to another of one and the same flow field, and $L$ should be locally defined. $\lambda$ is a moderate multiple of $\nu/a$, where $\nu$ is the kinematic viscosity and $a$ the sound speed. When $L$ may be taken as a length $d$ representative of the geometry of the boundaries, $\lambda/L$ is a moderate multiple of $M/R$, where $M$ is the Mach number and $R$ the Reynolds number based on $d$. Such rarefied gas flows, which occur at large enough values of the Knudsen number, entail finite $M$, and low $R$ and/or large $M$. 'Free molecule' flow is taken (admittedly still on the basis of insufficient experimental evidence) to be the regime where (roughly) $\lambda/L \approx M/R > 10$. For small $R$, the slip-flow regime is taken to lie roughly in $0.01 < \lambda/L \approx M/R < 0.1$. If $R$ is considered large enough for a boundary layer to be formed, in the layer the displacement thickness of the layer takes the place of $L$ in the Knudsen number. (Schaaf and Chambré, *op. cit.* on p. 99. Added October, 1959: for a more recent discussion with additional references see S. A. Schaaf, *Proc. 6th Midwestern Conf. on Fluid Mechanics*, Univ. of Texas, 1959, pp. 1–15.)

We do not here enter on a discussion of the physics of very condensed gases.

Leaving these questions aside, and complicated as the mathematics is for the discussion of the six fundamental equations mentioned, they apply only in what must be described as physically rather simple situations. Fortunately, these situations are of frequent occurrence, and the complicating effects, when they enter, may sometimes be small. However, they are also sometimes very important.

This section is intended only (i) to serve as a warning about the circumstances in which the basic equations need amplification or alteration, (ii) to give references to some of the literature which may serve as an introduction, (iii) to draw attention to situations where basic research is proceeding, but where more is still urgently needed. I make no attempt to discuss the subjects mentioned, nor to provide any kind of mathematical guiding lines in referring to the literature; I doubt if the time has yet come to attempt anything like that, and certainly I am not now qualified to do so. A discussion, *in extenso*, of the literature is impossible here, and would not be very helpful.

The first matter to which I allude is that of gas mixtures. The overall equation of continuity still applies, and so do the equations of momentum. However, we must also take into account the concentrations of the constituents. Gradients in concentration produce currents of the constituents, which are governed by diffusion-type equations. When the velocities of the constituents differ from the mean velocity because of varying diffusion rates, additional terms appear in the expression for the stress tensor; and in the equation for the total velocity of the fluid mixture an apparent stress also appears, which arises from the averaging, over the velocities of the constituents, of the nonlinear convection terms in the acceleration. The diffusion velocities also lead to an enthalpy transport, which appears in the energy equation. These are the main effects. Gradients of pressure and temperature may also set up diffusion currents. For a mixture in thermodynamic and concentration equilibrium the effects may be neglected; air is a mixture, but for the motions of air such as we have previously considered, the effects need not be considered. The most important applications (apart from condensation, dissociation, and ionization) are in the theory of combustion processes. Marble (*J. Ae. Sci.* 23 (1956) 462) has given a brief description, where other references will be found. (In particular see Karman, Fundamental Equations in Aerothermochemistry, *Proc. 2nd Agard Combustion Colloquium*, Liège, Belgium, 1955.) I should add that when exothermic chemical reactions are taking place, the heat evolved must appear in the equation of energy (or be taken into account in the definition of enthalpy); and that chemical reactions entail the presence of sources and sinks for the constituents, so the effect on the concentrations must, of course, be taken into account. (For an introduction to the thermodynamics of mixtures, with applications to dissociation and condensation, see Liepmann and Roshko, *Elements of Gasdynamics*.)

We have previously mentioned the effects of the "lags" in the adjustment of the gas molecule to equilibrium conditions. When the relaxation time is small in comparison with a characteristic time of the macroscopic process considered, the effect may be taken into account by the introduction of a "transfer" coefficient, such as a coefficient of viscosity or heat conduction; otherwise, if we use macroscopic equations, we must use equations of irreversible thermodynamics, with additions to the number of state variables; these equations contain "relaxation" equations involving a time derivative. The complete picture here is complicated. However, the "lag" in the translational modes is taken up by the use of the ordinary (shear) coefficient of viscosity and the coefficient of heat conduction; the "lag" in the rotational modes is imperfectly understood, but probably with the insertion of a bulk viscosity it would be only in extreme circumstances that anything further is required. We pass over the adjustment of the electronic states. The relaxation time of the vibrational modes is very much longer than that for the translational and rotational modes, and considerable work has been done on the consequences of this lag. The magnitude of the effect depends approximately on the product of the relaxation time and the vibrational heat capacity, and hence is biggest for those gases, like $CO_2$, for which both are fairly large. On the simplest theory, the relaxation equation equates the time rate-of-change of the internal energy for the vibrational degrees of freedom to its difference from its equilibrium value, divided by a relaxation time. (A similar treatment could be used for any internal degree of freedom.) Most of the investigations are concerned with the attenuation of sound waves, and with shock waves. Most investigations on shock waves assume an initial shock discontinuity, across which the energy content of the active (or fast) degrees of freedom instantaneously takes its new equilibrium value, while the inert (or slow) degrees of freedom retain their incident energy; this discontinuity is followed by a region in which adjustment to thermodynamic equilibrium takes place. (L. Talbot has recently computed a shock structure using the Navier-Stokes equations and the more complicated relaxation equations of Wang-Chang and Uhlenbeck, and concludes that the use of the simpler model is justified. *Univ. Calif. Inst. Eng. Res.*, Tech. Rep. He-150-145 (1957). There is a small, but not entirely negligible, effect in steady flow without shock waves. When, in a gas flow, appreciable changes of temperature of an element of the gas take place in times of the order of the relaxation time, the energies in the vibrational modes do not have their equilibrium values. The lags of the vibrational modes behind their equilibrium energy values cause an irreversible energy interchange within the gas. Consequently the entropy rises, and the available energy falls. Thus in flow in a supersonic wind-tunnel, for example, the effect will lead to lower measured temperatures than would be obtained without it; for flow past an obstacle it leads to a drag even in subsonic inviscid flow without shocks; the effect was measured by Kantrowitz by measuring the diminution of pressure in a gas allowed to expand isentropically into a uniform stream, and then made to impinge on a specially shaped impact tube. Reference for an introduction to all these effects may be made to Gunn, *Reports and Memoranda of the Aeronautical Council*, London, No. 2338 (1946, published 1952). The first application to shock waves seems to have been made by Bethe and Teller, Deviations from Thermal Equilibrium in Shock Waves, *Ballistic Research Lab. Report, Aberdeen Proving Ground*, Rep. X 117 (1940), and to general (steady) flows by Kantrowitz, *J. Chem. Phys.* **14** (1946), p. 150. See also the paper by Lighthill in the G. I. Taylor Memorial Volume, previously cited (p. 92). Further references may be found in the publications mentioned.

At sufficiently high temperatures (say about 3000°K for oxygen at pressures of about 1 atmosphere) molecular dissociation becomes important for gases with diatomic and polyatomic molecules. At still higher temperatures (say above 10,000°K for atomic oxygen at pressures of about 1 atmosphere), ionization of a monatomic gas begins to become important. For theoretical discussions of the thermodynamics of a diatomic gas in dissociation equilibrium, reference may be made to Liepmann and Roshko, *Elements of Gasdynamics*, and to Lighthill, *J. Fluid Mechs.*, **2** (1957), pp. 1–32. The former contains some mention of the theory for an ionizing monatomic gas; for further details, and in particular for the Saha formula for the degree of ionization, see the reference there given (A. von Engel, *Ionized Gases*, Oxford Univ. Press, 1955) and R. H. Fowler, *Statistical Mechanics*, Cambridge Univ. Press, 1936. On the use of the Saha equation (and also on the computation of real gas effects on shock discontinuities, referred to below), see R. A. Alpher, *J. Fluid Mechs.* **2** (1957), pp. 123–126.

The degree both of dissociation and ionization at any temperature depends on the pressure; the temperature at which dissociation (or ionization) becomes important decreases as the pressure decreases. For some gases (e.g. nitrogen) appreciable ionization begins before dissociation is complete. In a dissociating gas, energy is required to provide the heat of dissociation, i.e., in any process in the gas part of the heat produced goes into breaking the molecular bond, so temperatures are less than would be calculated in the absence of dissociation. A similar remark is true for ionization.

Tables of the thermodynamic properties of air, considered as a mixture of nitrogen and oxygen, have been prepared (see, for example, Hilsenrath and Beckett, Thermodynamic Properties of Argon-Free Air (0.78847 $N_2$, 0.21153 $O_2$) to 15,000°K, *National Bureau of Standards Report* 399, 1955, and the reference to Gilmore in Liepmann and Roshko, p. 352), and applied to the calculation of normal shock discontinuities, with the air behind the shock assumed in dissociation equilibrium (see, for example, M. F. Romig, *J. Ae. Sci.* **23** (1956), pp. 185, 186). The results for the temperature and density ratios, $T_2/T_1$ and $\rho_2/\rho_1$, depend on the pressure $p_1$ of the incident stream; the pressure ratio $p_2/p_1$ does not, and is not much affected by dissociation, even for quite strong shocks; $T_2/T_1$ is much reduced, and $\rho_2/\rho_1$ much increased, by the dissociation, and more for smaller than for larger pressures. It should be added that strong-shock formulae for a diatomic gas, with certain approximations, are given by Lighthill (*op. cit.*).

We should note that new chemical components will be formed in air at high temperatures. In various temperature ranges (considering only the changes in the oxygen and nitrogen content, and neglecting the presence of argon, etc.) we may expect to find $N_2$, $O_2$, $O$, $N$, $NO$, $NO_2$, $N_2O$, free electrons, $O^+$, $N^+$, $NO^+$, $N_2^+$, $O_2^+$, $O^-$, $O_2^-$. Some of these will be found in small quantities only, but relatively small concentrations may in some circumstances have large effects.

When dissociation equilibrium and local thermodynamic equilibrium are applied strictly at all times, particle (molecular, atomic, ionic, electronic) transport phenomena and radiative heat transfer are neglected. The relaxation times for dissociation, ionization, and recombination, and all molecular time lags are taken as negligibly small compared with the characteristic time of the macroscopic processes, and the theory then applies to the flow of an ideal gas (with shock discontinuities), but not to flow inside boundary layers or shock waves.

Effects of viscosity, heat conduction, etc. have been omitted so far. The next step is to allow for radiative heat transfer and for transport phenomena on a quasi-equilibrium basis, assuming all relaxation times in the gas short compared with a characteristic time of the macroscopic processes. Expressions must be found for the stress tensor, heat conduction vector, electric current vector, and radiative heat transfer. (The last will affect the equation of energy only; radiation pressure may be neglected.) A dissociating diatomic gas (for example) is a mixture of two "gases" at least, composed of atoms and molecules, and variations of temperature and pressure produce varying concentrations; this should be taken into account; for example, the stress tensor, even if taken to depend only on first powers of gradients, will include a term depending on the temperature gradient. Equations for concentration currents will also be involved. Theoretical guidance is required on the values of coefficients of viscosity, heat conduction, diffusion, etc., as well as specific heats and Prandtl number, at high temperatures under varying conditions of pressure and composition. I know of no published satisfactory physical and mathematical theory. Fundamental physical data are not sufficiently known, but nevertheless a theory establishing the correct macroscopic equations is required; the constants may then be found experimentally.

In addition, a non-equilibrium theory will be required, which will involve the rates of dissociation, ionization, and recombination. (For mixtures such as air, it will involve the reaction rates for the formation and decomposition of the chemical constituents present.) It will also involve the heat produced in exothermic reactions, and a further study of radiation effects (emission and absorption, and scattering). Mechanics, physics, and physical chemistry must all be combined in the theory. These considerations will enter, for example, in discussions of shock structure (including the flow behind shocks), and of the connection between the temperature of a wall and the recombination of gas atoms on it and in its neighborhood.

For further discussion from the aerodynamicist's point of view, and references, see A. Hertzberg, *Jet Propulsion*, **26** (1956), p. 549; J. G. Logan, *J. Ae. Sci.* **23** (1956), p. 123; H. H. Kurzweg and R. E. Wilson, *Aero. Eng. Rev.*, Dec. 1956, p. 32; W. H. Dorrance, *Aero. Eng. Rev.*, Jan. 1957, p. 26.

Finally, we remark that when a gas is completely ionized, we pass to another regime, in which mathematical discussion again becomes immediately profitable. We conclude with a simple example of a calculation in such a regime.

CHAPTER 13

# Longitudinal (Electrostatic) Waves in an Ionized Gas

We conclude with an example of the calculation of a wave motion in an ionized gas, when the rarefaction is too high to allow the use of an electrical conductivity independent of the frequency. The inclusion of this example, with the treatment of Boltzmann's equation given below, was suggested to me by Dr. Max Krook, of the Harvard Observatory, to whom I here express my thanks.

For an introduction to wave motion in an ionized gas, reference may be made to L. Spitzer, Jr., *Physics of Fully Ionized Gases*, Interscience, New York-London, 1956, Chap. 4.

We consider a purely longitudinal wave, the direction of propagation being along the axis of $x_1$, which we denote simply by $x$. To obtain a simple example, we consider only the case in which the oscillation is sufficiently rapid to allow the oscillatory movement of the positive ions to be neglected, because of their large inertia compared with that of the electrons. (For hydrogen, the mass of the ion is about 2000 times that of the electron; the ratio is higher for other gases.) The oscillations of the electrons lead to accumulations of electric charge, and produce a self-excited electric field. External electric and magnetic fields are assumed absent; the electric field and the current may be taken along the axis of $x$; and it is easy to see that the self-excited magnetic field is zero. All quantities are functions of $x$ and $t$ only. Also the effect of collisions in producing an ordinary scalar or a tensor conductivity, or anything in the nature of a shear stress, is neglected. Effects of external forces, such as gravity, are also neglected.

Again we must decide which system of units to use. Spitzer uses electromagnetic units (except for the charge of the electron). We shall continue to use the *MKSQ* system, introduced in

Section 3.[18] In this system, let $-q$ be the charge of the electron. Let $n$ be the number density of electrons — the number of electrons per unit volume. The gas is supposed neutral in the undisturbed state; if $n_0$ is the undisturbed value of $n$, $n_0$ is also the undisturbed number density of ions, and, with our assumptions, the number density of ions continues to be $n_0$; we have used the convention that an ion that carries a charge $mq$ is to be counted $m$ times.

If $\rho_e$ is the electric charge density,

(259) $$\rho_e = q(n_0 - n).$$

Refer to Maxwell's equations, (56). Since now $\mathbf{D} = \varepsilon_0 \mathbf{E}$, from the third of these equations

(260) $$\partial E/\partial x = \rho_e/\varepsilon_0 = q(n_0 - n)/\varepsilon_0.$$

From the second and third of equations (56) we have the equation for the conservation of charge,

(261) $$\partial j/\partial x = -\partial \rho_e/\partial t = q\, \partial n/\partial t,$$

where $j$ is the electric current density. We require also the macroscopic equation connecting $j$, $E$, and $n$, but we use only the linearized form, with squares and products of the disturbance neglected. With no magnetic field, and, as explained, the effect of collisions in producing shear stresses, or an ordinary scalar, or tensor, conductivity, neglected, this equation reduces simply to

(262) $$m\, \partial j/\partial t = n_0 q^2 E + \lambda k T q \partial n/\partial x,$$

where $m$ is the mass of an electron, $k$ is Boltzmann's constant, and $T$ is the undisturbed electron temperature, so, since there is equilibrium in the undisturbed state, $T$ is the temperature of the gas in the undisturbed state. $\lambda$ is a constant whose value depends on the particular process under study. For the one-dimensional compression waves that are to be studied here, the best value of $\lambda$ is 3. This equation follows from the derivation

---

[18] For a discussion of units, see Sommerfeld's *Electrodynamics*. Conversion tables are given by Stratton, *Electromagnetic Theory*, McGraw-Hill, New York, 1941. Note the units of mass, length, force, and energy in the *MKSQ* system.

given by Spitzer, and also, with the assumptions stated, from the more general and more nearly complete (unpublished) discussion by Krook.

The equations set out are all that is required for the simple macroscopic treatment. Note that we have passed from the one extreme, where we were concerned with equations determining macroscopic velocity components and the usual thermodynamic variables of state, right to the other extreme. Apart from the disturbance, $n - n_0$, in the number density, our equations now concern only the electrical quantities $E$ and $j$. In intermediate cases, everything must be taken into account.

If we now assume a wave in which $n - n_0$, $E$, and $j$ vary as $\exp[i(\omega t - \kappa x)]$, we find at once that

(263) $$\omega^2 = \omega_p^2 + 3\kappa^2 a^2,$$

where

(264) $$\omega_p^2 = n_0 q^2/\varepsilon_0 m, \quad a^2 = kT/m.$$

$\omega_p$ is called the plasma frequency. The wave velocity is given by

(265) $$V^2 = \omega^2/\kappa^2 = \omega_p^2/\kappa^2 + 3a^2.$$

The medium is dispersive, but with this approximation there is no attenuation.

Now consider a microscopic treatment, starting from Boltzmann's equation for the velocity-distribution function of the electrons. (An introduction to the Boltzmann equation may be found in the appendix to Spitzer's book.) For a more general mathematical discussion along the lines of the theory given below, see L. Landau, *J. Phys. U.S.S.R.* **10** (1946), pp. 25–34.

If **v** (with components $v_1$, $v_2$, $v_3$) is the electron velocity, the velocity-distribution function in the undisturbed, equilibrium, state is

(266) $$f_0 = n_0 \, (m/2\pi kT)^{\frac{3}{2}} \exp(-mv^2/2kT).$$

In the disturbed state, let

(267) $$f = f_0(1 + \phi).$$

Then $\phi$ is a function of $t$, $x$, and the $v_i$. In Boltzmann's equation, the force $-qE$, arising from the charge concentrations of the particles themselves, is inserted as an external force; this takes into account the forces on a particle arising from other particles whose distances are greater than a certain length (the Debye shielding distance); but otherwise the effects of collisions, and in particular of close collisions, are neglected. The force $-qE$ is along the $x$-axis, and the equation for $\phi$ is

$$(268) \qquad f_0\left(\frac{\partial \phi}{\partial t} + v_1 \frac{\partial \phi}{\partial x}\right) - \frac{q}{m} E \frac{\partial}{\partial v_1}\{f_0(1 + \phi)\} = 0.$$

Since $\partial f_0/\partial v_1 = -(mv_1/kT)f_0$, this equation, when linearized, becomes

$$(269) \qquad \frac{\partial \phi}{\partial t} + v_1 \frac{\partial \phi}{\partial x} + \frac{qv_1}{kT} E = 0.$$

If now $\phi$, $E$ and $n - n_0$ vary as $\exp[i(\omega t - \kappa x)]$, from (260) and (269)

$$E = -\frac{iq}{\varepsilon_0 \kappa}(n - n_0), \qquad \phi = \frac{iqv_1}{kT} \frac{E}{\omega - \kappa v_1}$$

$$= \frac{q^2}{\varepsilon_0 kT\kappa} \frac{v_1}{(\omega - \kappa v_1)}(n - n_0).$$

But

$$n_0 = \iiint f_0 \, dv_1 dv_2 dv_3, \qquad n = \iiint f \, dv_1 dv_2 dv_3$$

i.e.,

$$n - n_0 = \iiint f_0 \phi \, dv_1 dv_2 dv_3$$

$$(270) \qquad = n_0(n - n_0) \frac{q^2}{\varepsilon_0 kT\kappa} \left(\frac{m}{2\pi kT}\right)^{\frac{1}{2}} \int_{-\infty}^{\infty} \frac{v_1}{\omega - \kappa v_1} \exp(-mv_1^2/2kT) dv_1$$

$$\iint_{-\infty}^{\infty} \frac{m}{2\pi kT} \exp\left[-\frac{m(v_2^2 + v_3^2)}{2kT}\right] dv_2 dv_3$$

$$= n_0(n - n_0) \frac{q^2}{\varepsilon_0 kT\kappa} \left(\frac{m}{2\pi kT}\right)^{\frac{1}{2}} \int_{-\infty}^{\infty} \frac{v_1}{\omega - \kappa v_1} \exp(-mv_1^2/2kT) dv_1.$$

Substitute $\omega_p$ and $a$ from (264), and put $u = v_1/a$. Then

(271) $$\frac{1}{(2\pi)^{\frac{1}{2}}}\frac{\omega_p^2}{\kappa^2 a^2}\int_{-\infty}^{\infty}\frac{\exp(-\tfrac{1}{2}u^2)u\,du}{\zeta - u} = 1,$$

where

(272) $$\zeta = \omega/\kappa a.$$

(271) is the required equation connecting $\omega$ and $\kappa$. If $\zeta$ is real, and the integral is taken along the real axis, it is an improper integral. But in fact the wave is attenuated. This we shall verify; it was shown to be the case for any initial-value problem by Landau (*op. cit.*). If we have an initial-value problem, with $\kappa$ given and real, the imaginary part of $\omega$ must be positive. We assume $\kappa$ positive, so Im $(\zeta) > 0$. Similarly, for a boundary-value problem, with $\omega$ given and real, Im $(\kappa) < 0$. In this case we assume $\omega > 0$, so again Im $(\zeta) > 0$. We therefore take

$$0 < \arg \zeta < \pi.$$

If $\zeta$ approaches a real value, and the integral is taken along the real axis, the contour must be indented so that the pole at $u = \zeta$ lies above it; arg $u$ will increase from $-\pi$ to $0$ on the contour.

Write

(273) $$\int_{-\infty}^{\infty}\frac{\exp(-\tfrac{1}{2}u^2)u\,du}{\zeta - u} = -\int_{-\infty}^{\infty}\exp(-\tfrac{1}{2}u^2)du - \zeta F(\zeta)$$
$$= -(2\pi)^{\frac{1}{2}} - \zeta F(\zeta),$$

where

(274) $$F(\zeta) = \int_{-\infty}^{\infty}\frac{\exp(-\tfrac{1}{2}u^2)}{u - \zeta}du.$$

Let $D_n(z)$ be Weber's parabolic cylinder function of order $n$ (Whittaker and Watson, *Modern Analysis*, Section 16.5), and erfc $z$ the complementary error function, so that

(275) $$D_{-1}(z) = (\pi/2)^{\frac{1}{2}}\exp(\tfrac{1}{4}z^2)\,\text{erfc}\,(z/\sqrt{2}).$$

It is easily seen that, for large $|\zeta|$, $F(\zeta)$ has the same asymptotic expansion as $i(2\pi)^{\frac{1}{2}}\exp(-\tfrac{1}{4}\zeta^2)D_{-1}(-i\zeta)$. We may prove the identity of the two functions in several ways. One method is as follows.

$$F''(\zeta) + \zeta F' + F = \int_{-\infty}^{\infty} \frac{\exp(-\tfrac{1}{2}u^2)}{(u-\zeta)^3}\{2 + u(u-\zeta)\}du$$

$$= -\int_{-\infty}^{\infty} \frac{d}{du}\left[\frac{\exp(-\tfrac{1}{2}u^2)}{(u-\zeta)^2}\right]du = 0.$$

Hence $F$ satisfies the equation of which $\exp(-\tfrac{1}{4}\zeta^2)D_0(\zeta)$ and $\exp(-\tfrac{1}{4}\zeta^2)D_{-1}(-i\zeta)$ are two independent solutions, so

$$F(\zeta) = \exp(-\tfrac{1}{4}\zeta^2)\{c_1 D_{-1}(-i\zeta) + c_2 D_0(\zeta)\}.$$

The values of the constants $c_1$ and $c_2$ are found by comparing the values of $F$ and $F'$ at $\zeta = 0$. It is easily seen that $F(0) = \pi i$. To find $F'(0)$, substitute $t = \tfrac{1}{2}u^2$ in the integral for $F'(\zeta)$; the contour becomes the contour for Hankel's integral for the gamma function (Whittaker and Watson, Section 12.22); and on the contour $-2\pi \leq \arg t \leq 0$. The result is that $F'(0) = -(2\pi)^{\frac{1}{2}}$. Hence $c_1 = i(2\pi)^{\frac{1}{2}}$, $c_2 = 0$, and

(276)
$$\begin{aligned}F(\zeta) &= i(2\pi)^{\frac{1}{2}} \exp(-\tfrac{1}{4}\zeta^2) D_{-1}(-i\zeta) \\ &= i\pi \exp(-\tfrac{1}{2}\zeta^2) \operatorname{erfc}(\zeta/2^{\frac{1}{2}}i).\end{aligned}$$

The dispersion equation is

(277) $\quad i\zeta \exp(-\tfrac{1}{4}\zeta^2) D_{-1}(-i\zeta) + 1 + \kappa^2 a^2/\omega_p^2 = 0.$

When $\zeta$ is small, $\arg \zeta$ is near $\tfrac{1}{2}\pi$. Since $0 \leq \arg \zeta \leq \pi$, $-\tfrac{1}{2}\pi \leq \arg(-i\zeta) \leq \tfrac{1}{2}\pi$, and the asymptotic expansion of $D_{-1}(-i\zeta)$ for large $|\zeta|$ is

(278) $\quad D_{-1}(-i\zeta) \sim -\dfrac{\exp(\tfrac{1}{4}\zeta^2)}{i\zeta}\left\{1 + \dfrac{1}{\zeta^2} + \dfrac{1.3}{\zeta^4} + \ldots\right\}.$

For large $|\zeta|$ the dispersion equation becomes approximately

(279) $\quad 1/\zeta^2 + 1.3/\zeta^4 + \ldots = \kappa^2 a^2/\omega_p^2.$

If we solve by successive approximation, the first approximation is simply $\omega^2 = \omega_p^2$. For the next approximation we may take

(280) $\quad 1/\zeta^2 = \kappa^2 a^2/\omega^2 = \kappa^2 a^2/\omega_p^2 - 3/\zeta^4 = \kappa^2 a^2/\omega_p^2 - 3\kappa^2 a^2/\zeta^2 \omega_p^2,$

which gives

$$1/\omega_p^2 = 1/\omega^2 + 3\kappa^2 a^2/(\omega^2 \omega_p^2)$$

i.e.,

(263 bis) $$\omega^2 = \omega_p^2 + 3\kappa^2 a^2,$$

as in the simplest continuum theory above.

Other references are given by Spitzer. For later work, see Bhatnagar, Gross, and Krook, *Phys. Rev.* **94** (1954), p. 511 and Gross and Krook, *Phys. Rev.* **102** (1956), p. 593.

# APPENDIX I

# The Rate-of-Strain Components and the Vorticity

At time $t$ let $PQ$, $PR$ be two short straight lines of fluid particles, at right angles to each other, of lengths $\delta s_1$ and $\delta s_2$, respectively, and with direction cosines $\alpha_{1i}$ and $\alpha_{2i}$, respectively (Fig. 13).

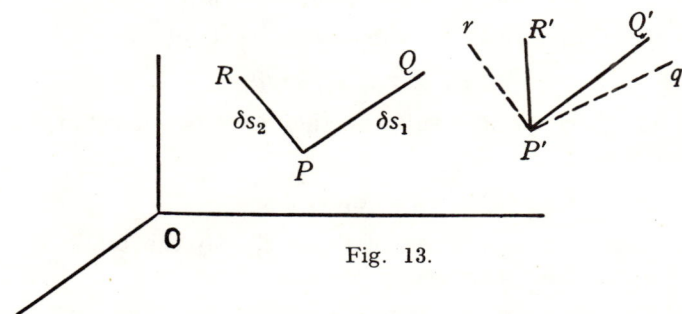

Fig. 13.

For the coordinates and velocity components at $P$, write $x_i$ and $v_i$; at $Q$, $x_i + \delta_1 x_i$, $v_i + \delta_1 v_i$; and at $R$, $x_i + \delta_2 x_i$, $v_i + \delta_2 v_i$. We retain only first-order terms in $\delta s_1$ and $\delta s_2$. If $\sigma$ may be either 1 or 2,

$$\delta_\sigma x_i = \alpha_{\sigma i} \delta s_\sigma, \qquad \delta_\sigma v_i = (\partial v_i/\partial x_j)\alpha_{\sigma j} \delta s_\sigma.$$

(When the *Greek* subscript $\sigma$ occurs twice in an expression, the sum is *not* to be taken over the possible values of the Greek subscript.) Also

$$\alpha_{1i}\alpha_{1i} = 1, \quad \alpha_{1i}\alpha_{2i} = 0.$$

Let the lines move with the fluid. After time $\delta t$, let $P$, $Q$, $R$ be at $P'$, $Q'$, $R'$, respectively. Then to the first order in $\delta t$, $P'$ is at $x_i + v_i \delta t$, $Q'$ and $R'$ at

$$x_i + v_i \delta t + \delta_\sigma x_i + \delta_\sigma v_i \delta t = x_i + v_i \delta t + \delta s_\sigma \{\alpha_{\sigma i} + \alpha_{\sigma j}(\partial v_i/\partial x_j)\delta t\}.$$

To this order $P'Q'$, $P'R'$ stay straight lines, and the components of the vectors **P'Q'**, **P'R'** are

$$\delta s_\sigma \{\alpha_{\sigma i} + \alpha_{\sigma j}(\partial v_i/\partial x_j)\delta t\}.$$

Now let $e'_{11}$ be twice the rate of extensional strain of $PQ$, so the length $P'Q'$ after time $\delta t$ is $\delta s_1(1 + \tfrac{1}{2}e'_{11}\delta t)$, and its square, to the first order in $\delta t$, is $(\delta s_1)^2(1 + e'_{11}\delta t)$. But $(\mathbf{P'Q'})^2$ is

$(\delta s_1)^2 \{\alpha_{1i} + \alpha_{1j}(\partial v_i/\partial x_j)\delta t\} \{\alpha_{1i} + \alpha_{1k}(\partial v_i/\partial x_k)\delta t\}$
$= (\delta s_1)^2 \{1 + \delta t(\alpha_{1i}\alpha_{1k}\, \partial v_i/\partial x_k + \alpha_{1i}\alpha_{1j}\, \partial v_i/\partial x_j)\}$
$= (\delta s_1)^2 \{1 + 2\delta t\, \alpha_{1i}\alpha_{1j}\, \partial v_i/\partial x_j\}.$

Hence
$$e'_{11}\delta t = 2\delta t\, \alpha_{1i}\alpha_{1j}\, \partial v_i/\partial x_j,$$
to the first order in $\delta t$. Divide by $\delta t$, and let $\delta t \to 0$.
$$e'_{11} = 2\alpha_{1i}\alpha_{1j}\, \partial v_i/\partial x_j.$$

Since $i$ and $j$ are dummy suffixes, they may be interchanged and half the sum taken, i.e.,
$$e'_{11} = \alpha_{1i}\alpha_{1j}\, e_{ij}.$$

Let the rate of shear be denoted by $e'_{12}$, so the angle $Q'P'R'$ is $\tfrac{1}{2}\pi - e'_{12}\delta t$; then the scalar product
$$\mathbf{P'Q'} \cdot \mathbf{P'R'} = \delta s_1 \delta s_2\, e'_{12}\delta t$$
to the first order in $\delta t$. From the expressions for the components this is equal to

$\delta s_1 \delta s_2 \{\alpha_{1i} + \alpha_{1j}(\partial v_i/\partial x_j)\delta t\} \{\alpha_{2i} + \alpha_{2k}(\partial v_i/\partial x_k)\delta t\}$
$= \delta s_1 \delta s_2\, \delta t \{\alpha_{1j}\alpha_{2i}\, \partial v_i/\partial x_j + \alpha_{1i}\alpha_{2k}\, \partial v_i/\partial x_k\}$
$= \delta s_1 \delta s_2\, \delta t \{\alpha_{1i}\alpha_{2j}\, \partial v_j/\partial x_i + \alpha_{1i}\alpha_{2j}\, \partial v_i/\partial x_j\}$
$= \delta s_1 \delta s_2\, \delta t\, \alpha_{1i}\alpha_{2j}\, e_{ij}.$

Hence
$$e'_{12} = \alpha_{1i}\alpha_{2j}\, e_{ij}.$$

**PQ** and **PR** are parallel to our previous axes of $x'_1$ and $x'_2$ (Section 1.6). We have therefore verified our previous (rather hasty) physical interpretation of $e'_{11}$ and $e'_{12}$, and have also recovered the transformation formulae.

We may go further. Let $\mathbf{P'q}$, $\mathbf{P'r}$ be parallel and equal to **PQ** and **PR** respectively. Let the angles $Q'P'r$, $R'P'q$ be $\tfrac{1}{2}\pi - a\delta t$, $\tfrac{1}{2}\pi - b\delta t$, respectively. Then

$$\mathbf{P'Q'} \cdot \mathbf{P'r} = \delta s_1 \delta s_2 \, a \delta t,$$

whereas, from the components, this is equal to

$$\delta s_1 \delta s_2 (\alpha_{1i} + \alpha_{1j}(\partial v_i/\partial x_j)\delta t)\, \alpha_{2i} = \alpha_{2i} \alpha_{1j}(\partial v_i/\partial x_j)\delta t\, \delta s_1 \delta s_2$$
$$= \alpha_{1i} \alpha_{2j}(\partial v_j/\partial x_i)\delta t\, \delta s_1 \delta s_2.$$

Hence

$$a = \alpha_{1i} \alpha_{2j}\, \partial v_j/\partial x_i.$$

$b$ is found by interchanging 1 and 2, and

$$b = \alpha_{1i} \alpha_{2j}\, \partial v_i/\partial x_j.$$

In the first place, $a + b = e'_{12}$. (Thus, although in general $\mathbf{P'Q'}$, $\mathbf{P'R'}$ are rotated out of the plane $qP'r$, this makes no first-order difference to the relation between the angles, as we can easily see geometrically.) Also

$$a - b = \alpha_{1i} \alpha_{2j} \xi_{ij} = \xi'_{12} = \omega'_3,$$

and we now have another physical interpretation of the vorticity. $\omega'_3$ is the difference of the rates at which $\mathbf{PQ}$ is rotated towards $\mathbf{PR}$ and at which $\mathbf{PR}$ is rotated towards $\mathbf{PQ}$ as the lines move with the fluid.

It is easily proved that if we take two line elements, $PQ$ and $PR$, in directions with direction cosines $\alpha_{1i}$ and $\alpha_{2i}$, as above, but if the angle between them is $\theta \neq \frac{1}{2}\pi$, then the rate of decrease, $\varepsilon$, of the angle between them is given by

$$\varepsilon \sin \theta = \alpha_{1i} \alpha_{2j} e_{ij} - \tfrac{1}{2}(e'_{11} + e'_{22}) \cos \theta,$$

where $\tfrac{1}{2}e'_{11}$, $\tfrac{1}{2}e'_{22}$ denote the rates of extensional strain along $PQ$ and $PR$ respectively.

Now let $y_i$ be general orthogonal coordinates, the square of the general line element being

$$(ds)^2 = dx_i dx_i = g_{ij} dy_i dy_j$$

(we shall not find it necessary to use superscripts), where $g_{ij} = 0$ if $i \neq j$, and when $i = j = \lambda$, we write

$$(g_{\lambda\lambda})^{\frac{1}{2}} = h_\lambda,$$

where $\lambda$ takes the values 1, 2, 3. (No summation with the Greek subscript $\lambda$.) Let $\alpha_{1\lambda}$ be the direction cosines of $PQ$ (as above) referred to the tangents at $P$ to the curves along which two of the $y_\lambda$ are constant, drawn in the sense in which the third is increasing. We shall compute the rate of extensional strain for $PQ$; the other computations may be carried out in a rather similar manner. The coordinates of $P$ are denoted by $y_\lambda$, those of $Q$ by $y_\lambda + \delta y_\lambda$, and $v_\lambda$ denotes the velocity component at $P$ at time $t$ in the direction in which $y_\lambda$ increases, the other two coordinates being constant. For a first-order calculation, the curvature of the curves along which two coordinates are constant may be neglected, but the variation of the $h_\lambda$ with position must be taken into account.

After a time $\delta t$, the displacement of $P$ has projections $v_\lambda \delta t$, so the coordinates of $P'$ are $y_\lambda + u_\lambda \delta t$, where $u_\lambda = v_\lambda / h_\lambda$. The coordinates of $Q'$ are $y_\lambda + u_\lambda \delta t + \delta y_\lambda + \delta u_\lambda \delta t$. If $\delta s$ is the length of $PQ$, $h_\lambda \delta y_\lambda = \alpha_{1\lambda} \delta s$, so if $\beta_{1\lambda} = \alpha_{1\lambda}/h_\lambda$,

$$\delta y_\lambda = \beta_{1\lambda} \delta s, \qquad \delta u_\lambda = \left(\frac{\partial u_\lambda}{\partial y_j}\right) \beta_{1j} \delta s,$$

and the coordinates of $Q'$ relative to $P'$ are $\delta s(\beta_{1\lambda} + \beta_{1j} \delta t \partial u_\lambda / \partial y_j)$. The $h_\lambda$ being the values at $P$, the values at $P'$ are

$$h_\lambda + \frac{\partial h_\lambda}{\partial y_j} u_j \delta t,$$

and the square of the length of $P'Q'$ is

$$(\delta s)^2 \sum_{\lambda=1}^{3} \left(h_\lambda + \frac{\partial h_\lambda}{\partial y_j} u_j \delta t\right)^2 \left(\beta_{1\lambda} + \beta_{1k} \frac{\partial u_\lambda}{\partial y_k} \delta t\right)^2$$

$$= (\delta s)^2 \sum_{\lambda=1}^{3} \left\{\alpha_{1\lambda}^2 + 2\delta t \left[\frac{\alpha_{1\lambda}^2}{h_\lambda} \frac{\partial h_\lambda}{\partial y_j} u_j + h_\lambda \alpha_{1\lambda} \beta_{1j} \frac{\partial u_\lambda}{\partial y_j}\right]\right\}.$$

If $\frac{1}{2} e'_{11}$ is the rate of extensional strain along $PQ$, this must be equal to $(\delta s)^2 (1 + e'_{11} \delta t)$, as before, so

$$e'_{11} = 2 \sum_{\lambda=1}^{3} \left(\frac{\alpha_{1\lambda}^2}{h_\lambda} \frac{\partial h_\lambda}{\partial y_j} u_j + h_\lambda \alpha_{1\lambda} \beta_{1j} \frac{\partial u_\lambda}{\partial y_j}\right).$$

This gives us

$$e'_{11} = \alpha_{1i}\alpha_{1j}e_{ij},$$

if

(A1)
$$e_{11} = 2\left\{\frac{\partial u_1}{\partial y_1} + \frac{u_j}{h_1}\frac{\partial h_1}{\partial y_j}\right\} = 2\left\{\frac{1}{h_1}\frac{\partial v_1}{\partial y_1} + \frac{v_2}{h_1 h_2}\frac{\partial h_1}{\partial y_2} + \frac{v_3}{h_3 h_1}\frac{\partial h_1}{\partial y_3}\right\},$$

$$e_{23} = \frac{h_3}{h_2}\frac{\partial u_3}{\partial y_2} + \frac{h_2}{h_3}\frac{\partial u_2}{\partial y_3} = \frac{h_3}{h_2}\frac{\partial}{\partial y_2}\left(\frac{v_3}{h_3}\right) + \frac{h_2}{h_3}\frac{\partial}{\partial y_3}\left(\frac{v_2}{h_2}\right),$$

and similarly for $e_{22}$, $e_{31}$, etc.

The rate of decrease of the angle between two line elements initially at an angle $\theta$ may be computed in the same way, and we verify that (as we expect) it is given by

$$\varepsilon \sin \theta = \alpha_{1i}\alpha_{2j}e_{ij} - \tfrac{1}{2}(e'_{11} + e'_{22})\cos \theta,$$

where the $e_{ij}$ are given by the same formulae as before (A1), and $\tfrac{1}{2}e'_{11}$, $\tfrac{1}{2}e'_{22}$ are the rates of extensional strain along the two line elements.

Thus (A1) provide formulae for the rate-of-strain components in general orthogonal coordinates.

In the case where the axes are not orthogonal, tensor calculus with the general metric should be used, following the treatment of Brillouin.[19] The above is included to bring out the physical picture of the method and results. Love[20] has also obtained the result by a use of elementary calculus similar to that above.

Note that since we are concerned with rates-of-strain, and divide by $\delta t$ and let $\delta t \to 0$, then as long as sufficient mathematical conditions on the existence and boundedness of the time derivatives of the functions involved are satisfied, no approximation is involved in working only to the first order in $\delta t$. In Hydrodynamics, we are concerned only with rates-of-strain, so, in contradistinction to Elasticity, we are concerned only with truly infinitesimal strains.

[19] See Brillouin, *Les Tenseurs*, Chap. 10, Section 6.
[20] See Love, *The Mathematical Theory of Elasticity*, Chap. 1, Section 20; see also Section 22C, and Note C.

# APPENDIX II

## Sources, Doublets, and Line Vortices in an Incompressible Fluid: Potential Flow

In three dimensions, if $R$ is distance from an origin $O$, and

(A2) $$\phi = -m/R,$$

$\phi$ is harmonic except at $O$, the only non-zero velocity component is the radial component $v_R = m/R^2$, and the volume flux out of any surface enclosing $O$ is the same for all such surfaces, and is $4\pi m$. If $m$ is positive, (A2) gives the velocity potential for a source of strength $m$. A negative source is called a sink.

The limit of a combination of two equal and opposite sources, $\pm m'$, at a distance $\delta s$ apart, as $\delta s \to 0$, $m' \to \infty$, and $m'\delta s$ remains finite and equal to $M$, is a doublet or dipole of strength $M$; the line along the limiting direction of $\delta \mathbf{s}$, from the sink to the source, is the axis of the doublet, whose strength is a directed quantity, or vector, $\mathbf{M}$. The velocity potential at a point $P$ for such a doublet at $O$ is

(A3a) $$\phi = -\mathbf{M} \cdot \nabla'(R^{-1}) = R^{-2}\mathbf{M} \cdot \nabla' R = -R^{-2}\mathbf{M} \cdot \nabla R$$
$$= -R^{-3}\mathbf{M} \cdot \mathbf{R} = -(M \cos \chi)/R^2,$$

where $\mathbf{OP} = \mathbf{R}$, $\nabla'$ is the gradient taken at $O$ and $\nabla$ the gradient at $P$, and $\chi$ is the angle between $\mathbf{OP}$ and the direction of $\mathbf{M}$.

In particular, for a doublet of strength $M$ at the origin of coordinates and in the negative direction of the axis of $x_1$, if $r$, $\vartheta$, $x$, are cylindrical polar coordinates such that $x_1 = r\cos\vartheta$, $x_2 = r\sin\vartheta$,

$$\phi = \frac{Mr\cos\vartheta}{(x_3^2 + r^2)^{\frac{3}{2}}},$$

and the velocity components $v_r$, $v_\vartheta$, $v_3$ are

$$v_r = \frac{M(x_3^2 - 2r^2)}{(x_3^2 + r^2)^{5/2}} \cos \vartheta, \quad v_\vartheta = -\frac{M \sin \vartheta}{(x_3^2 + r^2)^{3/2}},$$

(A3b)
$$v_3 = -\frac{3M r x_3 \cos \vartheta}{(x_3^2 + r^2)^{5/2}}.$$

These results are applied in Appendix III(a).

Higher-order singularities may be obtained by repeated differentiation, in the manner of Maxwell (*cf.* Lamb, Section 82; Maxwell, *Electricity and Magnetism*, Part 1, Chap. 9.).

The motion due to a source has axial symmetry about any line through the source. With spherical polar coordinates $R$, $\theta$, $\vartheta$, where $\theta$ is the angle between the axis of symmetry, as chosen, and the radius vector, Stokes's stream function for a source of strength $m$ on the axis is

(A4) $$\psi = m(1 - \cos \theta),$$

if $\psi = 0$ on the axis, or simply $\psi = -m \cos \theta + \text{constant}$. For a doublet of strength $M$ at the origin along the axis of symmetry,

(A5) $$\psi = (M/R) \sin^2 \theta,$$

and for a line of sources of uniform strength $\lambda$ per unit length along the axis, $\psi$ at a point $P$ is

$$\psi = \lambda (R_2 - R_1)$$

where $R_1$ and $R_2$ are the distances of $P$ from the beginning and end of the interval of the axis containing the sources.

In two dimensions, with $r$, $\theta$ as polar coordinates, and $z = re^{i\theta} = x_1 + ix_2$, if

$$w = \phi + i\psi = m \log z, \quad \phi = m \log r, \quad \psi = m\theta,$$

then $v_r = m/r$, $v_\theta = 0$, the flux out through any closed curve enclosing the origin is $2\pi m$, and we have a line source of strength $m$ through the origin. For a line source of strength $m$ through $z = z_0$, $w = m \log (z - z_0)$.

It is now easy to show that for a two-dimensional doublet (or line doublet) at the origin and of strength **M**, where the direction of **M** makes an angle $\alpha$ with the $x_1$-axis,

$w = -Me^{i\alpha}/z$, $\phi = -(M/r)\cos(\theta-\alpha)$, $\psi = (M/r)\sin(\theta-\alpha)$.

For such a doublet at $z = z_0$, $w = -Me^{i\alpha}/(z-z_0)$.

For a line vortex of strength $K$ along the $x_3$-axis, through the origin,

$$w = -(iK/2\pi)\log z, \quad \phi = K\theta/2\pi, \quad \psi = -(K/2\pi)\log r,$$

the circulation being $K$ in any circuit that encloses the origin, and zero in any circuit that does not. For a similar line vortex of strength $K$ through $z = z_0$, $w = -(iK/2\pi)\log(z-z_0)$.

$\phi$ is multiple-valued for a vortex, and $\psi$ is multiple-valued for a source.

APPENDIX III

# Some Remarks on the Irrotational Motion of an Incompressible Fluid

In the motion of an inviscid fluid of constant density acted on by conservative forces, diffused vorticity cannot occur. However, as pointed out in Section 8.1, after a (theoretically infinitely) long time, the limit of a motion of a viscous fluid, when the viscosity tends to zero, may be endowed with vorticity. An obvious example is a uniformly rotating fluid. Viscosity is needed to set up such motions, but the motion of a solid body in liquids so endowed with vorticity may be studied with viscosity neglected. We shall not enter into a discussion of such matters here. Some discussion may be found in Lamb, Section 159a, and other, later, interesting work may be found in the literature. Nor can we here pause to discuss such matters as secondary flow in pipe bends. Some flows with vortex sheets will be briefly mentioned, but there are other motions involving concentrated vorticity, including the celebrated Karman vortex-street, for which we must simply refer to the literature. (See, to start with, Lamb, Chap. VII Milne-Thomson, Chap. XIII, *Modern Developments*, Sections 242–247.)

(a) *Flows without discontinuities.*
*Sources and doublets for a body of revolution*

If we begin by considering motions which are strictly irrotational everywhere in the fluid, with **v** continuous, we are concerned with a simple boundary-value problem for harmonic functions; in two dimensions, $\phi + i\psi$ is a function of a complex variable. We do not here discuss the known solutions for spheres, ellipsoids, Rankine ovoids, and the like, or for circular and elliptic cylinders, or for Joukowski airfoils, or other solutions obtained by conformal transformations in two dimensions. All these have their place, and occasional uses, and are to be found fully discussed in the literature. For some comparisons of the theory with experiment in a real fluid, see *Modern Developments*, Chap. 1. With general shapes, numerical computations are necessary. We discuss only one case, which is of some practical importance, and raises questions of mathematical interest.

Consider flow past a body of revolution, which is assumed fixed in an otherwise undisturbed stream. With the axis of symmetry along the axis of $x_3$, suppose first that the undisturbed flow is $W$ along this axis. Now we know that for flow past a sphere the perturbation is that due to a doublet at the center, in the direction opposite to that of the undisturbed stream, and for some other bodies of revolution, it is that due to some other distribution of sources and sinks (of total strength zero) along the axis of symmetry. Without further consideration, let us try to find a distribution of sources, of strength $Wq(x_3)$ per unit length, which gives the required perturbation for the given body. Let the body intersect the $x_3$-axis at $x_3 = \pm l$. Denote by $r$ distance from the axis. For the uniform stream $W$, Stokes's stream function $\psi$ is $\tfrac{1}{2}Wr^2$. The sources must lie between $x_3 = \pm l$, and for the source distribution we must have

(A6) $$\int_{-l}^{l} q(\xi)d\xi = 0;$$

from Appendix II, the stream function for the sources alone is

$$-W \int_{-l}^{l} \frac{q(\xi)(x_3 - \xi)}{\{r^2 + (x_3 - \xi)^2\}^{\frac{1}{2}}} d\xi.$$

Let $r^2 = f(x_3)$ on the surface of the given body. The total value of the stream function is the sum of the two contributions, and the boundary condition is $\psi = 0$ on the body, so $q(\xi)$ must satisfy the integral equation

(A7) $$\int_{-l}^{l} \frac{q(\xi)(x_3 - \xi)}{\{f(x_3) + (x_3 - \xi)^2\}^{\frac{1}{2}}} d\xi = \tfrac{1}{2}f(x_3).$$

Approximate numerical solutions may certainly be obtained by taking $q(\xi)$ constant in each of a number of contiguous intervals, satisfying (A6), and also the equation (A7) at each of $n$ points on the boundary of a meridian section.

At large distances the source distribution may be shown to act like a doublet along the axis of $x_3$ of strength $W \int_{-l}^{l} q(\xi) \xi d\xi$ and the perturbation velocity potential

$$\phi \sim -\frac{W x_3}{R^3} \int_{-l}^{l} q(\xi) \xi d\xi$$

at large distances $R$ from the origin. $\phi$ is also the perturbation potential when the body moves with velocity $W$ along the $x_3$-axis in the negative direction.

Now we may easily construct flows symmetrical about the $x_3$-axis by superposing on a uniform flow $W$ the flow due to singularities other than sources on the axis — for example, symmetrically placed circular line vortices, and symmetrically placed rings of sources and sinks, of total strength zero. The streamline $\psi = 0$ for such a flow may be taken as the meridian section of a solid body placed in the flow $W$, and $\psi$ gives the disturbed flow; the singularities will be inside the body. Clearly in such a case the perturbation flow cannot be accurately represented by sources and sinks on the axis, for the velocity potential $\phi$ is harmonic, and cannot be continued inside the body without singularities being encountered off the axis. An accurate solution of (A7), with the condition (A6), presumably does not exist. But it does exist for a smooth contour having in common with the given smooth contour any number of points. This provides a good engineering solution; but the mathematics of this kind of situation does not appear to have been studied.

Suppose now that the undisturbed flow is $V$ at a (usually small) angle $\alpha$ to the axis of $x_3$. We may choose the axis of $x_1$ so that the undisturbed velocity is parallel to the $(x_3, x_1)$ plane. Then the undisturbed flow has components $W = V \cos \alpha$ and $U = V \sin \alpha$ parallel to the axes of $x_3$ and $x_1$ respectively; the fluid velocity may be found for each of these components separately, and added vectorially to give the fluid velocity in the stream $V$. (Pressures may not be found by simple addition, since the equation for the pressure is quadratic in the velocity.) The disturbance to the cross flow $U$ may be found, as Karman[21] pointed out, as the flow due to a distribution of doublets in the direction of the negative $x_1$-axis, supposed placed along the axis of $x_3$ between $x_3 = \pm l$. The boundary condition at the surface of the body is $v_r/v_3 = dr/dx_3$. If the doublets are of strength $Um(x_3)$ per unit length, then from the results of Appendix II, for the doublets alone

$$v_r = U \cos \theta \int_{-l}^{l} \frac{m(\xi)\{(x_3 - \xi)^2 - 2r^2\}}{\{(x_3 - \xi)^2 + r^2\}^{\frac{5}{2}}} d\xi,$$

$$v_3 = -3Ur \cos \theta \int_{-l}^{l} \frac{m(\xi)(x_3 - \xi)}{\{(x_3 - \xi)^2 + r^2\}^{\frac{5}{2}}} d\xi.$$

For the "undisturbed" flow $U$, $v_r = U \cos \theta$, $v_3 = 0$. The total velocity components for the cross flow are found by simple addition. Hence, if $r^2 = f(x_3)$ on the surface of the body, as before, the equation for $m(\xi)$ is

(A8) $$\int_{-l}^{l} \frac{m(\xi)\{(x_3 - \xi)^2 - 2f(x_3) + \frac{3}{2}(x_3 - \xi)f'(x_3)\}}{\{f(x_3) + (x_3 - \xi)^2\}^{\frac{5}{2}}} d\xi + 1 = 0.$$

At large distances the doublet distribution acts like a doublet in the negative direction of the $x_1$-axis, and

$$\phi \sim \frac{Ux_1}{R^3} \int_{-l}^{l} m(\xi) d\xi.$$

For further discussion of flows of incompressible fluids past bodies of revolution, reference may be made to A. H. Armstrong, *Reports and Memoranda of the Aeronautical Research Council*, London, No. 3020 (1954/1957).

(b) *Kinetic energy, impulse, forces*

If solid bodies are moving in a fluid of constant density, which is supposed otherwise unbounded and at rest at infinity, the kinetic energy of the fluid is

(A9) $$T = \tfrac{1}{2}\rho \int (\nabla \phi)^2 d\tau = -\tfrac{1}{2}\rho \int \phi \frac{\partial \phi}{\partial \nu} dS,$$

where the volume integral is over the space occupied by the

[21] *Aachener Abhandlungen*, No. 6, 1927; Collected Works, Vol. 2, p. 253.

fluid, and the surface integral is over the surfaces of the solids only, since the integral over any (infinitely large) boundary surface or "envelope" tends to zero when the surface goes to infinity in every direction.

If the space occupied by the fluid is simply connected, the fluid motion must, with the assumption of zero vorticity, be acyclic; if the space is not simply connected, we have still assumed the motion acyclic. (If there is circulation about an infinite cylinder, the kinetic energy per unit length is infinite.)

Consider the motion of a single body. For the calculation of $T$ it is convenient to compute $\phi$ for axes which coincide instantaneously with body axes. Let $U_1$, $U_2$, $U_3$ be the component velocities of the origin of the body axes, which is a fixed point in the body (usually, for convenience, the center of mass) and let $U_4$, $U_5$, $U_6$ be the component angular velocities of the body about the instantaneous position of those axes. Then we may write

$$\phi = \sum_{\sigma=1}^{6} U_\sigma \phi_\sigma$$

where the $\phi_\sigma$ are independent of the $U_\sigma$, and depend only on the configuration of the surface of the body relative to the axes. Thus $T$ is a homogeneous quadratic function of the $U_\sigma$, with 21 coefficients depending only on the configuration of the surface of the body.

The integrals, $\rho \int \nabla \phi \, d\tau$ and $\rho \int (\mathbf{x} \times \nabla \phi) d\tau$, for the momentum and angular momentum of the fluid are not convergent, so we cannot work with the limit of the momentum and angular momentum as any assumed bounding surface or envelope goes to infinity in every direction in a general manner. The impulse which would be needed to generate the motion from rest is therefore introduced. This impulse must both counteract the influence of the impulsive pressure $\tilde{\omega}$ on the surface, and generate the momentum $\mathbf{M}_S$ and the angular momentum $\mathbf{h}_S$ of the solid. The force and couple components of the impulse are therefore

(A10a) $$\mathbf{I} = \int_S \boldsymbol{\nu} \tilde{\omega} \, dS + \mathbf{M}_S,$$

and

(A10b) $$\mathbf{C} = \int_S (\mathbf{x} \times \mathbf{\nu}) \tilde{\omega} \, dS + \mathbf{h}_S,$$

respectively, with the integrals over the surface $S$ of the solid.

For the consideration of the rate of change of the impulse, we shall first use fixed axes, such that the equation for the pressure (with the body forces on the fluid eliminated in the manner explained in Section 2.4) is

$$p/\rho = -\tfrac{1}{2}\mathbf{v}^2 - \partial \phi/\partial t + F(t).$$

We introduce an enveloping surface $S'$, calculate the resultant force and couple on the combined system of solid and fluid inside $S'$ by using the equations of momentum and angular momentum, and let $S'$ go to infinity in all directions. We shall thus prove that the force and couple on the solid and infinite fluid together are equal to the rates of change of the force and couple constituents of the impulse. Since $\rho \mathbf{v} = -\boldsymbol{\nabla} \tilde{\omega}$, the momentum of the fluid between $S$ and $S'$ is

$$\int \rho \mathbf{v} \, d\tau = -\int \boldsymbol{\nabla} \tilde{\omega} \, d\tau = \int_S \boldsymbol{\nu} \tilde{\omega} \, dS - \int_{S'} \boldsymbol{\nu} \tilde{\omega} \, dS,$$

the volume integrals being now over the space between $S$ and $S'$. If $\mathbf{M}$ is the linear momentum of the solid and fluid together,

$$\mathbf{M} = \int \rho \mathbf{v} \, d\tau + \mathbf{M}_S = \mathbf{I} - \int_{S'} \boldsymbol{\nu} \tilde{\omega} \, dS.$$

If $\mathbf{F}$ is the force on the solid and the fluid apart from the pressure on $S'$, then

$$\mathbf{F} - \int_{S'} \boldsymbol{\nu} p \, dS = \frac{d\mathbf{M}}{dt} = \frac{d\mathbf{I}}{dt} - \frac{d}{dt}\int_{S'} \boldsymbol{\nu} \tilde{\omega} \, dS$$

$$= d\mathbf{I}/dt + \rho \int_{S'} \boldsymbol{\nu} \partial \phi / \partial t \, dS,$$

since $\tilde{\omega} = -\rho \phi$. Substitute for $p$, and let $S'$ tend to infinity. Since

$$\int_{S'} \boldsymbol{\nu} F(t) \, dS = 0,$$

and

$$\int_{S'} \boldsymbol{\nu} \mathbf{v}^2 \, dS \to 0,$$

we find that

(A11) $$\mathbf{F} = d\mathbf{I}/dt.$$

APPENDIX III 239

Similarly the resultant couple on the solid and fluid is $d\mathbf{C}/dt$. The time-derivatives are to be taken with fixed axes.

The body forces on the fluid having been eliminated, as in Section 2.4, the remaining external forces in fact act on the solid only; together with the resultant of the fluid pressures on the surface, they must produce the given motion of the solid and hence give rise to a force $d\mathbf{M}_S/dt$ and a couple $d\mathbf{h}_S/dt$. Therefore, if the fluid pressures on the surface have a resultant force $\mathbf{X}$ and couple $\mathbf{L}$, $d\mathbf{I}/dt + \mathbf{X} = d\mathbf{M}_S/dt$, and similarly for the couple, so

(A12) $$\mathbf{X} = \frac{d}{dt}\rho\int_S \mathbf{\nu}\phi dS, \quad \mathbf{L} = \frac{d}{dt}\rho\int (\mathbf{x}\times\mathbf{\nu})\phi dS$$

(on substituting $-\rho\phi$ for $\tilde{\omega}$).

The rates of change are with respect to fixed space axes. For axes rotating with angular velocity $\mathbf{\Omega}$ about their instantaneous position, and with the origin moving with velocity $\mathbf{U}$, the rates of change of the force and couple constituents of the impulse become

(A13) $d\mathbf{I}/dt + (\mathbf{\Omega}\times\mathbf{I})$ and $d\mathbf{C}/dt + (\mathbf{\Omega}\times\mathbf{C}) + (\mathbf{U}\times\mathbf{I})$,

respectively, and similarly for the right-hand sides of (A12). The formulae may be partly checked by applications to elementary problems.

We return to the calculation of the kinetic energy $T$ and the notation used there. After $T$ has been calculated in the manner specified, we may let the axes move with the body. By the ordinary methods of general dynamics, it may now be shown that $\mathbf{I}$ is the gradient with respect to $U_1, U_2, U_3$, and $\mathbf{C}$ the gradient with respect to $U_4, U_5$, and $U_6$, of the total kinetic energy of the whole system, solid and fluid together. Since $\mathbf{M}_S$ and $\mathbf{h}_S$ are given similarly in terms of the kinetic energy of the solid alone, it follows (with $\mathbf{\Omega} = U_4, U_5, U_6$) that

(A14) $$X_1 = -\left(\frac{d}{dt}\frac{\partial T}{\partial U_1} + U_5\frac{\partial T}{\partial U_3} - U_6\frac{\partial T}{\partial U_2}\right),$$
$$L_1 = -\left(\frac{d}{dt}\frac{\partial T}{\partial U_4} + U_5\frac{\partial T}{\partial U_6} - U_6\frac{\partial T}{\partial U_5} + U_2\frac{\partial T}{\partial U_3} - U_3\frac{\partial T}{\partial U_2}\right),$$

with similar formulae for $X_2$, $X_3$, $L_2$, $L_3$. (See Lamb, Chap. VI). The equivalence of (A12) (altered as explained in (A13)) and (A14) may be checked by quite elementary methods, by writing $\phi$ as a linear function of the $U$'s, as before, and computing the coefficients in $T$. Also (A14) follows at once from the principles of general dynamics, if we boldly assume that these principles may be applied with the infinite number of coordinates specifying the positions of the fluid particles treated as ignored coordinates, with the corresponding momenta zero.

If the solid body is moving with a constant velocity without rotation, the force **X** vanishes, but there is, in general, a couple, for which

$$(A15) \qquad L_1 = U_3 \, \partial T/\partial U_2 - U_2 \, \partial T/\partial U_3,$$

with similar expressions for $L_2$, $L_3$. We remark in passing that it can be shown that there are, for any solid body, three mutually perpendicular directions for uniform motions parallel to which the couple vanishes; these are the directions of "permanent translation." Only one such motion is stable; roughly speaking, we may say that a solid body which is to move without rotation will tend to set itself "broadside on" to the relative stream, a phenomenon which is easily observed (though the actual fluid motions are then far from potential motions).

The force **X** and couple **L** resulting from the fluid pressures on the surface, may, of course, be calculated by computing the pressure on the surface and evaluating the necessary surface integrals. In general, the motion is unsteady, and the pressure must be computed by the formula for moving axes. If, however, the solid body has a uniform velocity of translation, without rotation, the forces and couples are unaltered if we take the axes fixed in the body; the matter then becomes simple. The calculation is particularly simple and elegant, and capable of wide generalization, for two-dimensional motion past a cylinder. With this method, we may also include a circulation round the cylinder. If $w = \phi + i\psi$ for such a steady motion, and $z = x_1 + ix_2$, it is easy to show that the forces and the moment per unit length of the cylinder are given by

(A16)
$$X_1 - iX_2 = \tfrac{1}{2}\rho i \oint \left(\frac{dw}{dz}\right)^2 dz,$$
$$L_3 = -\tfrac{1}{2}\rho \operatorname{Re} \oint \left(\frac{dw}{dz}\right)^2 z\, dz,$$

where Re denotes "the real part of." The integrals are contour integrals round the boundary of the section of the cylinder by the $(x_1, x_2)$ plane. By Cauchy's theorem, the integrals may be taken round any contour enclosing this section, if there are no singularities in the fluid. Now, without the circulation, the disturbance is that due to a certain distribution of singularities inside the cylinder; at large distances the effect asymptotes that of a certain doublet at the origin. We may, for simplicity, take the axis of $x_1$ in the direction of the undisturbed velocity of the fluid relative to the cylinder; if this velocity is $U$, then for large $|z|$, with a circulation $-K$,

$$w = Uz + (iK/2\pi)\log z + (\alpha + i\beta)/z + O(z^{-2}),$$

with some constants $\alpha$ and $\beta$ depending on the section of the cylinder. If we now take the integrals round a large circle, whose radius is then made to tend to infinity, we find

(A17) $\quad X_1 = 0,\ X_2 = K\rho U,\ L_3 = -2\pi\rho\beta U.$

The origin of the "lift" force $K\rho U$ (the Kutta-Joukowski formula) is easily explained, since the circulation increases the velocity, and therefore decreases the pressure, on the side where $x_2$ is greater, and *vice versa* on the other side.

In general, $K$ is indeterminate according to inviscid fluid theory. However, for a section with a salient point there is a unique value of the circulation which gives finite velocity at that point; the velocity is infinite with any other circulation. For a flat-plate "airfoil" section of chord $c$, at an incidence $\alpha$ to the relative stream, if we neglect the singularity at the leading edge because in practice the section will be rounded off there, the condition for finite velocity at the trailing edge is $K = \pi U c \sin \alpha$, and the non-dimensional lift coefficient $C_L$, defined as $X_2/(\tfrac{1}{2}\rho U^2 c)$, is $2\pi \sin \alpha$. For an airfoil with thickness, the theoretical value for a finite or zero velocity at a sharp trailing edge is a little larger; for an airfoil with camber, $\alpha$ must be measured from the attitude of zero lift. Practically, $C_L$ increases linearly with $\alpha$ for sufficiently small $\alpha$, but the slope is less than the theoretical value because of the influence of the boundary layer and the wake.

The Kutta-Joukowski formula has been shown by G. I. Taylor[22] to be valid if the circuit round which the circulation is taken cuts across a wake behind the cylinder in which vorticity is present, provided that, when the circuit crosses the wake, it is *perpendicular to the undisturbed stream velocity*. This result is confirmed by experiment. The drag force is then, of course, different from zero, both theoretically and experimentally.

In the general calculation of forces for steady motion, we may allow for singularities, such as sources or sinks, in the fluid, by including the residues at the singularities as well as the effect of the asymptotic value of $w$ (Milne-Thomson, Sections 8.63, 8.83).

For formulae for the forces and moment per unit length on a cylinder in general motion in a fluid (two-dimensional motion) see Milne-Thomson, Sections 9.50–9.53.

For three-dimensional motions, the matter is not so simple. We are here concerned only with acyclic motions, since we are concerned with bodies such that the space outside is simply-connected. For steady motions, we may follow the same method as we should have used in two dimensions if we had set out that calculation in terms of real variables and harmonic functions. We may verify that $\mathbf{X} = 0$, and, after some calculation, we find that, if the velocity of the undisturbed fluid relative to the solid has components $-U_i$ ($i = 1, 2, 3$), and if, for large $R$,

(A18) $$\phi = -U_i x_i + A_i x_i/R^3 + O(R^{-3}),$$

then

[22] *Phil. Trans.* **A225** (1925), p. 238.

(A19) $$L_1 = 4\pi\rho(U_2 A_3 - U_3 A_2), \quad L_2 = 4\pi\rho(U_3 A_1 - U_1 A_3),$$
$$L_3 = 4\pi\rho(U_1 A_2 - U_2 A_1)$$

However, the easiest way to calculate the forces and couples in general is from the kinetic energy, by (A14). When the motion of the solid body is one of pure translation, there is a relation (involving the volume of the solid) between the expression for the kinetic energy and the expansion of $\phi$ for large $R$, which enables us to check (A19) from (A15), and to calculate the forces and couples in unsteady translational motion from (A14). The relation was found by different methods by Lamb and Taylor.[23] In (A18), $\phi' = \phi + U_i x_i$ is the potential of the motion when the body is considered as moving in otherwise undisturbed fluid with velocity components $U_i$. For this motion, let

(A20) $$2T/\rho = K_{ij} U_i U_j,$$

(with $K_{ij} = K_{ji}$). Then, if $V_0$ denotes the volume of the solid, Lamb shows that

$$4\pi\phi' = -V_0 U_i x_i/R^3 - U_i K_{ij} x_j/R^3 + O(R^{-3}).$$

i.e.,

(A21) $$\phi' = U_i \beta_{ij} x_j/R^3 + O(R^{-3}),$$

where

$$4\pi\beta_{ij} = 4\pi\beta_{ji} = -K_{ij} - V_0 \delta_{ij}.$$

Hence, with the $A_i$ as in (A18),

$$A_j = U_i \beta_{ij} = -(V_0 U_j + K_{ij} U_i)/4\pi,$$

and

(A22) $$2T/\rho = -V_0 U_i U_i - 4\pi A_i U_i,$$

which is Taylor's result. With this form for $T$, and $A_i = U_j \beta_{ij}$, with $\beta_{ij}$ symmetrical, (A19) is easily checked from (A15). Other results, for unsteady translations, are easily found. We have previously seen how to compute approximately the $A_i$ and $\beta_{ij}$ for any body of revolution.

[23] Lamb's *Hydrodynamics*, Section 121; G. I. Taylor, *Proc. Roy. Soc. (London)* **A120** (1928), p. 13.

The coefficients in $2T/\rho$ are called coefficients of added or hydrodynamic inertia; for a rectilinear motion along the axis of $x_1$, if $K_{12} = K_{13} = 0$, a body of mass $m$ moves as if its mass were increased to $m + \rho K_{11}$, which is called the virtual mass.

We must here take leave of this subject. Reference may be made both to Chap. 6 of Lamb's Hydrodynamics, for the dynamical theory of the motion of solids in liquids with irrotational flow, and to Landweber and Yih, *J. Fluid Mech.* **1** (1956), p. 319, and the references there given, for a discussion, with rotations included, of forces, moments, and added masses.

(c) *Flows with vortex sheets. Airfoil theory*

For steady flows of an unbounded fluid without vorticity the drag force is theoretically zero. The first theoretical deduction of a formula for a non-zero drag was provided by the theory of "free streamlines," developed, for flow past a plane flat plate, by Kirchhoff and Rayleigh according to the methods used by Helmholtz for two-dimensional jets, and extended by Levi-Civita and others to the case of curved rigid boundaries. According to this theory there is, in two-dimensional flow past a flat plate across the stream, for example, a mass of fluid at rest behind the plate, separated from the stream by two streamlines springing from the edges of the plate. The tangential velocity is discontinuous across these streamlines, which are traces of vortex sheets. These sheets reach to infinity; the fluid bounded by them and the rear of the plate is stagnant, at a constant pressure equal to that in the undisturbed stream at infinity; the velocity just outside the "free streamlines" is constant, and equal to that of the undisturbed stream. In this theory, therefore, the vortex sheet, which we may first suppose inserted at the solid boundary to bring the relative tangential velocity to zero, has separated from the surface, and become "free." When the boundary is curved, without sharp edges and infinite velocities in the classical theory of potential flow, the "free-streamline" solution is not unique, since the position at which the vortex sheet leaves the surface cannot be immediately specified. There is, however, only one solution in which the free streamlines have finite curvature at the position of separation, and when we consider the matter in connection with boundary-layer theory, there is little doubt that this solution must be the one chosen.

The results of the theory are largely in disagreement with the results of observation in real fluids of small viscosity. The relevant observed flows at large Reynolds numbers are unsteady; at large enough Reynolds numbers they are turbulent. Nevertheless, it is an important theoretical question whether these free-streamline flows would represent the limit of the steady flow of a viscous fluid as the viscosity tends to zero, the instability being neglected. Some remarks on this question are made in Section 8.1. For an attempt to provide a mathematical model with results more in agreement with observation in certain regimes, see Roshko, *J. Ae. Sci.* **22** (1955), p. 124.

Similar mathematical theories arise for jets, and in the study of cavitation. The theory, when suitably adjusted, has its greatest practical value in this last connection. Reference may be made to the *David Taylor Model Basin Reports* 842, 842A (1953), by Eisenberg. For the mathematical theory see Lamb and Milne-Thomson, and more particularly the recently published book, *Jets, Wakes, and Cavities*, by Birkhoff and Zarantonello.

Important applications of flows with vortex sheets are encountered in airfoil theory, for which reference may be made to Glauert, *Aerofoil and Airscrew Theory*, Cambridge Univ. Press, 1926; Milne-Thomson, *Theoretical Aerodynamics*, Macmillan, London, 1948; Kármán and Burgers, *Aerodynamic Theory* (ed. by W. F. Durand), Vol. 2, Springer, 1935; Robinson and Laurmann, *Wing Theory*, Cambridge Univ. Press, 1956. We shall here treat the density of the air as constant. For relative airfoil speeds whose square is small compared with the square of the speed of sound, the density variations are, in fact, small, and the approximation is a satisfactory first approximation. In order that there should be a lift on an airfoil, the pressure must be diminished, and the velocity increased, above it, and the contrary below it, from the values in a potential motion without circulation. This is accomplished by assuming a circulation round each section of the airfoil. But for an airfoil of finite span, the space outside is simply-connected. So there must be vorticity in the fluid.

Even for an airfoil of infinite span, if we suppose the motion to have started a long but finite time previously, it is still true that the circulation round an infinitely large circuit will be zero, and there will be vorticity in the fluid. For the steady motion of an airfoil of infinite span, we may suppose this vorticity concentrated in a single vortex which the airfoil has left behind it: the cast-off vortex. For an airfoil of finite span, there is still a cast-off vortex, but it is of finite length, and ends in the fluid. Vortex lines cannot begin or end in the fluid; moreover, there must be vortex lines in the fluid which are cut by any circuit round a section of the airfoil when the circuit is moved and deformed so as to shrink up to a point. Hence there are trailing vortex lines stretching from the airfoil to the cast-off vortex. If we now regard the cast-off vortex as being at an infinite distance, take axes fixed in the airfoil, and consider a steady motion, we may argue, either from Helmholtz's theorems, or from the fact that space is filled with streamlines on all of which the function $\mathfrak{U} = p/\rho + \tfrac{1}{2}\mathbf{v}^2$ (with body forces eliminated) has the same value, that the trailing vortex lines will lie along streamlines (Sections 4.1, 4.2). We now make a linearizing approximation, and take the trailing vortex lines along the undisturbed streamlines, i.e., in the direction of the undisturbed velocity $U$ of the stream relative to the airfoil, which we may take as the $x_3$-axis.

Suppose now that both the upper and lower surfaces of the airfoil lie near the plane $x_2 = 0$, and that the slopes of the surface are small. Near a rounded edge, this last assumption is violated, and the theory will not apply. Also assume that the disturbances produced in the stream are small. This assumption will not apply near a stagnation point, such as must occur at or near a rounded leading edge; nor will it apply where there is flow round a sharp edge. These deficiencies in a linearized theory are known to have serious effects only locally, and not seriously to affect the calculated over-all forces. Thus, if the equation of the airfoil surface is $x_2 = f(x_1, x_3)$ ($f_u$ on the upper surface and $f_l$ on the lower surface), then $f$ and its derivatives, $f_{(1)}$ and $f_{(3)}$, are assumed small. Let the velocity potential be $U(x_3 + \phi)$, so that $U\phi$ is the perturbation potential, and the derivatives of $\phi$ are assumed small. The boundary condition is

(A23) $$\nu_i \, \partial\phi/\partial x_i + \nu_3 = 0;$$

$\nu_1, \nu_3$ are small, and $|\nu_2|$ nearly 1. The terms $\nu_1 \, \partial\phi/\partial x_1$ and $\nu_3 \partial\phi/\partial x_3$ may be dropped, and $\nu_3$ taken as $-\nu_2 f_{(3)}$. The boundary condition becomes

(A24) $$\partial\phi/\partial x_2 = f_{(3)},$$

approximately. If now $2f_t$ is the airfoil thickness, and $f_c$ the ordinate of the mean surface, so that $f_u = f_c + f_t$ and $f_l = f_c - f_t$, we may divide $\phi$ into two parts, $\phi = \phi_1 + \phi_2$, each of which satisfies $\nabla^2\phi = 0$ off the vortex sheet, and where, on the upper and lower surfaces of the airfoil,

(A25) $$\begin{aligned}(\partial\phi_1/\partial x_2)_u &= -(\partial\phi_1/\partial x_2)_l = f_{t(3)} \\ (\partial\phi_2/\partial x_2)_u &= (\partial\phi_2/\partial x_2)_l = f_{c(3)}.\end{aligned}$$

Let $\Sigma$ be the projection of the mean wing surface on the plane $x_2 = 0$, $T$ the trailing vortex sheet, which is supposed to spring from the trailing edge of $\Sigma$, and $R$ the rest of the plane $x_2 = 0$. The boundary conditions may be applied on $\Sigma$. There is a discontinuity in $\partial\phi/\partial x_1$ across $T$; $\partial\phi/\partial x_2$ is continuous. It is easy to see that $\phi_1$ must be even and $\partial\phi_1/\partial x_2$ odd in $x_2$. Since $\partial\phi_1/\partial x_2$ is continuous outside $\Sigma$ on $x_2 = 0$, it is zero on $T$ and $R$. Hence $\partial\phi_1/\partial x_2$ is known at all points on the plane $x_2 = 0$, and the solution may be written down:

(A26) $$\phi_1 = -\frac{1}{2\pi}\iint_\Sigma \left(\frac{\partial\phi_1}{\partial x_2}\right)_{x_2=0} \frac{d\xi_1 d\xi_3}{[(\xi_1 - x_1)^2 + (\xi_3 - x_3)^2 + x_2^2]^{\frac{1}{2}}}$$

$\phi_2$ is odd and $\partial\phi_2/\partial x_2$ even in $x_2$, and the problem of determining $\phi_2$ is much harder. The Kutta-Joukowski condition must be satisfied; in linearized theory it takes the form that the component of the perturbation velocity normal to the mean surface must be finite at the trailing edge. From the continuity of the pressure across $T$, it is seen that $\partial\phi_2/\partial x_3$ must be continuous across $T$; but $\partial\phi_2/\partial x_3$ is odd, and therefore vanishes on $T$. On $T$, $\phi_2$ has the same value as at the point on the trailing edge of the airfoil on the same trailing vortex line. Also $\phi_2$ is continuous across $R$, and therefore vanishes on $R$. Thus, in addition to the Kutta-Joukowski condition, we have the following conditions on $\phi_2$: (i) $\partial\phi_2/\partial x_2$ given on $\Sigma$ (ii) $\phi_2 = 0$ on $R$ (iii) $\partial\phi_2/\partial x_3 = 0$ on $T$.

In determining $\phi_2$, the thickness of the airfoil may be neglected. The airfoil may be replaced by a "bound" vortex sheet, whose strength is the discontinuity in the tangential velocity on the two sides, and on which forces act to constrain it to keep the position of the airfoil's mean surface. We have now stated the problem of steady "lifting-surface" theory. We henceforth neglect the thickness.

If we now replace the cast-off vorticity, so as to think of the whole vortex sheet, bound vorticity, trailing vorticity, and cast-off vorticity, as finite, we may find formulae for the forces by considering the rates of change of the energy and impulse of a vortex sheet.

To consider the impulse, we return to Appendix III(b). We now also take into account impulsive forces, $\hat{\mathbf{F}}$ per unit mass, acting on the fluid, which may produce vorticity $\boldsymbol{\omega}$. But we suppose that outside some finite region, no matter how large, the motion is irrotational, with a velocity potential $\phi$, in which $\boldsymbol{\omega} = 0$, $\hat{\mathbf{F}} = 0$, and the impulsive pressure $\tilde{\omega} = -\rho\phi$. The envelope $S'$ is taken to lie in a region where these conditions are satisfied. The definitions of $\mathbf{I}$ and $\mathbf{C}$ in (A10) are altered by the addition of

$$\rho \int \hat{\mathbf{F}}\, d\tau \text{ and } \rho \int (\mathbf{x} \times \hat{\mathbf{F}})d\tau,$$

respectively. The pressure equation is unaltered on $S'$. But now $\rho\mathbf{v} = \rho\hat{\mathbf{F}} - \nabla\tilde{\omega}$, from the theory of a motion started impulsively. The expression for $\mathbf{M}$ is unaltered, and we prove as before that $\mathbf{F} = d\mathbf{I}/dt$, and the resultant couple on the fluid and solid is $d\mathbf{C}/dt$.

With no solids present, $\mathbf{M}_S = 0$, and

$$\mathbf{I} = \int \rho \mathbf{v} d\tau + \int_{S'} \boldsymbol{v}\bar{\omega} dS.$$

For a single vortex sheet, there is a velocity potential $\phi$, discontinuous across the sheet, and

(A27) $$\mathbf{I} = \int \rho \nabla \phi d\tau - \rho \int_{S'} \boldsymbol{v}\phi dS = -\rho \int_{S_1} \boldsymbol{v}\phi dS,$$

where $S_1$ is any surface containing the vortex sheet; it may be taken to coincide with both sides of the sheet and connecting surfaces at the edges. Similarly

(A28) $$\mathbf{C} = -\rho \int_{S_1} (\mathbf{x} \times \boldsymbol{v}) \phi dS.$$

Note that in these expressions $\phi$ is the complete potential, not a perturbation potential.

For a two-dimensional motion, the impulse, force, and moment are all taken per unit breadth. If, for example, we consider a pair of line vortices, of equal but opposite strength, the impulse is $-\rho \int \boldsymbol{v}\phi dS$ over both sides of a cut joining the intersections of the vortices with a plane perpendicular to them. The result is that $\mathbf{I}$ is in the plane of the motion, perpendicular to the line joining those intersections, and of magnitude $K\rho d$, where $K$ is the strength of the vortices, and $d$ their distance apart. $\mathbf{C}$ is perpendicular to the plane of the motion, of magnitude $K\rho d\bar{x}$, where $\bar{x}$ is the coordinate, parallel to the line joining the intersections, of the middle point of that line.

Now return to our airfoil theory, and let the airfoil move forward with velocity $U$, leaving the cast-off vortex at a fixed position behind. The lift force on the airfoil produces an equal and opposite reaction on the fluid, and this is the rate of change of the corresponding component of the impulse, which in our case is $-I_2$. In time $\delta t$, the length of the vortex sheet increases by $U\delta t$, so the lift force is given by

$$L = \rho U^2 \int v_2 \phi ds,$$

where $U\phi$ is now the perturbation potential, as before, and the integral is round a curve, in any plane $x_3 = $ constant behind the airfoil, enclosing the section of the trailing vortex sheet. The theory has been set out for an airfoil which is nearly plane; it may be extended at once to any "thin" body, which is near any given surface, and gives both components of the force normal to the direction of $U$; the force along the $x_1$-axis is found by replacing $\nu_2$ by $\nu_1$. In our case, $\nu_2 = 1$ on the upper surface of the sheet and $-1$ on the lower surface, and we may write

$$L = \rho U^2 \int (\phi_u - \phi_l) ds$$

over one side of the section of the trailing vortex sheet. Since $U(\phi_u - \phi_l)$ is the circulation $K$ round the corresponding section of the airfoil, this is exactly what we should expect, the integral of $K\rho U$ across the full span.

The drag force is most easily found by considering the rate of increase of the energy of the vortex system; it is

$$D = \tfrac{1}{2}\rho U^2 \int \left\{ \left(\frac{\partial \phi}{\partial x_1}\right)^2 + \left(\frac{\partial \phi}{\partial x_2}\right)^2 \right\} dS$$

over a plane $x_3 =$ constant far behind the airfoil, but with the cast-off vortex far behind it; in this expression we need to include in $\phi$ only the effect of the trailing vortex system. Thus

$$D = -\tfrac{1}{2}\rho U^2 \int \phi \frac{\partial \phi}{\partial \nu} ds,$$

where again the integral is around a curve enclosing the section of the trailing vortex sheet, or

$$D = -\tfrac{1}{2}\rho U^2 \int (\phi_u - \phi_l) \frac{\partial \phi}{\partial x_2} ds$$

over one side of that section.

Clearly (since $\nu_3 = 0$ on the sheet) we cannot find $D$ from the rate of change of the impulse according to this approximation. To find $D$ from the impulse it is necessary to allow for the displacement of the vortex sheet downwards, arising from the downwash due to the trailing vortices. When this is done, it is possible to recover the same formula for $D$.

These expressions give what is known as the induced drag,

arising from the increase of the kinetic energy of the trailing vortex wake. Drag forces arising from viscous and boundary layer effects must be added.

A different type of wake vorticity makes its appearance in the unsteady two-dimensional motion of an airfoil of infinite span. The circulation $K$ round the airfoil now varies with the time, and in each time interval $dt$ vorticity with total strength (circulation) equal to $-\dot{K}dt$ is fed into the wake vortex sheet from the trailing edge. The mechanism for this comes from the fact that at the trailing edge the pressure is different on the two sides of the airfoil, but this discontinuity must disappear as soon as we move off the airfoil surface. The theory of the impulse has been applied to the two-dimensional unsteady motion of an airfoil by Kármán and Sears[24].

For further developments of airfoil theory see the works cited on p. 243.

[24] *J. Ae. Sci.* **5** (1938), p. 379. See also Sears, *J. Franklin Inst.* **230** (1940), p. 95.

# APPENDIX IV

# Formulae for Shock Waves in Perfect Gases with Constant Specific Heats

## General equations for perfect gases with constant specific heats

*Equation of state*

$$p/\rho T = \Re$$

*Specific heats*

$$c_p = \gamma \Re/(\gamma - 1), \quad c_v = \Re/(\gamma - 1)$$

*Entropy*

$$S = c_p \log T - \Re \log p + \text{constant}$$
$$= c_v \log T - \Re \log \rho + \text{constant}$$
$$= c_v \log p - c_p \log \rho + \text{constant}$$

*Isentropic relations*

$$p/p_0 = (\rho/\rho_0)^\gamma; \quad T/T_0 = (\rho/\rho_0)^{\gamma-1} = (p/p_0)^{(\gamma-1)/\gamma}$$

*Acoustic velocity*

$$a^2 = \gamma \Re T = \gamma p/\rho.$$

*Mach number, M*

$$v = Ma = (\gamma \Re)^{\frac{1}{2}} M T^{\frac{1}{2}} = M(\gamma p/\rho)^{\frac{1}{2}}.$$
$$\rho v^2 = \gamma M^2 p.$$

*Enthalpy*

$$I = c_p T = \frac{\gamma}{\gamma - 1} \Re T = \frac{\gamma}{(\gamma - 1)} \frac{p}{\rho} = \frac{a^2}{\gamma - 1}$$

*Stagnation enthalpy*

$$I_0 = I + \tfrac{1}{2} v^2 = \frac{\gamma \Re T}{\gamma - 1} \left[ 1 + \tfrac{1}{2} (\gamma - 1) M^2 \right]$$

## Stagnation temperature and pressure

For a steady motion of an inviscid gas if variations in the potential energy $\Omega$ are neglected, $I_0$ is constant along a streamline, and is the stagnation enthalpy. If $\vartheta$ is the stagnation temperature, $I_0 = c_p \vartheta$. Hence

$$\vartheta = T[1 + \tfrac{1}{2}(\gamma - 1)M^2].$$

If $H$ is the stagnation pressure, or total head, i.e., the pressure in a gas brought to rest isentropically,

$$H/p = (\vartheta/T)^{\gamma/(\gamma-1)}.$$

## Critical and limiting velocities

With $\Omega$ neglected, and $I_0$ constant,

$$a^2/(\gamma - 1) + \tfrac{1}{2}v^2 = a_0^2/(\gamma - 1) = \tfrac{1}{2}v_G^2 = \tfrac{1}{2}(\gamma + 1)v_S^2/(\gamma - 1),$$

where $v_G$ is the limiting speed, $v_S$ the critical speed, and $a_0$ the acoustic speed at a stagnation point. For subsonic flow $0 < v < v_S$; for sonic flow, $v = v_S$; for supersonic flow $v_S < v < v_G$.

## Stationary shock waves

The subscript 1 refers to conditions upstream; the subscript 2 to conditions downstream. Let the velocities be $v_1$ and $v_2$ on the two sides, with directions that make angles $\alpha_1$ and $\alpha_2$ with the normal to the shock (Fig. 14). For a normal shock, $\alpha_1 = \alpha_2 = 0$. For an oblique shock $\alpha_1$ and $\alpha_2$ are different from zero.

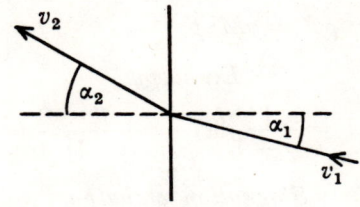

Fig. 14.

The governing equations are

APPENDIX IV 255

*Continuity*
$$\rho_1 v_1 \cos \alpha_1 = \rho_2 v_2 \cos \alpha_2,$$
i.e.
$$(M_1 p_1/T_1^{\frac{1}{2}}) \cos \alpha_1 = (M_2 p_2/T_2^{\frac{1}{2}}) \cos \alpha_2 \ (= \lambda, \text{ say})$$
Also
$$v_1 \sin \alpha_1 = v_2 \sin \alpha_2.$$

*Momentum*
$$p_1 - p_2 = \rho_2 v_2^2 \cos^2 \alpha_2 - \rho_1 v_1^2 \cos^2 \alpha_1$$
$$= \gamma p_2 M_2^2 \cos^2 \alpha_2 - \gamma p_1 M_1^2 \cos^2 \alpha_1$$
$$= \gamma \lambda^2 (T_2/p_2 - T_1/p_1).$$

*Energy*
$$I_1 - I_2 = \tfrac{1}{2}(v_2^2 \cos^2 \alpha_2 - v_1^2 \cos^2 \alpha_1) = \tfrac{1}{2}(v_2^2 - v_1^2)$$
i.e.,
$$T_1 - T_2 = \tfrac{1}{2}(\gamma - 1)(T_2 M_2^2 \cos^2 \alpha_2 - T_1 M_1^2 \cos^2 \alpha_1)$$
$$= \tfrac{1}{2}(\gamma - 1) \lambda^2 (T_2^2/p_2^2 - T_1^2/p_1^2)$$
$$\{= \tfrac{1}{2}(\gamma - 1)(T_2 M_2^2 - T_1 M_1^2)\}.$$

### Equations in terms of the pressure-ratio $p_2/p_1$

Divide the energy equation by the momentum equation
$$\frac{T_2 - T_1}{p_2 - p_1} = \frac{\gamma - 1}{2\gamma} \left(\frac{T_1}{p_1} + \frac{T_2}{p_2}\right)$$
Set

(A29)
$$p_2/p_1 = 1 + x$$

Then

(A30)
$$\frac{T_2}{T_1} = (1 + x) \left(1 + \frac{\gamma - 1}{2\gamma} x\right) \bigg/ \left(1 + \frac{\gamma + 1}{2\gamma} x\right).$$

Also

(A31a)
$$\frac{\rho_2}{\rho_1} = \frac{p_2}{p_1} \frac{T_1}{T_2} = \left(1 + \frac{\gamma + 1}{2\gamma} x\right) \bigg/ \left(1 + \frac{\gamma - 1}{2\gamma} x\right)$$

and

(A31b)
$$\frac{v_2 \cos \alpha_2}{v_1 \cos \alpha_1} = \frac{\rho_1}{\rho_2}.$$

Also

$$\frac{\rho_2 v_2^2 \cos^2 \alpha_2}{\rho_1 v_1^2 \cos^2 \alpha_1} = \frac{p_1}{p_2} = \frac{v_2 \cos \alpha_2}{v_1 \cos \alpha_1},$$

$$p_2 - p_1 = \rho_1 v_1^2 \cos^2 \alpha_1 \left(1 - \frac{v_2 \cos \alpha_2}{v_1 \cos \alpha_1}\right)$$

$$= \gamma p_1 M_1^2 \cos^2 \alpha_1 \frac{2x}{2\gamma + (\gamma + 1)x},$$

and (since $p_2/p_1 \neq 1$ unless $M_1 \cos \alpha_1 = 1$)

(A32) $$M_1^2 \cos^2 \alpha_1 - 1 = \frac{\gamma + 1}{2\gamma} x.$$

This last equation may be used to express $M_1^2 \cos^2 \alpha_1$ in terms of $p_2/p_1$, or *vice-versa*.

### Equations in terms of $M_1^2 \cos^2 \alpha_1$

Define $y$ by

(A33) $$M_1^2 \cos^2 \alpha_1 = 1 + y.$$

Then

(A34) $$\frac{p_2}{p_1} = 1 + x = \frac{2\gamma M_1^2 \cos^2 \alpha_1 - (\gamma - 1)}{\gamma + 1} = 1 + \frac{2\gamma}{\gamma + 1} y$$

(A35) $$\frac{v_2 \cos \alpha_2}{v_1 \cos \alpha_1} = \frac{(\gamma - 1) M_1^2 \cos^2 \alpha_1 + 2}{(\gamma + 1) M_1^2 \cos^2 \alpha_1} = \frac{\gamma + 1 + (\gamma - 1) y}{(\gamma + 1)(1 + y)}$$

(A36) $$\frac{\rho_2}{\rho_1} = \frac{(\gamma + 1) M_1^2 \cos^2 \alpha_1}{(\gamma - 1) M_1^2 \cos^2 \alpha_1 + 2} = \frac{(\gamma + 1)(1 + y)}{\gamma + 1 + (\gamma - 1)y}$$

$$\frac{M_2^2 \cos^2 \alpha_2}{M_1^2 \cos^2 \alpha_1} = \left(\frac{v_2 \cos \alpha_2}{v_1 \cos \alpha_1}\right)^2 \frac{\rho_2 p_1}{\rho_1 p_2} = \frac{p_1}{p_2} \frac{v_2 \cos \alpha_2}{v_1 \cos \alpha_1}$$

$$= \frac{1}{M_1^2 \cos^2 \alpha_1} \frac{(\gamma - 1) M_1^2 \cos^2 \alpha_1 + 2}{2\gamma M_1^2 \cos^2 \alpha_1 - (\gamma - 1)},$$

so

$$M_2^2 \cos^2 \alpha_2 = \frac{(\gamma - 1) M_1^2 \cos^2 \alpha_1 + 2}{2\gamma M_1^2 \cos^2 \alpha_1 - (\gamma - 1)}$$

(A37) $$= \frac{\gamma + 1 + (\gamma - 1)y}{\gamma + 1 + 2\gamma y}.$$

$$\frac{T_2}{T_1} = \frac{1 + \tfrac{1}{2}(\gamma - 1) M_1^2 \cos^2 \alpha_1}{1 + \tfrac{1}{2}(\gamma - 1) M_2^2 \cos^2 \alpha_2}$$

$$= \frac{\{2\gamma M_1^2 \cos^2 \alpha_1 - (\gamma - 1)\}\{(\gamma - 1) M_1^2 \cos^2 \alpha_1 + 2\}}{(\gamma + 1)^2 M_1^2 \cos^2 \alpha_1}$$

(A38)
$$= \frac{\{\gamma + 1 + 2\gamma y\}\{\gamma + 1 + (\gamma - 1)y\}}{(\gamma + 1)^2 (1 + y)}.$$

Since

$$v_1 v_2 \cos \alpha_1 \cos \alpha_2 = \frac{\rho_1}{\rho_2} v_1^2 \cos^2 \alpha_1 = \frac{2a_1^2 + (\gamma - 1) v_1^2 \cos^2 \alpha_1}{\gamma + 1}$$

$$= \frac{2a_1^2 + (\gamma - 1) v_1^2 - (\gamma - 1) v_1^2 \sin^2 \alpha_1}{\gamma + 1},$$

$$v_1 v_2 \cos \alpha_1 \cos \alpha_2 = v_S^2 - \frac{\gamma - 1}{\gamma + 1} v_1^2 \sin^2 \alpha_1,$$

and for a normal shock,

(A39)
$$v_1 v_2 = v_S^2.$$

## Equations for the stagnation pressure (or total head)

If viscosity, heat conduction, etc. are neglected except at the shock discontinuity, and variations in the potential energy $\Omega$ are neglected, in a steady flow the stagnation temperature along a streamline is constant throughout:

$$\vartheta_1 = T_1[1 + \tfrac{1}{2}(\gamma - 1) M_1^2] = T_2[1 + \tfrac{1}{2}(\gamma - 1) M_2^2] = \vartheta_2$$

In terms of $M_1^2$,

$$\frac{\vartheta_2}{T_2} = \frac{T_1}{T_2}[1 + \tfrac{1}{2}(\gamma - 1) M_1^2]$$

$$= \frac{(\gamma + 1)^2 M_1^2 \cos^2 \alpha_1 \{(\gamma - 1) M_1^2 + 2\}}{2\{2\gamma M_1^2 \cos^2 \alpha_1 - (\gamma - 1)\}\{(\gamma - 1) M_1^2 \cos^2 \alpha_1 + 2\}}.$$

In wind-tunnel flow the observed quantity is $H_2/p_1$ (or $p_1/H_2$). Now

$$\frac{H_2}{p_1} = \frac{H_2}{p_2} \frac{p_2}{p_1} = \left(\frac{\vartheta_2}{T_2}\right)^{\gamma/(\gamma-1)} \frac{p_2}{p_1}$$

(A40) $$= \left\{\frac{[M_1 \cos \alpha_1]^{2\gamma} [\tfrac{1}{2}(\gamma+1)]^{\gamma+1}}{\gamma M_1^2 \cos^2 \alpha_1 - \tfrac{1}{2}(\gamma-1)}\right\}^{1/(\gamma-1)}$$
$$\left\{\frac{(\gamma-1) M_1^2 + 2}{(\gamma-1) M_1^2 \cos^2 \alpha_1 + 2}\right\}^{\gamma/(\gamma-1)}.$$

Also
$$\frac{H_1}{p_1} = [1 + \tfrac{1}{2}(\gamma-1) M_1^2]^{\gamma/(\gamma-1)},$$

and

(A41) $$\frac{H_2}{H_1} = \left[\frac{[M_1 \cos \alpha_1]^{2\gamma} (\gamma+1)^{\gamma+1}}{\{2\gamma M_1^2 \cos^2 \alpha_1 - (\gamma-1)\}\{(\gamma-1) M_1^2 \cos^2 \alpha_1 + 2\}^\gamma}\right]^{1/(\gamma-1)}.$$

Since $\vartheta_2 = \vartheta_1$, equation (A41) also follows from

$$\frac{H_2}{H_1} = \frac{H_2}{p_2}\frac{p_2}{p_1}\frac{p_1}{H_1} = \left(\frac{\vartheta_2}{T_2}\right)^{\gamma/(\gamma-1)} \frac{p_2}{p_1}\left(\frac{T_1}{\vartheta_1}\right)^{\gamma/(\gamma-1)} = \left(\frac{T_1}{T_2}\right)^{\gamma/(\gamma-1)}\frac{p_2}{p_1}.$$

### Equations for the gain in entropy

The gain in entropy is given by

$$\frac{S_2 - S_1}{\Re} = \frac{\gamma}{\gamma-1} \log \frac{T_2}{T_1} - \log \frac{p_2}{p_1} = \log \frac{H_1}{H_2}.$$

Hence

$$\frac{(\gamma-1)(S_2 - S_1)}{\Re} = \log [2\gamma M_1^2 \cos^2 \alpha_1 - (\gamma-1)]$$
$$+ \gamma \log [(\gamma-1) M_1^2 \cos^2 \alpha_1 + 2] - \gamma \log (M_1^2 \cos^2 \alpha_1)$$
$$- (\gamma+1) \log (\gamma+1)$$

(A42) $$= \log \left[1 + \frac{2\gamma}{\gamma+1} y\right] + \gamma \log \left[1 + \frac{\gamma-1}{\gamma+1} y\right]$$
$$- \gamma \log (1 + y).$$

Denote the right-hand side by $f(y)$. Then

$$f'(y) = \frac{2\gamma(\gamma-1) y^2}{[1+y][\gamma+1+2\gamma y][\gamma+1+(\gamma-1) y]}.$$

$1 + y \,(= M_1^2 \cos^2 \alpha_1)$ is positive; $\gamma + 1 + 2\gamma y$ is positive, since

$p_2/p_1$ cannot be negative. Also $\gamma > 1$. Hence $f'(y)$ is positive. Also $f(0) = 0$. So $S_2 \gtreqless S_1$ according as $y \gtreqless 0$. By the second law of thermodynamics, $S_2 \geqq S_1$. Hence $y \geqq 0$, $M_1^2 \cos^2 \alpha_1 \geqq 1$. The sign of equality holds only if there is no discontinuity. With $y > 0$, it follows that $p_2 > p_1$, $\rho_2 > \rho_1$, $T_2 > T_1$, $v_2 \cos \alpha_2 < v_1 \cos \alpha_1$, $M_2^2 \cos^2 \alpha_2 < 1$. For a normal shock, $M_2 < 1$, and the flow always passes from supersonic to subsonic. The wave is always a compression wave, with $\rho_2 > \rho_1$, $p_2 > p_1$, $T_2 > T_1$.

The change in entropy is of the third order in $y$ (and therefore in the pressure-ratio increase, $p_2/p_1 - 1$). In fact, for sufficiently small $y$,

$$\frac{(\gamma - 1)(S_2 - S_1)}{\Re} = \sum_{n=1}^{\infty} \frac{(-1)^{n-1}}{n} \frac{\gamma}{(\gamma + 1)^n}$$
$$\{2^n \gamma^{n-1} - (\gamma + 1)^n + (\gamma - 1)^n\} y^n,$$

(A43)
$$\frac{S_2 - S_1}{\Re} = \frac{2\gamma}{(\gamma + 1)^2} \left\{ \frac{y^3}{3} - \frac{4\gamma}{\gamma + 1} \frac{y^4}{4} + \frac{(11\gamma^2 + 1)}{(\gamma + 1)^2} \frac{y^5}{5} \right.$$
$$\left. - \frac{2\gamma(13\gamma^2 + 3)}{(\gamma + 1)^3} \frac{y^6}{6} + \frac{(19\gamma^2 + 1)(3\gamma^2 + 1)}{(\gamma + 1)^4} \frac{y^7}{7} + \cdots \right\}.$$

Since, from (A32), $x = 2\gamma y/(\gamma + 1)$, the value of $S_2 - S_1$ is easily written down in terms of $x$.

### Formulae for $\gamma = 1.4$

For air at normal temperatures and pressures, we may take $\gamma = 1.4$. Then

$$\frac{p_2}{p_1} = 1 + \frac{7y}{6}, \quad \frac{\rho_2}{\rho_1} = \frac{v_1 \cos \alpha_1}{v_2 \cos \alpha_2} = \frac{1+y}{1+y/6},$$

$$\frac{T_2}{T_1} = \frac{\{1 + 7y/6\}\{1 + y/6\}}{1 + y},$$

$$M_2^2 \cos^2 \alpha_2 = \frac{1 + y/6}{1 + 7y/6},$$

$$\frac{H_2}{p_1} = \left(\frac{6}{5}\right)^{\frac{7}{2}} \frac{(1+y)^{\frac{7}{2}}}{(1 + 7y/6)^{\frac{5}{2}}} \left\{\frac{1 + (y + M_1^2 \sin^2 \alpha_1)/6}{1 + y/6}\right\}^{\frac{7}{2}},$$

$$\frac{H_1}{p_1} = \left(\frac{6}{5}\right)^{\frac{7}{2}} [1 + (y + M_1^2 \sin^2 \alpha_1)/6]^{\frac{7}{2}}, \quad \frac{H_2}{H_1} = \frac{(1+y)^{\frac{7}{2}}}{(1 + 7y/6)^{\frac{5}{2}}(1 + y/6)^{\frac{7}{2}}}.$$

## Deflection and downstream Mach number for an oblique shock

$M_2^2 \cos^2 \alpha_2$ is given by (A37). From $v_1 \sin \alpha_1 = v_2 \sin \alpha_2$, it follows that $M_1 a_1 \sin \alpha_1 = M_2 a_2 \sin \alpha_2$, so

$$M_2^2 \sin^2 \alpha_2 = \frac{a_1^2}{a_2^2} M_1^2 \sin^2 \alpha_1 = \frac{T_1}{T_2} M_1^2 \sin^2 \alpha_1$$

$$= \frac{(\gamma+1)^2(1+y)(M_1^2 - 1 - y)}{\{\gamma + 1 + 2\gamma y\}\{\gamma + 1 + (\gamma-1)y\}}$$

By addition

(A44)
$$M_2^2 = \frac{\{\gamma + 1 + (\gamma - 1)y\}^2 + (\gamma + 1)^2(1+y)(M_1^2 - 1 - y)}{\{\gamma + 1 + 2\gamma y\}\{\gamma + 1 + (\gamma - 1)y\}}$$
$$= \frac{(\gamma+1)^2 M_1^4 \cos^2 \alpha_1 - 4(\gamma M_1^2 \cos^2 \alpha_1 + 1)(M_1^2 \cos^2 \alpha_1 - 1)}{\{2\gamma M_1^2 \cos^2 \alpha_1 - (\gamma - 1)\}\{(\gamma - 1)M_1^2 \cos^2 \alpha_1 + 2\}}.$$

By division

$$\tan \alpha_2 = \frac{\gamma + 1}{\gamma + 1 + (\gamma - 1)y} \{(1+y)(M_1^2 - 1 - y)\}^{\frac{1}{2}}$$

$$= \tan \alpha_1 \frac{(\gamma+1)(1+y)}{\gamma + 1 + (\gamma - 1)y};$$

this also follows from

$$\frac{\tan \alpha_2}{\tan \alpha_1} = \frac{v_2 \sin \alpha_2}{v_2 \cos \alpha_2} \frac{v_1 \cos \alpha_1}{v_1 \sin \alpha_1} = \frac{v_1 \cos \alpha_1}{v_2 \cos \alpha_2},$$

together with equation (A35). The angle, $\delta$, through which the velocity is deflected through the shock wave, is

(A45)
$$\delta = \alpha_2 - \alpha_1,$$

and

(A46)
$$\tan \delta = \tan \alpha_1 \frac{y}{\frac{1}{2}(\gamma + 1)M_1^2 - y}$$
$$= \frac{M_1^2 \sin 2\alpha_1 - 2 \tan \alpha_1}{2 + M_1^2 (\gamma - \cos 2\alpha_1)}.$$

Since $M_1^2 \cos^2 \alpha_1 \geqq 1$, $y \geqq 0$. Let $\mu_1$ be the Mach angle upstream, defined as the acute angle for which

(A47) $$\sin \mu_1 = 1/M_1.$$

Then $\alpha_1$ lies between $\tfrac{1}{2}\pi - \mu_1$ (for a shock of zero strength) and 0 (for a normal shock). For a shock of zero strength, $y = 0$, the formulae give $M_2 = M_1$, $\delta = 0$, as they must. For a normal shock, $\alpha_1 = 0$, $y = M_1^2 - 1$, and equation (A44) reduces correctly to equation (A37); also $\delta$ is again zero.

If, for a given $M_1$, $\alpha_1$ decreases from $\tfrac{1}{2}\pi - \mu_1$ to 0, $M_2$ decreases monotonically from the supersonic value $M_1$ to the subsonic value for a normal shock. The flow behind the shock is just sonic, $M_2 = 1$, if $y$ is the positive root of the equation

(A48) $$[2\gamma/(\gamma + 1)]\, y^2 + y(3 - M_1^2) - M_1^2 + 1 = 0.$$

For values of $\alpha_1$ less than the value provided by (A48), the flow downstream of the shock is subsonic; for greater values of $\alpha_1$ it is supersonic.

Also, $\delta$ is zero for $\alpha_1 = \tfrac{1}{2}\pi - \mu_1$ and for $\alpha_1 = 0$, and is positive in between. It therefore has at least one maximum. It is easily shown to have just one maximum, for a given $M_1$, when $y$ is the positive root of

(A49) $$[2\gamma/(\gamma + 1)]\, y^2 + y(4 - M_1^2) - 2(M_1^2 - 1) = 0.$$

As $M_1 \to \infty$, both (A48) and (A49) yield $\cos^2 \alpha_1 = (\gamma + 1)/2\gamma$; as $M_1 \to 1$, $\alpha_1 \to 0$ in both cases. For intermediate values of $M_1$, the value of $\alpha_1$ from (A49) is a little less than the value from (A48). Also, for a given $\alpha_1$, $\tan \delta$ increases monotonically with $M_1$; $M_1 \geqq \sec \alpha_1$, and for $M_1 = \sec \alpha_1$, $\delta = 0$; the limiting value of $\tan \delta$ as $M_1 \to \infty$ is

$$\sin \alpha_1 \cos \alpha_1 / \{\tfrac{1}{2}(\gamma + 1) - \cos^2 \alpha_1\}.$$

Thus for a given $M_1$ and a given $\alpha_1$ (between 0 and $\tfrac{1}{2}\pi - \mu_1$), there is a unique $\delta$; for a given $\alpha_1$ in that range and a given $\delta$ in the range specified above, there is a unique $M_1$; for a given $M_1$ and a given $\delta$, there is no solution for $\alpha_1$ if the given $\delta$ exceeds the greatest possible value for the given $M_1$, as given by (A46) and (A49); if the given $\delta$ is less than that value, there are two possible values of $\alpha_1$, for the smaller of which the flow behind the shock is subsonic (strong shock) and for the larger of which (weak shock) the flow behind the shock is supersonic except when $\delta$ is so near the maximum for the given $M_1$ that this larger value of $\alpha_1$ lies in the small range of values between the values given by (A48) and (A49). (An illustrative diagram will be found in Liepmann and Roshko, *Elements of Gasdynamics*, Section 4.3.) For a given $M_1$, the range of values of $\delta$ for which the flow behind the weak shock is subsonic is very small indeed.

## Weak oblique shock waves. Hypersonic approximations for small deflections

For a weak shock, $y$ is small, $\alpha_1$ is near $\tfrac{1}{2}\pi - \mu_1$, the flow behind the shock is supersonic, and $\delta$ is small. Approximate expressions in terms of $\delta$ play an important role in gasdynamics and supersonic aerodynamics. In place of $M_1$, it is often convenient to use the Mach angle, defined in (A47). The expressions are obtained as follows.

From the definition of $y$ in (A33)

$$\cos \alpha_1 = (1 + y)^{\frac{1}{2}} \sin \mu_1.$$

Hence

$$\sin \alpha_1 = (1 - y \tan^2 \mu_1)^{\frac{1}{2}} \cos \mu_1,$$

and

$$\cot \alpha_1 = \tan \mu_1 (1 + \tfrac{1}{2} y \sec^2 \mu_1) + O(y^2).$$

From (A46), $\tan \delta$ and $\delta$ are now found correct to $O(y^2)$, and then $y$ in terms of $\delta$ correct to $O(\delta^2)$. Equations (A34), (A36), and (A38) now give $p_2/p_1$, $\rho_2/\rho_1$, and $T_2/T_1$ to $O(\delta^2)$. Equations (A43) show that $S_2 - S_1$ is of the third order in $\delta$. Also $v_2/v_1$ is found from

$$v_2^2 = v_2^2 \sin^2 \alpha_2 + v_2^2 \cos^2 \alpha_2 = v_1^2 \sin^2 \alpha_1 + (\rho_1/\rho_2)^2 v_1^2 \cos^2 \alpha_1,$$

i.e.,

$$(v_2/v_1)^2 = 1 + \{(\rho_1/\rho_2)^2 - 1\} \cos^2 \alpha_1.$$

The first-order results are

$$y = \delta(\gamma + 1) \operatorname{cosec} 2\mu_1,$$
$$\alpha_1 = \tfrac{1}{2}\pi - \mu_1 - \tfrac{1}{4}\delta(\gamma + 1) \sec^2 \mu_1,$$
$$p_2/p_1 - 1 = 2\delta\gamma \operatorname{cosec} 2\mu_1, \quad \rho_2/\rho_1 - 1 = 2\delta \operatorname{cosec} 2\mu_1,$$
(A50) $\quad T_2/T_1 - 1 = 2\delta(\gamma - 1) \operatorname{cosec} 2\mu_1, \quad v_2/v_1 - 1 = -\delta \tan \mu_1.$

For the second-order terms (in $\delta^2$), reference may be made to Illingworth, *Modern Developments: High Speed Flow*, Chap. IV. In particular

$$p_2/p_1 - 1 = 2\delta\gamma \operatorname{cosec} 2\mu_1 + \delta^2 \gamma (\gamma + \cos^2 2\mu_1)$$
$$\operatorname{cosec}^2 2\mu_1 \sec^2 \mu_1 + O(\delta^3).$$

If we define a non-dimensional pressure coefficient, $C_p$, by

$$C_p = \frac{p_2 - p_1}{\tfrac{1}{2}\rho v_1^2} = 2\frac{p_2 - p_1}{\gamma M_1^2 p_1} = \frac{2}{\gamma}\left(\frac{p_2}{p_1} - 1\right) \sin^2 \mu_1,$$

then

(A51) $\quad C_p = 2\delta \tan \mu_1 + \tfrac{1}{2}\delta^2(\gamma + \cos^2 2\mu_1) \sec^4 \mu_1 + O(\delta^3).$

If $\delta$ is small, then from (A46) $y$ is small unless $\tan \alpha_1$ is small or $M_1^2$ is large. For the "weak shock," $\tan \alpha_1$ cannot be small. But if $M_1^2$ is large, $\delta$ may be small without $y$ being small. [Moreover, even if we assume $y$ small, the expansions of which the first terms are given in (A50) are not valid unless $M_1\delta$ is small; for large $M_1$, $\mu_1$ is small, and the second-order result for $y$, for example, is

$$y = \delta(\gamma + 1) \operatorname{cosec} 2\mu_1 + \tfrac{1}{2}\delta^2(\gamma + 1)(\gamma + \cos^2 2\mu_1)$$
$$\operatorname{cosec}^2 2\mu_1 \sec^2 \mu_1 + O(\delta^3).]$$

For the "weak shock" for a given, small, $\delta$ and a given, large, $M_1$, $\alpha_1$ is near to $\tfrac{1}{2}\pi$. We set $\tfrac{1}{2}\pi - \alpha_1 = \alpha'$, and calculate $y$ without assuming it small. $y + 1$ is approximately equal to $M_1^2 \alpha'^2$. From (A46), with $\delta$ and $\alpha'$ small, $y$ is approximately equal to $\tfrac{1}{2}\alpha'(\gamma + 1)M_1^2\delta$. Hence we have the approximation that $\alpha'$ is the positive root of

$$\alpha'^2 - \tfrac{1}{2}\alpha'(\gamma + 1)\delta - 1/M_1^2 = 0,$$

so

(A52) $$\alpha' = \frac{\gamma+1}{4}\delta + \left\{\left(\frac{\gamma+1}{4}\right)^2 \delta^2 + \frac{1}{M_1^2}\right\}^{\frac{1}{2}}$$

and

(A53) $$y = \frac{(\gamma+1)^2}{8} M_1^2 \delta^2 + \frac{\gamma+1}{2} M_1 \delta \left\{\left(\frac{\gamma+1}{4}\right)^2 M_1^2 \delta^2 + 1\right\}^{\frac{1}{2}},$$

(A54) $$C_p = \delta^2 \left[\frac{\gamma+1}{2} + \left\{\left(\frac{\gamma+1}{2}\right)^2 + \frac{4}{M_1^2 \delta^2}\right\}^{\frac{1}{2}}\right].$$

This is the hypersonic approximation for the "weak" shock for a given, small, $\delta$ and a given, large, $M_1$. $M_1\delta$ may be neither very large nor very small, and there is no generally satisfactory manner of approximating in a simple way to the square root. $y$ is no longer necessarily small. From (A44) the Mach number, $M_2$, after the shock is of order either $M_1$ or $\delta^{-1}$, and is still large.

## References

For further discussions of shock waves, reference may be made to the texts previously cited; Liepmann and Roshko, *Elements of Gasdynamics*; *Modern Developments*: *High Speed Flow*; Courant and Friedrichs, *Supersonic Flow and Shock Waves*. For more elaborate tables and graphs, see N.A.C.A. Tech. Rep. No. 1135 (1953), Equations, Tables and Charts for Compressible Flow, by the Ames Research Staff, and two volumes prepared on behalf of the (British) Aeronautical Research Council by the Compressible Flow Tables Panel (Chairman: L. Rosenhead) and published by the Oxford University Press: *A Selection of Tables for Use in Calculations of Compressible Airflow* (1952), and *A Selection of Graphs for Use in Calculations of Compressible Airflow* (1954). References to other tables and graphs are contained in the literature. The important and interesting subjects of shock-wave intersections and reflections, and spherical and cylindrical waves (blast waves) must regretfully be omitted.

# APPENDIX V

# Formulae for a Prandtl-Meyer Expansion in a Perfect Gas with Constant Specific Heats

For ease of reference, certain formulae which have appeared elsewhere in (Section 4.1 and Appendix IV) are included below. The formulae apply to the steady motion of an inviscid perfect gas with constant specific heats, and hold along a streamline, and throughout any region of constant entropy and irrotational motion.

The subscript zero is used for conditions at a stagnation point, and the subscript $s$ for conditions when the local Mach number is unity. As before, $v_G$ is the limiting velocity, and $v_s$ the critical velocity. (See equation (93).) Let

(A55) $$\tau = \frac{v^2}{v_G^2} = \frac{\gamma-1}{2}\frac{v^2}{a_0^2} = \tau_s \frac{v^2}{v_s^2}$$

where $\tau_s$ is the value of $\tau$ when $v = v_s$,

(A56) $$\tau_s = v_s^2/v_G^2 = (\gamma - 1)/(\gamma + 1).$$

($\tau_s = \tfrac{1}{6}$ for $\gamma = 1.4$.)

With variations in $\Omega$ neglected,

$$I_0 = a^2/(\gamma-1) + \tfrac{1}{2}v^2 = a^2/(\gamma-1) + a_0^2\tau/(\gamma-1) = a_0^2/(\gamma-1)$$
$$= \tfrac{1}{2}v_G^2 = v_s^2/2\tau_s = \{a^2/(\gamma-1)\}\{1 + \tfrac{1}{2}(\gamma-1)M^2\}$$
(A57) $$= \{\gamma\Re T/(\gamma-1)\}\{1 + \tfrac{1}{2}(\gamma-1)M^2\} = \gamma\Re T_0/(\gamma-1).$$

($T_0$ was denoted by $\vartheta$ in Appendix IV.) Also $a^2/a_0^2 = T/T_0$. Hence

(A58) $$T/T_0 = (\rho/\rho_0)^{\gamma-1} = (p/p_0)^{(\gamma-1)/\gamma} = 1 - \tau$$
$$= [1 + \tfrac{1}{2}(\gamma-1)M^2]^{-1}$$

Also

(A59) $$M^2 = \frac{v^2}{a^2} = \frac{2\tau}{(\gamma-1)(1-\tau)} = \frac{(1-\tau_s)\tau}{\tau_s(1-\tau)}$$
$$M^2 - 1 = \frac{\tau/\tau_s - 1}{1 - \tau}$$

Consider now supersonic flow round a convex corner. Such a deflection must be expansive, and cannot be accomplished by a shock wave (which gives the flow in a concave corner). See p. 82. It is, then, taken to be an isentropic expansion. The region of influence of the corner will be bounded by the characteristic curves through it; in a uniform flow the characteristic curves are straight lines making an angle $\mu = \sin^{-1}(1/M)$ with the streamlines, so the expansive turning of the flow from a uniform flow in one direction to a uniform flow in another direction is accomplished between radii vectors through the corner, making angles $\mu_1 = \sin^{-1}(1/M_1)$ and $\mu_2 = \sin^{-1}(1/M_2)$, respectively, with the streamlines in the uniform flows upstream and downstream of the deflection. In the problem of pure convex or expansive turning, without consideration of other boundaries, there is no characteristic length to define a scale in the problem, so, if $(r, \theta)$ are plane polar coordinates, with the corner as origin, the dependent variables can depend only on the angle $\theta$ and not on the distance $r$ from the corner. The result from these considerations is a centered simple wave, or Prandtl-Meyer expansion. If B is the corner (Fig. 15), the fluid flows along the wall AB,

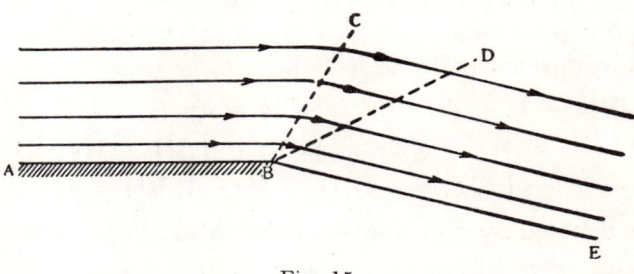

Fig. 15

expands between BC and BD, and then flows parallel to BE. BC and BD are characteristic curves, or Mach lines; the angle ABC $= \pi - \mu_1$, and the angle DBE $= \mu_2$; the disturbance does not reach the region ABC, and the uniform flow in DBE may be joined to the flow in the disturbed region CBD when BD is reached. BE may be a solid boundary, or the wall may end at

B and BE may be a free surface, with the gas expanding to a lower pressure in the space below ABD than in the space above ABC.

The flow in CBD must be irrotational, from the general theories of Section 4. However, the flow is two-dimensional, with the velocity components $v_r$, $v_\theta$, the pressure $p$, density $\rho$, etc., all independent of $r$, and it is easily proved, as follows (without recourse to general theorems), that such a motion must be irrotational. For a two-dimensional motion, the only possible non-zero component of vorticity is the component, $\omega_z$, perpendicular to the plane of the motion; with $p$ and $\mathbf{v}^2$ independent of $r$, the equation of steady motion gives at once that $(\mathbf{v} \times \boldsymbol{\omega})_r = 0$, i.e., $v_\theta \omega_z = 0$, so, since $v_\theta \neq 0$, $\omega_z = 0$. There is therefore a velocity potential $\phi$, which must be of the form $\phi = rf(\theta)$, and

$$v_r = f(\theta), \quad v_\theta = f'(\theta), \quad \mathbf{v}^2 = f^2 + f'^2.$$

The equation of steady two-dimensional motion,

$$a^2 \nabla^2 \phi - \boldsymbol{\nabla}\phi \cdot \boldsymbol{\nabla}\tfrac{1}{2}\mathbf{v}^2 = 0,$$

then gives at once

$$r^{-1}(f + f'')(a^2 - f'^2) = 0.$$

$f + f'' = 0$ would mean $d\mathbf{v}^2/d\theta = 0$, and $\mathbf{v}^2$ would be a constant. From the constancy of $I_0$, it follows that $a^2$, $p$, $\rho$, $M$, $\mu$, etc. would all be constant in such a flow. Also it is easily seen that the flow direction would be constant, and this solution simply represents a uniform flow. Hence we have $f'^2 = a^2$; with the sense in which $\theta$ increases taken from BC to BD, $v_\theta$ is positive, and $v_\theta = a$. The streamlines make an angle $\mu$ with the radii vectores, which form one family of characteristic curves, or Mach lines.

Since $v_\theta = a$,

$$I_0 = v_G^2/2 = v_\theta^2[(\gamma - 1)^{-1} + \tfrac{1}{2}] + \tfrac{1}{2}v_r^2 = v_\theta^2/2\tau_s + \tfrac{1}{2}v_r^2.$$

Hence

$$f' = \tau_s^{\frac{1}{2}}(v_G^2 - f^2)^{\frac{1}{2}}$$

which is a differential equation for $f$ as a function of $\theta$. If the origin for $\theta$ is suitably chosen,

(A60) $\quad f = v_r = v_G \sin(\tau_s^{\frac{1}{2}}\theta), \quad f' = v_\theta = a = v_G \tau_s^{\frac{1}{2}} \cos(\tau_s^{\frac{1}{2}}\theta)$
$\quad\quad\quad = v_s \cos(\tau_s^{\frac{1}{2}}\theta)$

$\mathbf{v}^2$ increases as $\theta$ increases, and $\theta$ is measured from a position in an assumed backward continuation of the turning, expanding flow at which $v = a = v_s$, $v_r = 0$ — i.e., $\theta$ is measured from the sonic position in an assumed analytical continuation of the turning flow backward from BC.

Since the streamlines make angles $\mu$ with the radii vectores,
$$v_r = v \cos \mu, \quad v_\theta = v \sin \mu,$$
and
$$v_r/v_\theta = \cot \mu = \tau_s^{-\frac{1}{2}} \tan(\tau_s^{\frac{1}{2}}\theta),$$
i.e.,
(A61) $\quad\quad\quad \tan \tau_s^{\frac{1}{2}}\theta = \tau_s^{\frac{1}{2}} \cot \mu.$

This equation connects $\theta$ and $\mu$. In particular, it gives the value of $\theta$ (measured from the assumed sonic position) to be assigned on BC, where $\mu$ has the known value $\mu_1$.

Since the streamlines make angles $\mu$ with the radii vectores, the angle turned through by the flow is the change in $\theta + \mu$. At the assumed sonic position, $\theta = 0$, $\mu = \frac{1}{2}\pi$. If we measure from that position, the flow deflection is the change in

$\omega = \theta + \mu - \frac{1}{2}\pi = \tau_s^{-\frac{1}{2}} \tan^{-1}\{\tau_s(M^2-1)\}^{\frac{1}{2}} - \tan^{-1}(M^2-1)^{\frac{1}{2}}$
(A62) $= \tau_s^{-\frac{1}{2}} \tan^{-1}[(\tau-\tau_s)/(1-\tau)]^{\frac{1}{2}} - \tan^{-1}[(\tau/\tau_s-1)/(1-\tau)]^{\frac{1}{2}}$

$\omega$ is known as the Prandtl-Meyer function. It plays an important role both in the theory of characteristics, and in hodograph theory. It is tabulated against $M$, $\mu$, or $\tau$. For a given value of one of these parameters at BC, the value of $\omega$ there is found. For deflection through a given angle, the value of $\omega$, increased by this angle, gives the value on BD, and the values of $M$, $\mu$, and $\tau$ in the flow downstream of BD are then found, again from the tables. The values of $M$, $\mu$, and $\tau$ may be similarly found at any stage of the deflection, and the corresponding values of $T/T_0$, $\rho/\rho_0$, and $p/p_0$ are given by (A58).

Note also that

$$\tau = v^2/v_G^2 = \sin^2(\tau_s^{\frac{1}{2}}\theta) + \tau_s \cos^2(\tau_s^{\frac{1}{2}}\theta)$$

(A63)
$$= 1 - \frac{2}{\gamma + 1}\cos^2(\tau_s^{\frac{1}{2}}\theta),$$

and

(A64) $\quad T/T_0 = (\rho/\rho_0)^{\gamma-1} = (p/p_0)^{(\gamma-1)/\gamma} = [2/(\gamma + 1)]\cos^2(\tau_s^{\frac{1}{2}}\theta).$

For expansion round a corner from an initial to a final pressure, with a given stagnation pressure, we therefore have, from the values of the pressure, the initial and final values of $M$, $\tau$, $\mu$, and $\theta$, and therefore also the deflection. (Of course if the initial values of $p$ and $M$ are given, they determine $p_0$.) For an assumed expansion into a vacuum, in the final state $p = 0$, $v = v_G$, $v_\theta = 0$, $\tau_s^{\frac{1}{2}}\theta = \frac{1}{2}\pi$ ($\theta = 220.454°$ approximately, for $\gamma = 1.4$), and $\mu = 0$. The maximum deflection from the sonic condition would be $\frac{1}{2}\pi(\tau_s^{-\frac{1}{2}} - 1)$ ($= 130.454°$ for $\gamma = 1.4$). Any deflection will be less than this. Note that even for a jet expanding round a corner into a vacuum, the flow would not fill the space below the wall AB.

Along any radius vector, the velocity direction and magnitude, the pressure, density, temperature, etc., are all constant, and the turning and expansion may be joined to a uniform stream after any deflection less than the maximum.

The equations of the streamlines and of the second family of Mach lines are easily calculated, and are found to be

(A65) $\quad\quad\quad r = \text{constant } \{\sec(\tau_s^{\frac{1}{2}}\theta)\}^{1/\tau_s}$

for the streamlines, and

(A66) $\quad\quad\quad r = \text{constant } \{\sec(\tau_s^{\frac{1}{2}}\theta)\}^{1/2\tau_s} \{\csc(\tau_s^{\frac{1}{2}}\theta)\}^{\frac{1}{2}}$

for the second family of Mach lines.

Finally, note that, for a small deflection, the pressure, density, temperature, and velocity ratios may be expanded in powers of the deflection, and that, correct to the second order in the deflection, the results are the same as for deflection by an oblique shock wave, due allowance being made for the sign of the deflection. This was to be expected, since to the second order the flow through a shock wave may be treated as isentropic. In particular, the pressure coefficient is still given by equation (A51) of Appendix IV. All this applies, however, only if $M_1$ is not large. For if $M_1$ is large, a small deflection by a shock does not necessarily imply that $M_1^2 \cos^2 \alpha_1 - 1$ or $p_2/p_1 - 1$ is small, and the entropy change is not negligible. We may carry out a hypersonic approximation to a Prandtl-Meyer expansion. From (A62), with $M_1$ and $M_2$ large, it is easy, by approximating to the inverse tangents, to see that the deflection is approximated by

$$\delta = [2/(\gamma - 1)](M_1^{-1} - M_2^{-1}).$$

From (A58), $p_2/p_1 - 1$ is found in terms of $M_1$ and $M_2$, and approximated for large $M_1$ and $M_2$. Then $M_2$ is eliminated. The result is that $C_p/\delta^2$ is still a function of $M_1 \delta$, but now

(A67)
$$C_p = \frac{p_2 - p_1}{\tfrac{1}{2}\rho v_1^2} = \frac{2}{\gamma M_1^2}\left(\frac{p_2}{p_1} - 1\right)$$
$$= \frac{2}{\gamma M_1^2}\{[1 - \tfrac{1}{2}(\gamma - 1)M_1 \delta]^{2\gamma/(\gamma-1)} - 1\}.$$

On this hypersonic approximation, the pressure $p_2$ falls to zero when $\delta = 2/[(\gamma - 1)M_1]$.

# BOOK LIST

LAMB, H., *Hydrodynamics*, 6th ed., Cambridge University Press, 1932, or Dover Publications Reprint (paperback).

PRANDTL, L., and TIETJENS, O. G., Vol. 1. *Fundamentals of Hydro- and Aeromechanics* (translated by L. Rosenhead); Vol. 2. *Applied Hydro- and Aeromechanics* (translated by J. P. den Hartog), McGraw-Hill, New York, 1934.

SOMMERFELD, A., *Lectures on Theoretical Physics*, Vol. 2. *Mechanics of Deformable Bodies* (translated by G. Kuerti), Academic Press, New York, 1950; Vol. 3, *Electrodynamics* (translated by E. G. Ramberg), Academic Press, New York, 1952.

MILNE-THOMSON, L. M., *Theoretical Hydrodynamics*, Macmillan, London, 1949.

*Modern Developments in Fluid Dynamics* (ed. by S. Goldstein), Oxford University Press, 1938, 2 vols.

SCHLICHTING, H., *Boundary Layer Theory* (translated by J. Kestin), McGraw-Hill, New York, 1955.

LIEPMANN, H. W., and ROSHKO, A., *Elements of Gasdynamics*, Wiley, New York, 1957.

*Modern Developments in Fluid Dynamics: High Speed Flow* (ed. by L. Howarth), Oxford University Press, 1953, 2 vols.

*General Theory of High Speed Aerodynamics* (ed. by W. R. Sears: Vol. 6 of *High Speed Aerodynamics and Jet Propulsion*), Princeton University Press, 1954.

COURANT, R. and FRIEDRICHS, K. O., *Supersonic Flow and Shock Waves*, Interscience, New York–London, 1948.

MISES, R. VON, GEIRINGER, H., and LUDFORD, G. S. S., *Mathematical Theory of Compressible Fluid Flow*, Academic Press, New York, 1958.

## Monographs

COWLING, T. G., *Magnetohydrodynamics*, Interscience, New York–London, 1957.

BATCHELOR, G. K., *The Theory of Homogeneous Turbulence*, Cambridge University Press, 1953.

WARD, G. N., *Linearized Theory of Steady High-Speed Flow*, Cambridge University Press, 1955.

LIN, C. C., *The Theory of Hydrodynamic Stability*, Cambridge University Press, 1955.

TOWNSEND, A. A., *The Structure of Turbulent Shear Flow*, Cambridge University Press, 1956.

FRANKL, F. I. and KARPOVICH, E. A., *Gas Dynamics of Thin Bodies* (translated by M. D. Friedman), Interscience, New York–London, 1953.

SPITZER, L., JR., *Physics of Fully Ionized Gases*, Interscience New York–London, 1956.

### SOME ADDITIONAL BOOKS

LANDAU, L. D., and LIFSHITZ, E. M., *Fluid Mechanics*, Pergamon Press, London, Paris, Frankfurt and Addison-Wesley, Reading, Mass., 1959.

*Fundamentals of Gas Dynamics* (ed. by H. W. Emmons: Vol. 3 of *High Speed Aerodynamics and Jet Propulsion*), Princeton University Press, 1958.

*Theory of Laminar Flows* (ed. by F. K. Moore: Vol. 4 of *High Speed Aerodynamics and Jet Propulsion*), Princeton University Press, 1964.

*Laminar Boundary Layers* (ed. by L. Rosenhead), Oxford University Press, 1963.

HINZE, J. O., *Turbulence*, McGraw-Hill, New York, Toronto, and London, 1959.

CHANDRASEKHAR, S., *Hydrodynamic and Hydromagnetic Stability*, Oxford University Press, 1961.

BATCHELOR, G. K., *An Introduction to Fluid Dynamics*, Cambridge University Press, 1967.

VAN DYKE, M., *Perturbation Methods in Fluid Mechanics*, Academic Press, New York and London, 1964.

# Some Problems of Magneto-Gasdynamics

SPECIAL LECTURES BY J. M. BURGERS

## Introduction

In the preceding sections of these lectures on fluid mechanics Professor Goldstein has given the basic equations for the flow of fluids subjected to electromagnetic fields. I have thought that it might be worthwhile to consider a few problems appearing in such cases in some detail.

The first problem to be treated is connected with Eq. (81), at the end of Section 3.5, "Equations for $\mathbf{B}$ and $\rho_e$." This equation describes the effect of the motion of the medium, supposed to be conductive for electricity, on the magnetic flux. In Section 4.3 and in Section 5.3 it has been pointed out that there is an analogy between the behavior of the magnetic flux and that of vorticity. (To be precise the analogy requires that the coefficient of the electric conductivity and the coefficient of the viscosity both are constants, since otherwise questions will arise concerning the behavior of their derivatives.) I will discuss a nontrivial example of the application of Eq. (81), in order to illustrate some methods of approximation useful in magneto-gasdynamics.

The second contribution refers to the "tensorial character" of the electric conductivity when a magnetic field is present, a question which was briefly mentioned by Professor Goldstein at the end of Section 3.1.

## 1. Basic equations

Since the equations to be applied have been given in Professor Goldstein's chapters, only the most important of them need be repeated here. We shall make use of the approximation described in Section 3.5, in which free charges and displacement currents are neglected. This can be done when there are no high-frequency oscillations and when the speed of the medium is small compared with the speed of light.

Equation (58) for the electric current then simplifies to:

(1) $$\mathbf{J} = \sigma\{\mathbf{E} + (\mathbf{v} \times \mathbf{B})\},$$

and Eqs. (56) give:

(2) $$\begin{aligned}\nabla \times \mathbf{E} &= -\partial \mathbf{B}/\partial t \\ \nabla \times \mathbf{B} &= \mu \mathbf{J} \\ \nabla \cdot \mathbf{E} &= 0 \\ \nabla \cdot \mathbf{B} &= 0\end{aligned}$$

Use is made of the relations $\mathbf{D} = \varepsilon \mathbf{E}$ and $\mathbf{B} = \mu \mathbf{H}$, and it is assumed that the dielectric coefficient $\varepsilon$ and the magnetic permeability $\mu$ are constants.

We shall come later to the equations of motion for the fluid.

It is convenient to introduce the vector potential $\mathbf{A}$ of the electromagnetic field and the electrostatic scalar potential $\phi$, in such a way that:

(3a) $$\mathbf{B} = \nabla \times \mathbf{A}$$

(3b) $$\nabla \cdot \mathbf{A} = 0$$

(3c) $$\mathbf{E} = -\partial \mathbf{A}/\partial t - \nabla \phi$$

We then obtain:

(3d) $$\mathbf{J} = \sigma\{-\partial \mathbf{A}/\partial t - \nabla \phi + \mathbf{v} \times (\nabla \times \mathbf{A})\}$$

Substitution into the equation $\nabla \times \mathbf{B} = \mu \mathbf{J}$ gives (after some reduction):

(4) $$\partial \mathbf{A}/\partial t - \mathbf{v} \times (\nabla \times \mathbf{A}) + \nabla \phi = \kappa \Delta \mathbf{A}$$

where $\kappa = (\mu \sigma)^{-1}$, while $\Delta = \nabla^2$ denotes the Laplace operator. This equation must be used in conjunction with the condition:

(3b) $$\nabla \cdot \mathbf{A} = 0$$

Application of the operator $\nabla \cdot$ to Eq. (4) gives the result:

(5) $$\Delta\phi = \mathbf{v} \cdot (\Delta\mathbf{A}) + (\nabla \times \mathbf{v}) \cdot (\nabla \times \mathbf{A})$$

When written in the components, Eq. (4) represents three linear equations for the 4 unknowns $A_1$, $A_2$, $A_3$ and $\phi$, provided it is assumed that the components of the velocity of the fluid, $v_1$, $v_2$, $v_3$, have been given. Either Eq. (3b) or Eq. (5) represents a fourth equation for the 4 unknowns. As will be evident from a comparison with Eq. (3d), Eq. (5) must ensure the continuity condition for the electric current: $\nabla \cdot \mathbf{J} = 0$.

We apply these equations to the case of a plane shock wave, with its front perpendicular to the $x$-axis and propagating itself with a speed $V$ in the direction of that axis, in a gas originally at rest and having constant density, constant pressure and constant temperature, it being supposed that there is a magnetic dipole at the origin with its magnetic field around it. The gas is assumed to be conductive and the conductivity will be treated as very high. We shall approach the case in various steps. In the first step we neglect the reaction of the magnetic forces on the motion of the gas and treat the gas motion as given, with a constant value of $V$ and a constant and uniform gas speed $v$ behind the shock front in the direction of the $x$-axis ($v_1 = v$; $v_2 = v_3 = 0$). Moreover, in this first step the conductivity will be taken to be infinite. In the second step we keep the velocity field of the gas the same, but introduce a small resistivity of the gas. In the third step we shall give some attention to the effect of the magnetic field on the motion of the gas.

When we leave aside a constant factor, the dipole field has the components:

(6) $$B_1 = -3xz/r^5, \quad B_2 = -3yz/r^5, \quad B_3 = (r^2 - 3z^2)/r^5$$

The sign is chosen such that $B_3$ is positive in the $x$, $y$-plane (equatorial plane of the dipole); this requires that the dipole has its negative pole directed upward. The corresponding components of the vector potential are:

(7) $$A_1 = y/r^3, \quad A_2 = -x/r^3; \quad A_3 = 0$$

These components satisfy the conditions $\nabla \cdot \mathbf{A} = 0$ and

$$\Delta A_1 = \Delta A_2 = \Delta A_3 = 0$$

We suppose that the shock front is coming from the left (that is, from the side of the negative $x$) and is moving towards the origin, so that $V$ and $v$ are positive. It is convenient to assume that the front reaches the origin at $t = 0$; the position of the shock front at any time is then given by $x = Vt$. Since the region of interest in the following considerations is to the left of the origin, we shall have to do with negative values of $x$ and of $t$.

## 2. Equations for the vector potential $A$ when the conductivity of the gas is infinite

With $\kappa = 0$ and $v_1 = v$, $v_2 = v_3 = 0$ in the domain behind the shock front, Eqs. (4), (3b), and (5) take the forms:

(8)
$$\partial A_1/\partial t = -\partial \phi/\partial x$$
$$\partial A_2/\partial t + v(\partial A_2/\partial x) = v(\partial A_1/\partial y) - \partial \phi/\partial y$$
$$\partial A_3/\partial t + v(\partial A_3/\partial x) = v(\partial A_1/\partial z) - \partial \phi/\partial z$$

(9)
$$\partial A_1/\partial x + \partial A_2/\partial y + \partial A_3/\partial z = 0$$

(10)
$$\Delta \phi = v \Delta A_1$$

It is easily seen that these equations can be satisfied with $\phi = 0$. In that case the first equation of (8) shows that $A_1$ will retain the value given in (7); since this value satisfies the condition $\Delta A_1 = 0$, Eq. (10) gives $\Delta \phi = 0$, and as there is no reason to expect a singularity of $\phi$ in the domain behind the shock front, we come back to $\phi = 0$.

With $\phi = 0$ and $A_1 = y/r^3$, the right-hand sides of the second and third equations of (8) are known. These equations can be integrated along a line for which $dx/dt = v$; $dy/dt = dz/dt = 0$. Let us draw such a line through the point $Q(x, y, z, t)$ where $A_2$ and $A_3$ must be found (compare Fig. 1, giving the $x$, $t$-plane), and let us follow it in the direction of negative $t$ until, at $Q'$, it

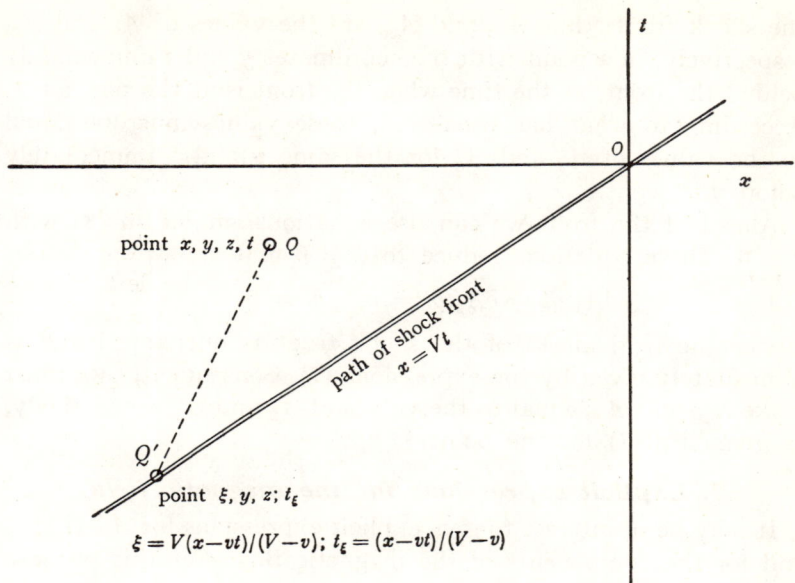

Fig. 1. $x$, $t$-diagram for a shock wave of constant speed $V$. $Q'O$ is the path of the shock front $x = Vt$; $QQ'$ is the line $dx/dt = v$; $dy/dt = dz/dt = 0$.

intersects with the line $x = Vt$ describing the path of the shock front. The $x$-coordinate of the point of intersection has the value:

(11) $$\xi = V(x - vt)/(V - v)$$

($\xi$ will be negative, like $x$ and $t$).

The expressions for $A_2$ and $A_3$ behind the shock front then become:

(12)
$$A_2 = \int_\xi^x (\partial A_1/\partial y)dx + A_{20}$$
$$A_3 = \int_\xi^x (\partial A_1/\partial z)dx + A_{30}$$

In order to determine the integration constants $A_{20}$, $A_{30}$, we observe that the values of $A_1, A_2, A_3$ must be continuous throughout the whole field if we wish to prevent infinite values of the magnetic flux. Now $\xi = x$ when the point $x$ is chosen immediately behind the shock front; thus $A_{20}$ and $A_{30}$ are the values of $A_2$ and $A_3$, respectively, in a point with the coordinates $y$ and $z$ immediately behind the front, at the time when the front is at the position $\xi$. According to what has been said, these values must be equal to the values of $A_2$ and $A_3$ for the same $y$ and $z$ immediately before the front.

Ahead of the front we can use equations similar to (8) with $v = 0$. These equations reduce to $\phi = 0$ and

(13) $\qquad \partial A_1/\partial t = \partial A_2/\partial t = \partial A_3/\partial t = 0$

Hence the field ahead of the shock front is unchanged and is immediately given by the expressions (7). Consequently we must make $A_{20}$ and $A_{30}$ equal to the values of $A_2$ and $A_3$, respectively, as given by (7) for the point $\xi$, $y$, $z$.

### 3. Explicit expressions for the magnetic field

It may be of interest to give explicit expressions for $A_1$, $A_2$, $A_3$ and for the components of the magnetic flux. For this purpose we introduce the following auxiliary functions:

(14a) $\qquad \begin{aligned} G_2(X) &= (1 + X/R)y/(y^2 + z^2) \\ G_3(X) &= (1 + X/R)z/(y^2 + z^2) \end{aligned}$

where $X$ is a quantity to be specified in various cases, while in every case $R^2 = X^2 + y^2 + z^2$ (also $X$ will be negative in the following formulas). Two other functions to be used are:

(14b) $\qquad \begin{aligned} F_2(X) &= -\partial G_2/\partial y = \partial G_3/\partial z + X/R^3 \\ F_3(X) &= -\partial G_2/\partial z = -\partial G_3/\partial y \end{aligned}$

These functions satisfy the relations:

$$\partial F_2/\partial X = -(R^2 - 3y^2)/R^5; \quad \partial F_3/\partial X = 3yz/R^5;$$
$$\partial F_2/\partial y + \partial F_3/\partial z = -3Xy/R^5; \quad \partial(F_2 - X/R^3)/\partial y + \partial F_3/\partial z = 0$$

We can now write in the wake of the shock wave $(x < Vt)$:

(15a) $\qquad \begin{aligned} A_1 &= y/r^3 \\ A_2 &= -F_2(x) + F_2(\xi) - \xi/r_1^3 \\ A_3 &= -F_3(x) + F_3(\xi) \end{aligned}$

and ahead of the shock wave $(x > Vt)$:

(15b)
$$A_1 = y/r^3$$
$$A_2 = -x/r^3$$
$$A_3 = 0$$

In these expressions $\xi$ is the quantity given by (11), while $r_1^2 = \xi^2 + y^2 + z^2$. At the shock front we have $x = Vt$ and at the same time $\xi = Vt$; thus $\xi = x$ and $r_1 = r$. Hence there is no discontinuity of the components of $A$ at the shock front. It can also be checked without difficulty that the divergence of $A$ is zero.

By differentiation we can obtain the values of the components of the magnetic flux $B_1$, $B_2$, $B_3$ (it must be kept in mind that $\xi$ is a function of $x$, given by (11), and that partial differentiation with respect to $x$ supposes that $t$ remains constant). One finds for $x < Vt$:

(16a)
$$B_1 = -3\xi z/r_1^5$$
$$B_2 = -[V/(V-v)]3yz/r_1^5$$
$$B_3 = [V/(V-v)](r_1^2 - 3z^2)/r_1^5$$

and for $x > Vt$:

(16b)
$$B_1 = -3xz/r^5$$
$$B_2 = -3yz/r^5$$
$$B_3 = (r^2 - 3z^2)/r^5$$

In the field behind the shock front the value of $B_1$ is simply the value transported from the point $\xi$, $y$, $z$. The values of $B_2$ and $B_3$ appear to be increased in the ratio $V/(V-v)$, which is the ratio in which also the density of the gas increases.

The values of $B_2$ and $B_3$ are discontinuous at the shock front. This has the consequence that when we calculate the components of the electric current, it is found that the shock front carries a concentrated current sheet. Moreover there is a system of currents in the domain behind the shock front. When the current system has been found, it is not difficult to recalculate the magnetic field from the current system. Both the continuous currents in the region behind the shock front, and the concentrated current sheet at the shock front produce fields ahead of the shock front as well as behind the shock front. It is found that the fields produced ahead of the shock front exactly cancel each other, so that here we are left with the original magnetic field, as it was given by (6). The fields produced in the wake of the shock front, when added together and combined with the original magnetic field, lead us back to the results given in (15a), (16a).

## 4. Case where the gas ahead of the shock front is non-conducting

We shall see in Section 5 that as soon as we introduce a (small) finite resistivity of the gas, so that the conductivity is no longer infinite, the concentrated current sheet becomes replaced by a system of strong currents in a thin layer just *ahead* of the shock front.

This has a curious consequence when we consider a case in which the gas ahead of the shock front should have no conductivity at all, while the gas behind the shock front has a very high (infinite) conductivity due to the fact that the passage of the shock produces a high temperature, with consequent high degree of ionization. Although formally the solution obtained above would hold, it is probable that it must be rejected from a physical point of view since the strong currents ahead of the shock front cannot exist in this case. Thus in the present case we must doubt the presence of a concentrated current sheet.

It is possible to find a solution which makes both the components of $A$ itself and their derivatives continuous at the shock front. The latter circumstance entails that there are no discontinuities in the components of the magnetic flux, and thus there are no infinitely concentrated electric currents. This solution is given behind the shock front $(x < Vt)$ by the formulas:

(17a)
$$A_1 = y/r^3$$
$$A_2 = -F_2(x) + \alpha\{F_2(\xi) - \xi/r_1^3\}$$
$$A_3 = -F_3(x) + \alpha F_3(\xi)$$

and ahead of the shock front $(x > Vt)$ by:

(17b)
$$A_1 = y/r^3$$
$$A_2 = -x/r^3 - (1-\alpha)\{F_2(\bar{x}) - \bar{x}/\bar{r}^3\}$$
$$A_3 = -(1-\alpha)F_3(\bar{x})$$

with the coefficient $\alpha$ equal to:

(18) $$\alpha = 2(V-v)/(2V-v)$$

while

$$\bar{x} = 2Vt - x; \quad \bar{r}^2 = \bar{x}^2 + y^2 + z^2$$

It will be seen that the values of $A_1$, $A_2$, $A_3$ ahead of the front now are different from their original values (7). This means that the current system obtained behind the shock front sends a magnetic field ahead of the front which is no longer compensated by a current sheet at the front. We may consider $\alpha$ as a coefficient of transmission, and $1 - \alpha = v/(2V-v)$ as a coefficient of reflection (note that, in consequence of the infinite value of $\kappa$ ahead of the front, Eqs. (13) do not hold there).

The components of the magnetic flux are given by the expressions, for $x < Vt$:

$$B_1 = -[2(V-v)/(2V-v)]3\xi z/r_1^5,$$
$$B_2 = -[2V/(2V-v)]3yz/r_1^5,$$
$$B_3 = +[2V/(2V-v)](r_1^2 - 3z^2)/r_1^5;$$

and for $x > Vt$:

$$B_1 = -3xz/r^5 + [v/(2V-v)]3\bar{x}z/\bar{r}^5,$$
$$B_2 = -3yz/r^5 - [v/(2V-v)]3yz/\bar{r}^5,$$
$$B_3 = (r^2 - 3z^2)/r^5 + [v/(2V-v)](\bar{r}^2 - 3z^2)/\bar{r}^5.$$

## 5. Influence of a small value of the resistivity of the gas

With a finite value of the resistivity of the gas Eqs. (8) must be replaced by the following ones:

(19)
$$\partial A_1/\partial t = -\partial \phi/\partial x + \kappa \Delta A_1$$
$$\partial A_2/\partial t + v(\partial A_2/\partial x) = v(\partial A_1/\partial y) - \partial \phi/\partial y + \kappa \Delta A_2$$
$$\partial A_3/\partial t + v(\partial A_3/\partial x) = v(\partial A_1/\partial z) - \partial \phi/\partial z + \kappa \Delta A_3$$

They can be used both behind and ahead of the shock front; in the latter case we have to put $v = 0$. It is not necessary to assume that the value of $\kappa$ is the same in the two domains, but within each domain $\kappa$ will be considered to be a constant. Again we can find a solution with $\phi = 0$ and $A_1 = y/r^3$, making $\Delta A_1 = 0$. The second and third equations of the system (19) now become linear equations of the second order for $A_2$ and $A_3$. It can be expected that the derivatives of $A_2$ and $A_3$ will assume large values near the shock front only, and moreover, that only the $x$-derivatives will become large. This makes it possible to substitute $\kappa(\partial^2 A_2/\partial x^2)$, $\kappa(\partial^2 A_3/\partial x^2)$ for $\kappa\Delta A_2$, $\kappa\Delta A_3$, respectively. We thus have two equations referring only to the $x$, $t$-plane.

The solution of these equations can be obtained by standard methods. We must treat the region ahead of the shock path $x = Vt$ separately from that behind the shock path, and in each case we must introduce two distributions of source functions of unknown strengths along the line $x = Vt$, one for the calculation of $A_2$ and another for the calculation of $A_3$. There are thus 4 unknown source functions along this line. Their values are obtained by writing down the equations expressing that $A_2$, $A_3$, $\partial A_2/\partial x$ and $\partial A_3/\partial x$ must be continuous at this line (attention is needed in calculating the values of various quantities). The calculations had been published in Technical Note BN–102 of June 1957 of the Institute for Fluid Dynamics and Applied Mathematics of the University of Maryland (copies no longer available). A simplified deduction of the result can be given as follows, taking the case of $A_2$ as an example. We use expressions (15a), (15b) given above, which satisfy Eqs. (19) with an error of order $\kappa$. They are continuous at the shock front, but their $x$-derivatives are discontinuous. To get rid of the discontinuity without disturbing the continuity of $A_2$ itself to more than a term of order $\kappa$, we add on one side of the shock an amount of the form

(*) $$a_2 = \kappa\alpha \exp[\beta(x - Vt)/\kappa],$$

where $\alpha$ and $\beta$ are supposed to be of normal order of magnitude. At the shock front $a_2$ is of order $\kappa$, which we consider as negligible; while its $x$-derivative is $\alpha\beta$, which has a finite value that can be adjusted so as to balance the jump in the value of $\partial A_2/\partial x$, resulting from the use of (15a), (15b). Now to satisfy Eqs. (19) in terms of the highest order (finite terms without $\kappa$), we need

$$-\beta(V - v) = \beta^2,$$

from which $\beta = -(V - v)$. It follows that the expression (∗) can be applied only at the side where $x - Vt > 0$, that is, in front of the shock. It then vanishes rapidly with increasing values of $x$. If used on the other side, one would obtain a divergent expression when going away further and further from the shock front. Thus it is found that the rapid change of the magnetic field, represented by the term (∗), must be located *in front* of the shock and it is here that we find the concentrated current system. This proves the assertion made in Section 4. As $v = 0$ in this region, we have $\beta = -V$, and the following result is obtained for the $B$'s:

(20)
$$\begin{aligned} B_1 &= B_{10} + \text{terms of order } \kappa \\ B_2 &= B_{20}[1 + \{v/(V - v)\} \exp{(-V\eta/\kappa)}] \\ B_3 &= B_{30}[1 + \{v/(V - v)\} \exp{(-V\eta/\kappa)}] \end{aligned}$$

Here $\eta$ stands for $x - Vt$, while $B_{10}$, etc. represent the values of the $B$'s given by Eqs. (6) or (16b) for $x = \xi$. At a distance ahead of the front of the order of a few times $\kappa/V$ the flux components have the values of the original field as given by Eqs. (6); while just at the front, where $\eta = 0$, the values agree with those given in Eqs. (16a) for a point on the front.

## 6. More accurate results for the region just ahead of the shock front, when account is taken of the reaction of the magnetic forces on the motion of the gas

In the preceding calculations it had been assumed that the motion of the gas was given; no attention has been paid to the influence of the magnetic forces on this motion. The results obtained make us expect that the most important influence will occur in the narrow region ahead of the shock front where concentrated electric currents are found. In order to investigate the main aspects of the flow in this narrow region we can make use of the theory developed for stationary one-dimensional shock waves in a magnetic field.

We assume that a plane shock wave is situated at the origin (in the $y$, $z$-plane) and we take the $x$-axis in the direction of the flow relative to the shock front. The gas speed is now measured with respect to the shock front (considered as stationary); it will be denoted by $u$, which takes the value $V$ on one side of the front, and $V - v$ on the other side (downstream), changing rapidly in a narrow domain. We take the $y$-axis in the direction of the magnetic flux component $B = (B_2^2 + B_3^2)^{\frac{1}{2}}$ parallel to the front. The electric field strength $E$ and the electric current $J$ will then be along the $z$-axis. Having regard to the signs involved in the vector formulas, Eqs. (1) and (2) reduce to

(21)
$$J = \sigma(E + uB)$$
$$\partial E/\partial x = \partial B/\partial t$$
$$\partial B/\partial x = \mu J$$

To these equations we must join the equations of motion for the gas, completed with terms representing the effect of the electromagnetic forces. Thus we have the equation of continuity:

(22) $$\partial \rho/\partial t + u(\partial \rho/\partial x) = -\rho(\partial u/\partial x)$$

the equation of momentum:

(23) $$\rho\{\partial u/\partial t + u(\partial u/\partial x)\} = -\partial(p - f)/\partial x - JB$$

($f$ is the viscous traction connected with $\partial u/\partial x$; $-JB$ is the ponderomotive force exerted by the magnetic field on an element of volume carrying an electric current); and the equation of energy:[1]

(24) $$\rho\{\partial U/\partial t + u(\partial U/\partial x)\} = -(p - f)(\partial u/\partial x) - \partial q/\partial x + J^2/\sigma$$

(where $q$ is the heat flow through conduction and $J^2/\sigma$ is the Jcule heat produced per unit volume in unit time).

We further have the equations of state (for an ideal gas), for the pressure: $p = RT\rho$, and for the internal energy per unit mass:

$$U = RT/(\gamma - 1)$$

[1] S. I. Pai, *Phys. Review* **105** (1957), pp. 1424–1426.

Since we suppose the field to be stationary, all partial derivatives with respect to $t$ drop out. The second equation of (21) then gives

$$E = \text{constant}$$

Elimination of $J$ between the first and third equations of (21) leads to

(25) $$uB - \kappa(\partial B/\partial x) = -E = \text{constant}$$

It is not necessary that $\kappa$ shall have a constant value, provided it remains small. We can neglect the term $\kappa(\partial B/\partial x)$ at distances upstream and downstream from the shock of an order large compared with $\kappa/u$ and thus we may write:

(25a) $$u_\text{I} B_\text{I} = u_\text{II} B_\text{II} = -E$$

where the subscripts I, II refer to the regions far upstream and downstream from the shock front, respectively. It should be kept in mind that the relation (25a) will not hold for $B_\text{I}$ when the domain upstream of the shock front should be nonconducting, for this implies an infinite value of $\kappa$; we come back to this point at the end of Section 8. For the present we keep to the assumption that the gas is highly conductive already upstream of the shock front. As regards signs, both $u$ and $B$ are positive quantities in our treatment; thus the constant must be positive and $E$ is a negative quantity (directed along the negative $z$-axis). It is expected that $B$ will not decrease when we move with the gas; hence we must require:

$$\partial B/\partial x \geqq 0$$

and thus in the domain where (25) holds:

$$uB \geqq u_\text{I} B_\text{I}.$$

## 7. Hugoniot equations for a shock in a magnetic field

When the partial derivatives with respect to $t$ are omitted, we can integrate Eqs. (22) and (23) with respect to $x$ (in the case of Eq. (23) we make use of the last relation of 21). We can also integrate the equation which is obtained after adding (23) multiplied by $u$ to (24). The results of the integrations are:

(26) $$\rho u = \rho_I u_I$$

(27) $$\rho u^2 + p - f + B^2/2\mu = \rho_I u_I^2 + p_I + B_I^2/2\mu$$

(28) $$\tfrac{1}{2}u^2 + \gamma RT/(\gamma - 1) - f/\rho + q/\rho_I u_I + BB_I/\mu\rho_I =$$
$$= \tfrac{1}{2}u_I^2 + \gamma RT_I/(\gamma - 1) + B_I^2/\mu\rho_I$$

(the last equation has been divided by $\rho_I u_I$).

In applying these equations we will assume that the viscosity and heat conductivity of the gas are much smaller than the resistivity coefficient $\kappa$. This may not always be the case. However, when the assumption holds, certain simplifications are possible, since we can exclude from consideration a narrow transition region of a thickness of the order of the viscosity, where $u$, $p$, and $T$ suffer rapid changes connected with large values of $f$ and $q$. Within this narrow region the behavior of the gas is governed wholly by the thermodynamic relations of ordinary shock wave theory; the magnetic field $B$ practically does not vary in this region and the forces derived from the magnetic field are insignificant in comparison with the pressure gradient. Outside of this narrow region we can neglect $f$ and $q$; the equations then allow us to investigate the effect of the ponderomotive force unhampered by effects of viscosity etc. With this simplification, Eqs. (26), (27), and (28) may be denoted as the Hugoniot equations for our problem (although the name Hugoniot equations is usually reserved for the equations relating the end states I and II only).

It is convenient to introduce dimensionless variables as follows:

(29)
$$u = zu_I; \quad \rho = \rho_I/z; \quad RT = \theta u_I^2;$$
$$RT_I = \theta_I u_I^2; \quad p = RT\rho = (\theta/z)\rho_I u_I^2;$$
$$B = (2\mu\rho_I u_I^2)^{\frac{1}{2}}h; \quad B_I = (2\mu\rho_I u_I^2)^{\frac{1}{2}}h_I$$

Equations (27) and (28) then lead to the following results:

$$\theta + z^2 + h^2 z = (1 + \theta_I + h_I^2)z;$$
$$[\gamma/(\gamma - 1)]\theta + \tfrac{1}{2}z^2 + 2hh_I = [\gamma/(\gamma - 1)]\theta_I + \tfrac{1}{2} + 2h_I^2$$

Elimination of $\theta$ gives a relation between $z$ and $h$:

(30) $\frac{1}{2}(\gamma + 1)z^2 - \gamma(1 + \theta_I + h_I^2 - h^2)z + \frac{1}{2}(\gamma - 1) + \gamma\theta_I -$
$\quad\quad -2(\gamma - 1)(h_I h - h_I^2) = 0$

This can still further be simplified if we write:

(31) $\begin{cases} \theta_I = (\gamma + 1)z_0/2\gamma - (\gamma - 1)/2\gamma; \\ h^2 = (\gamma + 1)b^2/2\gamma; \quad h_I^2 = (\gamma + 1)b_I^2/2\gamma; \\ k = 2(\gamma - 1)/\gamma \quad (<1). \end{cases}$

We then obtain:

(32) $\quad z^2 - (1 + z_0 + b_I^2 - b^2)z + z_0 - kb_I(b - b_I) = 0.$

It may be useful to observe that the quantity $\theta_I$, defined by the fourth of Eqs. (29), is equal to $1/(\gamma M_I^2)$ where $M_I$ is the ordinary Mach number (not taking account of magnetic effects) for the flow far ahead of the shock front. It is implicit in our assumptions that $M_I > 1$; hence we have $\theta_I < 1/\gamma$. Looking at the definition of $z_0$ given in (31) it follows that $z_0$ will be between the limits 1 and $(\gamma - 1)/(\gamma + 1)$. When the magnetic field is absent, Eq. (32) turns into a simple quadratic equation for $z$, having the roots $z = 1$ and $z = z_0$; we thus see that $z_0$ determines the speed which would be obtained behind an ordinary shock when no magnetic field was present.

## 8. Discussion of Equation (32)

Equation (32) is an algebraic equation in finite terms giving a relation between the velocity of the gas and the intensity of the magnetic field. It must hold throughout the flow, both ahead of and behind the shock front, with the exception of the very narrow region at the front itself where viscous forces and heat conductivity are the main operating agents. The equation must be used in conjunction with the differential equation (25) which connects distances along the $x$-axis with the velocity of the gas and the strength of the magnetic field. Written in terms of the dimensionless variables, Eq. (25) becomes:

(33) $\quad\quad\quad\quad db/dx = u_I(zb - b_I)/\kappa$

with the condition $zb \geqq b_I$.

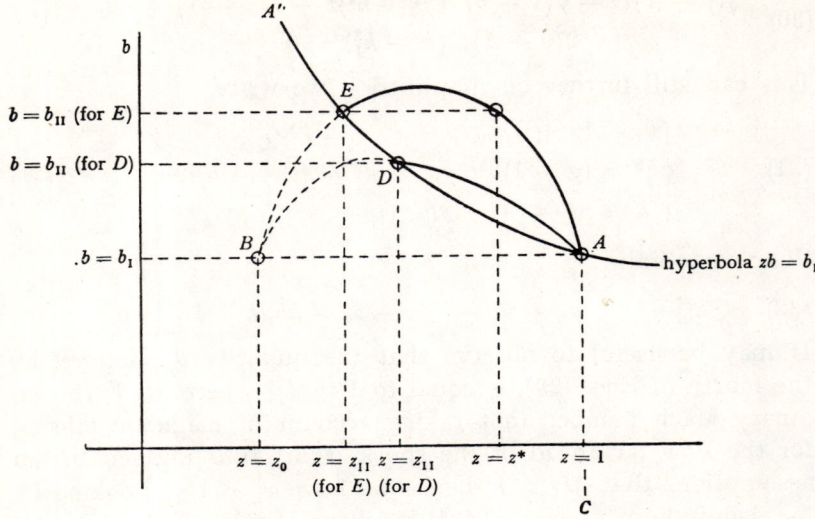

Fig. 2. Relation between $z$ and $b$ according to Eq. (32) or Eq. (34). In the diagram have been marked the points $A$ ($z = 1$; $b = b_I$), representing the state of the field far ahead of the front; and $B$ ($z = z_0$; $b = b_I$), which would be reached in a shock without change of the magnetic field strength. The curve $AA'$ is the hyperbola $zb = b_I$. The curves $AD$ and $AE$ give two possible forms for the upper branch of the function $b(z)$ as determined by (34); the curve $AD$ represents a case where $b$ does not pass through a maximum, while the curve $AE$ gives a case with a maximum for $b$. In both cases the intersection with the hyperbola $AA'$ determines the point $z_{II}$, $b_{II}$. The point $z = z^*$ on the curve $AE$ has also been marked. The point $C$ gives the second (negative) root for $b$ when $z = 1$; the lower branch of the function $b(z)$ passing through this point has not been indicated. Neither has the hyperbola $b = kb_I/2z$ been indicated; it may pass over $B$ or below $B$ depending on the value of $k/2z_0$.

It is convenient to consider $z$ as an independent variable in Eq. (32) and to solve for $b$. This gives:

(34) $\quad b = kb_I/2z \pm \{(1 - k/2z)^2 b_I^2 + [(1 - z)(z - z_0)]/z\}^{\frac{1}{2}}$

For $z = 1$ we find two values (compare Fig. 2): (*1*) $b = b_I$, which is the starting point $A$ (state of the field ahead of the shock wave); (*2*) $b = -(1 - k)b_I$, a point $C$ on the second branch of the curve; since this value is negative, this point has no physical meaning in our problem.

From (32) we can calculate the values of $db/dz$ and $d(zb)/dz$. It is found that the first quantity is negative at the point $z = 1$, $b = b_\mathrm{I}$, which means that a decreasing velocity is accompanied by an increasing magnetic field. The condition $zb \geq b_\mathrm{I}$ will be satisfied if $zb$ rises above $b_\mathrm{I}$; this requires $d(zb)/dz$ to be negative. This condition will be satisfied at the point $z = 1$, $b = b_\mathrm{I}$ provided:

$$b_\mathrm{I}^2 < \tfrac{1}{2}\gamma(1 - z_0)$$

a formula which can be transformed into a condition for $u_\mathrm{I}$:

(35) $$u_\mathrm{I}^2 > a_\mathrm{I}^2 + B_\mathrm{I}^2/\mu\rho_\mathrm{I} = (a_\mathrm{eff})_\mathrm{I}^2$$

Here $a_\mathrm{I}$ is the ordinary sound speed in the gas far upstream of the shock, while $(a_\mathrm{eff})_\mathrm{I}$ is an *effective sound speed* in that domain, which takes account of the influence of the magnetic field on the propagation of sound waves. We thus find that a solution can be obtained if the velocity of the gas upstream of the shock is supersonic with respect to the effective sound speed. This condition will certainly be satisfied far away from the dipole, where the magnetic field is weak. It is probable that when the shock wave approaches the domain where the magnetic field is strong, the condition will remain satisfied in consequence of an automatic adjustment of $u_\mathrm{I}$. (It should be kept in mind that $u_\mathrm{I}$ is the speed of the gas ahead of the front, relative to the shock front; in our previous picture this is the speed $V$ of the shock with respect to the gas at rest. The assertion is that the speed of the shock front automatically increases when the shock penetrates into a region where the magnetic field is stronger, and in this way adjusts itself so as always to satisfy the condition $V > (a_\mathrm{eff})_\mathrm{I}$.)

Formula (34) shows that there are two real values for $b$ so long as $1 > z > z_0$. The value with the negative sign before the radical, which has no physical meaning at the point $z = 1$, always remains below the hyperbola $zb = \tfrac{1}{2}kb_\mathrm{I}$, and thus also below the hyperbola $zb = b_\mathrm{I}$. Hence it is only the branch with the positive sign before the radical which can be used. This branch intersects the hyperbola $zb = b_\mathrm{I}$ in three points; one is $z = 1$ (the starting point); one point gives a negative value for $z$ and thus has no meaning for us; there remains a root $z_\mathrm{II}$ which is of importance. It is found that $z_\mathrm{II} > z_0$; also $z_\mathrm{II} < 1$ when condition (35) is satisfied.

It follows that the transition process occurring in the neighborhood of the shock front brings us from the state $z = 1$, $b = b_\mathrm{I}$ to the state determined by $z = z_\mathrm{II}$, with a corresponding value of $b_\mathrm{II}$. Upstream from the state I and downstream from the state II no further changes occur: here $db/dx = 0$ and also $dz/dx = 0$.

Having regard to the form of the curve defined by (32) it appears that two cases still must be distinguished, depending on the circumstance whether the curve (the branch corresponding to the positive root in 34) has or has not a maximum between $z = 1$ and $z = z_\mathrm{II}$. In the latter case (no maximum) we have $db/dz < 0$ along the entire curve. A continuous transition of velocity and magnetic field strength then is possible without the appearance of a real thermodynamic shock (sudden decrease of the velocity not accompanied by a change of the magnetic field). On the other hand when there is a maximum, it is highly improbable that the transition will follow the entire curve in such a way that $b$ first would increase to the maximum and then decrease again. In order to prevent this we follow the curve until the value $b = b_\mathrm{II}$ is reached for the first time; let this be for a value $z^*$, which is larger than $z_\mathrm{II}$. Now a thermodynamic shock can take place, bringing a sudden decrease of the velocity so that $z$ abruptly changes from the value $z^*$ to the value $z_\mathrm{II}$, while the magnetic field stays at the value determined by $b_\mathrm{II}$.

Whether the first or the second case will occur, depends on the value of $b_\mathrm{I}$. The second case certainly presents itself when the magnetic field is weak, since it must occur when there is no magnetic field at all.

A few words may be added concerning the case where the gas ahead of the shock wave would have zero conductivity. As has been remarked already earlier, we cannot say that $u_\mathrm{I} B_\mathrm{I} = -E$ in this case. It appears that Eq. (27) still holds, but in Eq. (28) $BB_\mathrm{I}/\mu\rho_\mathrm{I}$ must be replaced by $-EB/\mu\rho_\mathrm{I} u_\mathrm{I}$, and $B_\mathrm{I}^2/\mu\rho_\mathrm{I}$ by $-EB_\mathrm{I}/\mu\rho_\mathrm{I} u_\mathrm{I}$. We can keep to the definition of $h_\mathrm{I}$ as given at the end of Eqs. (29), but we must also introduce a new dimensionless quantity $\epsilon$ defined by

$$E = -\varepsilon u_\mathrm{I}(2\mu\rho_\mathrm{I} u_\mathrm{I}^2)^{\frac{1}{2}}$$

The final result is that Eq. (32) is replaced by:

(36)     $z^2 - (1 + z_0 + b_I^2 - b^2)z + z_0 - k\varepsilon^*(b - b_I) = 0$

with
$$\varepsilon^* = \varepsilon[2\gamma/(\gamma + 1)]^{\frac{1}{2}}$$

Since there are no electric currents ahead of the shock front, the magnetic field cannot change here; hence ahead of the shock $b$ retains the value $b_I$. In the very narrow domain of the shock itself, the magnetic field cannot change appreciably and magnetic forces are insignificant in comparison with the pressure gradient. The only solution to which Eq. (36) leads with $b = b_I$ is $z = z_0$, the value obtained from the ordinary nonmagnetic shock theory. It can be proved that no further change of the magnetic field and of the state of flow will occur with gradients of the order $\kappa^{-1}$ in the region behind the shock front.

## 9. Final remarks on the shock wave problem

Many further questions must be analyzed before the problem of the penetration of a shock wave into a magnetic field can be brought to a solution. Although we have given some attention to the effect of the magnetic field on the velocity of the gas near the shock front, this has been done on the simplifying assumption that the shock wave had a constant speed $V$. Actually both the flow of the gas and the propagation of the shock wave are affected by the magnetic forces. Since the equations are nonlinear, a full solution cannot be given, but it is possible to deduce some information concerning what will happen to the part of the shock front close to the $x$-axis (which part will remain normal to the $x$-axis), by approximating the situation through the use of a "model equation". For instance, we may consider the following system of equations (where $\alpha^2 = 1/\rho$):

$$\partial u/\partial t + u(\partial u/\partial x) + \alpha^2 B(\partial B/\partial x) = F(x);$$
$$\partial B/\partial t + u(\partial B/\partial x) + B(\partial B/\partial x) = 0.$$

The first equation is an equation of motion for an imaginary gas with constant density, not subjected to pressure forces, but only to a ponderomotive force $-JB = -B(\partial B/\partial x)$, due to the magnetic field; and to an imposed external force $\rho F(x)$. The second equation, for the magnetic field, follows from (21) if $\sigma = \infty$. The direction of $B$ is parallel to the $y$-axis, as before. We consider the domain $x \geqq 0$, $t \geqq 0$, and assume that at $t = 0$ we have $u = 0$ everywhere; $B = fx$, where $f$ (like $\alpha$) is a constant. We take $F = \alpha^2 f^2 x$, so as to balance the ponderomotive force of the magnetic field at $t = 0$. It is then assumed that from $t = 0$ onward a finite velocity $v$ is imposed upon the gas at $x = 0$. This will produce gas flow and wave motion, starting with a shock wave running into the gas (starting from the origin at $t = 0$). The history of the field can be described by means of the Riemannian equations

$$\{(\partial/\partial t) + (u \pm \alpha B)(\partial/\partial x)\}(u \pm \alpha B) = F.$$

When $v$ is small compared with the Alfvén velocity $\alpha B$ (which is the velocity of propagation of small disturbances of the magnetic field), an approximate solution can be constructed, giving the position of the shock front as

$$X = (v/\alpha f) \sinh z,$$

with shock front speed

$$V = dX/dt = (v/2) \exp z,$$

where $z$ is an auxiliary variable connected with $t$ by

$$\alpha f t = z + \tfrac{1}{2}[1 - \exp(-2z)].$$

The magnetic field strength immediately in front of the shock is $B_\mathrm{I} = fX = (v/\alpha) \sinh z$; and immediately behind the front, $B_\mathrm{II} = (v/2\alpha) \exp z$. The gas velocity in front of the shock is zero; that immediately behind it is $u = (v/2) \exp(-z)$. Evidently $B_\mathrm{II} > B_\mathrm{I}$, and

$$V > \alpha B_\mathrm{I} \quad \text{(locally supersonic speed);}$$
$$V - u = \alpha B_\mathrm{I} < \alpha B_\mathrm{II} \quad \text{(locally subsonic flow).}$$

We may consider this as an instance of a general rule: when a shock penetrates into a region of increasing magnetic field strength, the shock speed increases as a result of the increased "resilience" of the field, in such a way that it is always larger than the local effective sound speed, so that condition (35) remains satisfied. At the same time reflected waves appear; the speed of the gas behind the shock wave becomes smaller and the shock wave itself becomes weaker.

The field actually is always three-dimensional, and the parts of the shock front not close to the $x$-axis will be affected in a way different from those near the $x$-axis. The shock front consequently will not remain plane.

An interesting problem is whether a stationary flow pattern around the dipole may result sometime after the passage of the shock wave, assuming that this is followed by gas flow of uniform speed. It can be expected that in this case there will be found a domain around the dipole, where the magnetic field is so strong that it is impenetrable to the flow. This problem has obtained great interest in space science, as there actually exists such a domain around the Earth, into which the solar wind cannot penetrate. There are already numerous observations by satellites, which have given information concerning the shape of the so-called *magnetosphere*, and theoretical calculations by various authors have given results close to the observed form. Outside the magnetosphere the magnetic field of the Earth is swept away (although in the interplanetary space smaller magnetic forces are found due to the field of the Sun). Since the flow cannot penetrate into the magnetosphere, the latter can be compared to a solid body, and there is found an aerodynamic shock wave in front of it.

Space does not permit to pursue this subject further, and the reader is referred to the literature.[2]

## 10. The calculation of the electric current in an ionized gas subjected simultaneously to an electric and a magnetic field

At the end of Section 3.1 of his lectures Professor Goldstein

[2] Some attention to these problems had been given in the Technical Note BN–102 mentioned earlier; and in a contribution to "Magnetohydrodynamics", ed. by R. K. M. Landshoff (Stanford University Press, 1957). The subject has also been considered in a contribution on "The transportation of magnetic lines of force by a highly conducting fluid", in J. Soc. Industr. Appl. Math. **13**, pp. 189–199 and references (1965). This paper mentions publications on the Earth's magnetosphere by various authors, mainly in the J. Geophys. Res. until 1964.

has pointed out that when an ionized gas is subjected simultaneously to an electric and a magnetic field, the diffusion of the electrons and the ions will not be wholly in the direction of the electric field, since there may appear a drift velocity at right angles to the electric and magnetic fields. This will influence the direction of the electric current produced; consequently the behavior of the gas cannot be described with the aid of a scalar conductivity coefficient, but some form of tensor relation will result.

In the following pages we deduce expressions for the components of the electric current resulting from the motion of the electrons and the ions. It will be assumed that the gas has uniform density, temperature, and pressure, so that there is no diffusion due to pressure or temperature gradients. The treatment parallels that presented by S. Chapman and T. G. Cowling in *The Mathematical Theory of Non-uniform Gases*, Cambridge, 1939, Sections 18.4–18.42 (pp. 329–336), which, however, only refers to a completely ionized gas; and also that of T. G. Cowling, in *Magnetohydrodynamics*, Interscience Publishers, Inc., New York, 1957, Chapter 6 (pp. 99–108). It has been attempted to present the deduction of the equations as straightforward as possible, starting from the basic equation of diffusion theory, and to give the final equations in a simpler form than is found in these books. The deeper points connected with the precise calculation of the collision cross-sections involved are left aside; it should be observed that also Chapman and Cowling, in the sections quoted, content themselves with the formal introduction of "collision times."

## 11. Basic formulas of the theory of diffusion

When the electric field is described by the vector $\mathbf{E}$ and the magnetic field by the flux $\mathbf{B}$, the force $\mathbf{f}_s$ acting on a single molecule carrying an electric charge $\mathbf{e}_s$ and having the individual velocity $\boldsymbol{\xi}_s$, is given by:

(37) $$\mathbf{f}_s = \mathbf{e}_s\{\mathbf{E} + (\boldsymbol{\xi}_s \times \mathbf{B})\}$$

When there are $N_s$ molecules of type $s$ per unit volume, the mean force on these molecules per unit volume has the value:

(38) $$\mathbf{F}_s = N_s \mathbf{e}_s\{\mathbf{E} + (\mathbf{u} \times \mathbf{B}) + (\mathbf{w}_s \times \mathbf{B})\}$$

The quantities $\mathbf{u}$ and $\mathbf{w}_s$ appearing here are defined in the following lines.

We write $\mathbf{u}_s$ for the mean value of $\boldsymbol{\xi}$, taken over all the molecules of type $s$ in the element of volume considered. A mean mass velocity $\mathbf{u}$ is defined by means of the equation:

(39) $$\rho \mathbf{u} = \sum_s \rho_s \mathbf{u}_s$$

where $\rho_s = N_s m_s =$ the density of the molecules of type $s$ per unit volume, while $\rho = \sum_s \rho_s$ is the total density of the gas; finally we write:

(40) $$\mathbf{u}_s = \mathbf{u} + \mathbf{w}_s$$

so that $\mathbf{w}_s$ will represent the *mean diffusion velocity* of the molecules of type $s$ with respect to the mean mass flow. It follows from these definitions that

$$\sum_s \rho_s \mathbf{w}_s = 0$$

The "driving force" for the diffusion of the molecules of type $s$ is not $\mathbf{F}_s$, but is given by:[3]

(41) $$\boldsymbol{\psi}_s = \mathbf{F}_s - (\rho_s/\rho)\mathbf{F}$$

where

$$\mathbf{F} = \sum_s \mathbf{F}_s$$

[3] When the composition of the gas mixture is not uniform, so that there are gradients in the number density of the constituents and pressure gradients, the expression for the "driving force" must be amplified by terms depending on the pressure gradients in the following way:

$$\boldsymbol{\psi}_s = \mathbf{F}_s - (\rho_s/\rho)\mathbf{F} - \nabla p_s + (\rho_s/\rho)\nabla p$$

($p_s$ = partial pressure of the gas labeled $s$; $p = \Sigma_s p_s$ = total pressure). The quantity so defined is equivalent to $-p\mathbf{d}$ in the notation applied by Chapman and Cowling, *op. cit.*, p. 140, formula 7 (compare also the "Notes added in 1951," p. 408, formula 1). The same notation is applied by J. O. Hirschfelder in Section D "The Transport Properties of Gases and Gaseous Mixtures" of "Thermodynamics and Physics of Matter," edited by F. D. Rossini, *High Speed Aerodynamics and Jet Propulsion*, Vol. I, Princeton, 1955, p. 345, formula 2–3.

The second term represents a correction necessary in order to eliminate the effect of the forces $\mathbf{F}_s$ upon the mass flow. It will be seen that

$$\sum_s \psi_s = 0$$

The diffusion equations can now be written:[4]

(42) $$\psi_s = \sum_t K_{st}(\mathbf{w}_s - \mathbf{w}_t)$$

where the coefficients $K_{st}$ determine the resistances connected with the relative motions of the various constituents of the gas mixture. We have $K_{st} = K_{ts}$.

Evidently all the intricacies of the molecular theory of diffusion go into the values of the coefficients $K_{st}$. In order to relate these coefficients to collision cross-sections, we write $S_{st}(g)$ for the collision cross-section referring to the collision of molecules $s$ and $t$ with relative velocity $g$. We introduce the mean collision cross-section appropriate for transfer of momentum:

(43) $$Z_{st} = Z_{ts} = (\pi^{\frac{3}{2}}\alpha^3)^{-1} \int_0^\infty 4\pi g^2 \exp(-g^2/\alpha^2) \cdot (g^3/\alpha^3) S_{st}(g) dg$$

where

$$\alpha = (2kT/\mu)^{\frac{1}{2}},$$

with

$$\mu = m_s m_t/(m_s + m_t).$$

[4] These equations are equivalent with Eqs. (2–10) given by J. O. Hirschfelder et al. in the book mentioned in the preceding reference, when the terms referring to thermal diffusion due to a temperature gradient are omitted. The coefficient $K_{st}$ used in (42) of our text takes the following form in the notation used by Chapman and Cowling and by Hirschfelder:

$$K_{st} = N_s N_t p/N^2 D_{st}$$

where $D_{st}$ is the coefficient of diffusion for a mixture of the gases $s$ and $t$. This is most easily seen from Eq. 7, p. 144, of Chapman and Cowling's book. These authors consider diffusion in a binary mixture; the formulas, however, can be used for an arbitrary mixture, provided we keep in mind that $N = \Sigma_s N_s$ is the total number of molecules per unit volume, and that $p$ is the total gas pressure, so that $p = NkT$.

The value of $K_{st}$ is then given by:[5]

(44) $$K_{st} = \tfrac{2}{3}\alpha\mu N_s N_t Z_{st}$$

It is possible to write:

(44a) $$K_{st} = \mu N_s/\tau_{st}$$

and we may call $\tau_{st}$ the "mean collision time" for collisions of a molecule $s$ with all the molecules $t$ around it. We then have:

(44b) $$(\tau_{st})^{-1} = \tfrac{2}{3}\alpha N_t Z_{st}$$

It should be observed that $\tau_{st} \neq \tau_{ts}$.

## 12. Application to an ionized gas

We consider a gas consisting of atoms of a single species, of mass $m$, which partly are dissociated into electrons of mass $m_e$ (very small in comparison with $m$) and positive ions of a mass $m - m_e$. We suppose that the ions have unit positive charge, $+e$, the electrons having unit negative charge, $-e$, so that the numbers of the positive ions and the electrons must be the same. The numbers per unit of volume of neutral atoms, positive ions and electrons can be represented by

---

[5] The "first approximation" to the value of $D_{st}$ is given by Chapman and Cowling in the following form on p. 165, Eq. 1:

(1) $$D_{st} = 3E/2Nm_0 = 3kT(m_s + m_t)/16Nm_s m_t \Omega_{st}^{(1)}(1)$$

when use is made of the definitions of $E$ (p. 164) and of the quantities $m_0$, $M_1$, $M_2$ introduced by Chapman and Cowling (p. 54). Referring to p. 157, Eqs. 2 and 3, for the definition of the quantities $\phi_{st}^{(1)}$ and $\Omega_{st}^{(1)}(1)$, we find that in our notation:

$$\phi_{st}^{(1)} = (g/2\pi)S_{st};$$
$$\Omega_{st}^{(1)}(1) = (\alpha/8)Z_{st}$$

With these results it is easy to verify our formula (44) for $K_{st}$. Apart from the factor 2/3 it represents the product of an average thermal velocity $\alpha$, depending on the mean mass $\mu$, into that mean mass and the product of the number densities $N_s$, $N_t$, and the mean collision cross-section $Z_{st}$.

The circumstance that the weight factor in Eq. (43) for the mean collision cross-section $Z_{st}$ is the product of the Maxwellian distribution function for the mean relative velocity $g$ into some power of $g/\alpha$, is connected with the laws governing the transfer of momentum in collisions.

$$(1-\beta)N,\ \beta N,\ \beta N$$

respectively, where $\beta$ is the degree of ionization.

The electromagnetic forces per unit volume on the three species of particles are given by:

neutral atoms: $\quad\mathbf{F}_1 = 0;$

positive ions: $\quad\mathbf{F}_2 = \beta Ne\ \{\mathbf{E} + (\mathbf{w}_2 \times \mathbf{B})\};$

electrons: $\quad\mathbf{F}_3 = -\beta Ne\ \{\mathbf{E} + (\mathbf{w}_3 \times \mathbf{B})\}$

We have omitted the term with the mass velocity $\mathbf{u}$; it can be supposed, either that there is no mass velocity, or else that the $\mathbf{E}$ used above stands for $\mathbf{E} + \mathbf{u} \times \mathbf{B}$.

We have:

$$\mathbf{F} = \mathbf{F}_1 + \mathbf{F}_2 + \mathbf{F}_3 = \beta Ne(\mathbf{w} \times \mathbf{B})$$

where we have written:

$$\mathbf{w} = \mathbf{w}_2 - \mathbf{w}_3$$

With sufficient approximation the following expressions can be used for the densities of the three components:

atoms: $\quad\rho_1 = (1-\beta)Nm;$

positive ions: $\quad\rho_2 = \beta N(m - m_e) \cong \beta Nm;$

electrons: $\quad\rho_3 = \beta Nm_e$

summing up to the total density $\rho = Nm$.

The values of the "driving forces" to the same approximation become:

(45)
$$\begin{aligned}\psi_1 &= -(1-\beta)\mathbf{F} = -\beta(1-\beta)Ne(\mathbf{w} \times \mathbf{B}) \\ \psi_2 &= \beta Ne\{\mathbf{E} + (\mathbf{w}_2 \times \mathbf{B})\} - \beta^2 Ne(\mathbf{w} \times \mathbf{B}) \\ \psi_3 &= -\beta Ne\{\mathbf{E} + (\mathbf{w}_3 \times \mathbf{B})\}\end{aligned}$$

making

$$\psi_1 + \psi_2 + \psi_3 = 0$$

We now must solve the equations:

(46)
$$\begin{aligned}\psi_1 &= K_{12}(\mathbf{w}_1 - \mathbf{w}_2) + K_{13}(\mathbf{w}_1 - \mathbf{w}_3) \\ \psi_2 &= K_{12}(\mathbf{w}_2 - \mathbf{w}_1) + K_{23}(\mathbf{w}_2 - \mathbf{w}_3) \\ \psi_3 &= K_{13}(\mathbf{w}_3 - \mathbf{w}_1) + K_{23}(\mathbf{w}_3 - \mathbf{w}_2)\end{aligned}$$

together with:

(47) $$\rho_1 \mathbf{w}_1 + \rho_2 \mathbf{w}_2 + \rho_3 \mathbf{w}_3 = 0$$

It is convenient to solve for $\mathbf{w}$, provisionally keeping the cross product $\mathbf{w} \times \mathbf{B}$ in the result, as if it were a known quantity. We then obtain:

(48) $$\mathbf{w} = \xi\{\mathbf{E} - \eta(\mathbf{w} \times \mathbf{B})\} + \beta^2(1-\beta)^2 \frac{N^2 e^2}{Q}(\mathbf{w} \times \mathbf{B}) \times \mathbf{B}$$

where the following abbreviations have been used:

(49a) $\quad Q = K_{12}K_{13} + K_{12}K_{23} + K_{13}K_{23}$

(49b) $\quad \xi = \beta N e (K_{12} + K_{13})/Q;$

(49c) $\quad \eta = (K_{12} - K_{13} + 2\beta K_{13})/(K_{12} + K_{13})$

Instead of solving the vector equation in a formal way, we introduce a rectangular system of coordinates with the $x$-axis in the direction of the magnetic flux $\mathbf{B}$. The components of the magnetic flux then have the values:

$$B, 0, 0$$

those of the cross product $\mathbf{w} \times \mathbf{B}$:

$$0, w_z B, -w_y B$$

and those of the product $(\mathbf{w} \times \mathbf{B}) \times \mathbf{B}$:

$$0, -w_y B^2, -w_z B^2$$

With the further abbreviation:

(49d) $$\zeta = \beta^2(1-\beta)^2 N^2 e^2 B^2/Q$$

the following equations are obtained:

(50) $$\begin{aligned} w_x &= \xi E_x \\ (1+\zeta)w_y &= \xi(E_y - \eta w_z B) \\ (1+\zeta)w_z &= \xi(E_z + \eta w_y B) \end{aligned}$$

from which:

(51) $$\begin{aligned} w_y &= \xi[(1+\zeta)E_y - \eta\xi B E_z]/[(1+\zeta)^2 + \eta^2 \xi^2 B^2] \\ w_z &= \xi[(1+\zeta)E_z + \eta\xi B E_y]/[(1+\zeta)^2 + \eta^2 \xi^2 B^2] \end{aligned}$$

The resulting electric current is given by

(52) $$\mathbf{i} = \beta N(e\mathbf{w}_2 - e\mathbf{w}_3) = \beta Ne\mathbf{w}$$

Hence the equation for the $x$-component takes the form:
(53) $$i_x = \sigma_0 E_x$$
where $\sigma_0$ represents the conductivity in so far as it is not influenced by the magnetic field, with the value:

(54) $$\sigma_0 = \beta Ne\xi = \beta^2 N^2 e^2 (K_{12} + K_{13})/Q$$

## 13. Discussion and simplification of the formulas obtained

The values of the coefficients $K_{12}$, $K_{13}$, $K_{23}$ can formally be derived from Eq. (44). We introduce "mean collision times" by means of formulas analogous to Eq. (44b). In order to avoid the difficulty which is due to the nonsymmetric character of the $\tau_{st}$, leading to 6 different quantities, we shall be content with three "mean collision times", viz.

$\tau_i$: mean collision time for a positive ion, colliding with a neutral atom;

$\tau_e$: mean collision time for an electron, colliding with a neutral atom;

$\tau$: mean collision time for a positive ion colliding with an electron, or for an electron colliding with a positive ion (in this case the symmetry exists).

With this choice the quantity $N_s$ in Eq. (44a) in all three cases will have the value $\beta N$. We thus arrive at:

$$K_{12} = \tfrac{2}{3}[2kT/(m/2)]^{\frac{1}{2}}(m/2)\beta(1-\beta)N^2 Z_{12} = (m/2)\beta N(\tau_i)^{-1}$$
(55) $$K_{13} = \tfrac{2}{3}(2kT/m_e)^{\frac{1}{2}}m_e\beta(1-\beta)N^2 Z_{13} = m_e\beta N(\tau_e)^{-1}$$
$$K_{23} = \tfrac{2}{3}(2kT/m_e)^{\frac{1}{2}}m_e\beta^2 N^2 Z_{23} = m_e\beta N(\tau)^{-1}$$

Evidently $\tau_i$ and $\tau_e$ will become infinite when $\beta = 1$ (gas completely ionized and neutral atoms completely absent).

The precise calculation of the mean collision cross-sections $Z_{12}$, $Z_{13}$ and $Z_{23}$ involves great difficulties. The cross-section $Z_{23}$ in particular depends on the effect of the Coulomb attraction between positive ions and electrons; since this attraction is a long range force, the value of $Z_{23}$ is dependent on the number of particles per unit volume. On the other hand, the cross-sections $Z_{12}$ and $Z_{13}$ refer to collisions of one charged particle with a neutral particle; although multipole effects (induction effects) may play a part, it is probable that these cross-sections still will be closely related to the diameters of the neutral particle. It seems reasonable therefore to expect that $Z_{12}$ and $Z_{13}$ will not differ in order of magnitude. If we now compare the expressions for $K_{12}$ and $K_{13}$ it will be seen that the factors occurring in them are either the same, or of the same order of magnitude, with the exception of the masses $m$ and $m_e$. We consequently may expect that $K_{13}$ will be much smaller than $K_{12}$, approximately in the ratio $(m_e/m)^{\frac{1}{2}}$. This result makes it possible to neglect $K_{13}$ in the combinations $K_{12} + K_{13}$ and $K_{12} - K_{13}$. We then obtain:

(56a) $\qquad Q \cong K_{12}(K_{13} + K_{23})$

(56b) $\qquad \xi \cong \beta N e/(K_{13} + K_{23})$

(56c) $\qquad \eta \cong 1$

(56d) $\qquad \zeta \cong \beta^2 (1-\beta)^2 N^2 e^2 B^2/\{K_{12}(K_{13} + K_{23})\}$

(56e) $\qquad \sigma_0 \cong \beta^2 N^2 e^2/(K_{13} + K_{23})$

A further simplification is obtained by introducing the mean collision times $\tau_i$, $\tau_e$ and $\tau$ into the formulas. We combine the latter two into

(57) $\qquad \tau_m = \tau \tau_e/(\tau + \tau_e)$

which represents an average collision time for collisions between electrons and either positive ions (to which refers $\tau$) or neutral atoms (to which refers $\tau_e$). In the case of complete ionization ($\beta = 1$), which makes $\tau_e$ infinite, the value of $\tau_m$ is equal to $\tau$.

We also introduce:

$$\omega = eB/m_e$$

which is the angular velocity for the spiraling motion of an electron around the lines of magnetic flux. We can then write:

(58) $\qquad \sigma_0 = \beta N (e^2/m_e) \tau_m$

(59a) $\qquad \eta \xi B = \omega \tau_m$

(59c) $\qquad \zeta = (1-\beta)^2 (2m_e/m) \omega^2 \tau_i \tau_m$

and obtain:

(60)
$$i_x = \sigma_0 E_x$$
$$i_y = \sigma_0[(1+\zeta)E_y - \omega\tau_m E_z]/[(1+\zeta)^2 + \omega^2\tau_m^2]$$
$$i_z = \sigma_0[(1+\zeta)E_z + \omega\tau_m E_y]/[(1+\zeta)^2 + \omega^2\tau_m^2].$$

These results have a simpler form than the equations given by Cowling (*op. cit.* p. 107, Eqs. 6–22). It will be seen that the relation between the current strength and the electric field in the direction of the magnetic field is not affected by the presence of the latter, and that it depends on the degree of ionization and on the average collision time $\tau_m$, defined by (57). The components transverse to the direction of the magnetic field suffer the influence of the latter in two ways:

(*1*) through the coefficient $\omega\tau_m$, depending on the product of the spiral frequency of the electrons and the average collision time $\tau_m$; this coefficient becomes large when the electron can describe many cycles between two successive collisions;

(*2*) through the coefficient $\zeta$, which also depends on the mean collision time $\tau_i$ for the positive ions. This coefficient vanishes when the gas is completely ionized (it is true that $\tau_i$ becomes infinite in that case, but only as $(1-\beta)^{-1}$; hence $(1-\beta)^2\tau_i$ goes to zero). As Cowling observes, *op. cit.*, p. 107, 108, the effect represented by this coefficient indicates the presence of a dissipation mechanism in a partially ionized gas, which is absent in a wholly ionized gas.

Equations (60) illustrate what is meant by the "tensorial" character of the electric conductivity in the presence of a magnetic field.[6]

---

[6] Since this was written the subject of resistance coefficients, diffusion and electric conductivity has been treated in detail in my book "Flow Equations for Composite Gases" (Academic Press Inc., New York, 1969), in which also the problem of the cross-sections for electrostatic interaction between charged particles has been discussed. Results for a partially ionized gas are given in Sections 38 (pp. 180–190) and 41 (pp. 205–208). It has been checked that the results there obtained agree with Eqs. (60) and (48) given above (there are some differences in notation, and Gaussian units for electric and magnetic quantities have been used in the book). The text in the book at the same time gives attention to the features which lead to thermal diffusion. For their effect on the resistance coefficients for a partially ionized gas in a magnetic field, see A. C. Pipkin, The Physics of Fluids **4** (1961), pp. 154–158.

# Author Index

Abraham, M., 24, 55
Adams, Mac C., 183, 193
Alden, H. L., 138, 140
Alpher, R. A., 211
Ames research staff, 264
Armstrong, A. H., 236
Batchelor, G. K., 61, 93, 134, 168, 169, 271, 272
Becker, R., 24
Beckett, C. W., 212
Bethe, H. A., 211
Bhatnagar, P. L., 221
Birkhoff, G., 245
Bond, R., 122
Brillouin, L., 227
Bryson, A. E., 183
Burgers, J. M., 245, 273, 302
Burnett, D., 98
Carrier, G. F., 144, 145
Chaplygin, S., 177
Chambré P. L., 99, 100, 208
Chandrasekhar, S., 272
Chapman, D. R., 156
Chapman, S., 92, 294, 295, 296, 297
Chester, W., 182
Chu, Boa-Teh, 48
Cole, J. D., 182
Coles, D., 168
Compressible Flow Tables Panel, (British) Aeronautical Research Council, 264
Courant, R., 78, 79, 86, 171, 264, 271
Cowan, G. R., 114
Cowling, T. G., 92, 271, 294, 295, 296, 297, 302
Crocco, L., 70, 162
Davies, R. M., 93
Dijkstra, D., 145
Dorrance, W. H., 214
Durand, W. F., 245
Eisenberg, P., 245
Elliot, D., 122
Emmons, H. W., 272

Engel, A. von, 211

Falkner, V. M., 128
Finn, R., 178
Flugge, W., 183
Fowler, R. H., 95, 211
Fox, H., 144
Frankl, F. I., 171, 272
Friedrichs, K. O., 78, 79, 86, 171, 264, 271
Geiringer, H., 271
Gilbarg, D., 112, 178
Gilmore, F. R., 212
Glauert, H., 245
Glauert, M. B., 122
Goldstein, S., 61, 128, 137, 138, 141, 271, 273, 293
Grad, H., 99, 112
Greene, E. F., 114
Gross, E. P., 221
Guggenheim, E. A., 95
Gunn, J. C., 211

Hantzsche, W., 163
Hardy, G. H., 129
Hartree, D. R., 128, 129
Hayes, W. D., 182
Helmboltz, H. von, 244
Hertzberg, A., 214
Hilsenrath, J., 212
Hinze, J. I., 272
Hirschfelder, J. O., 295, 296
Horning, D. F., 114
Howarth, L., 39, 70, 92, 103, 171, 175, 271
Illingworth, C. R., 45, 110, 112, 159, 162, 163, 263, 264
Imai, I., 133, 144, 145, 179
Jones, R. T., 144
Kantrowitz, A., 211
Kaplun, S., 110, 127, 144
Karman, Th. von, 171, 209, 236, 245, 251
Karpovich, E. A., 171, 272
Kawaguti, M., 133

Kelly, H. R., 122
Kennard, E. H., 99
Kirchoff, G., 244
Kochin, N. E., 138, 141
Kopal, Z., 181
Krook, M., 215, 217, 221
Kuo, Y. H., 126
Kurzweg, H. H., 214

Lagerstrom, P. A., 110, 144
Lamb, H., 4, 6, 18, 25, 31, 45, 73, 74, 107, 177, 197, 199, 233, 240, 243, 244, 245, 271
Landau, L. D., 217, 219, 272
Landshoff, R. K. M., 293
Landweber, L., 244
Laurmann, J. A., 245
Lees, L., 148
Levi-Civita, T., 244
Lewy, H., 144
Libby, P. A., 144
Liepmann, H. W., 93, 111, 112, 171, 175, 209, 211, 262, 264, 271
Lifshitz, E. M., 272
Lighthill, M. J., 92, 97, 112, 114, 122, 166, 171, 177, 199, 211, 212
Lin, C. C., 145, 169, 272
Logan, J. G., 214
Love, A. E. H., 227
Ludford, G. S. S., 176, 271

Mangler, W., 122, 130, 152
Marble, F. E., 209
Mills, R. H., 130
Milne-Thomson, L. M., 25, 233, 242, 245, 271
Mises, R. von, 154, 271
Møller, C., 56
Moore, F. K., 115, 272
Morawetz, C. S., 180
Morgan, A. J. A., 110
Murray, J. D., 144, 145

Pai, S. I., 122, 284
Paolucci, D., 112
Pauli, W., 55
Pearson, J. R. A., 110
Pipkin, A. C., 302
Prandtl, L., 4, 73, 271
Prigogine, I., 97

Probstein, R. F., 122, 182
Proudman, I., 110

Rayleigh, Lord, 83, 244
Riemann, B., 83
Robinson, A., 245
Romig, M. F., 212
Rosenhead, L., 97, 115, 264, 272
Roshko, A., 93, 111, 112, 171, 175, 209, 211, 245, 262, 264, 271
Rubesin, M. W., 156

Seban, R. A., 122
Schaaf, S. A., 99, 100, 144, 208
Schiffer, M., 144
Schlichting, H., 105, 107, 115, 271
Schubauer, G. B., 168
Sears, W. R., 115, 116, 171, 183, 193, 251, 271
Sherman, F. S., 114
Skan, S. W., 128
Sommerfeld, A., 24, 31, 47, 52, 216, 271
Spitzer, L., Jr., 215, 217, 221, 272
Squire, H. B., 105, 133, 147
Stewartson, K., 122, 129, 159
Stratton, J. A., 216

Talbot, L., 114, 210
Taylor, G. I., 169, 242, 243
Teller, E., 211
Tietjens, O. G., 4, 271
Townsend, A. A., 168, 169, 272

Uhlenbeck, G. E., 210

Van de Vooren, A. I., 145
Van Driest, E. R., 168
Van Dyke, M., 137, 144, 145, 272

Wang-Chang, C. S., 210
Ward, G. N., 171, 183, 193, 201, 271
Watson, G. N., 193, 219, 220
Wendt, H., 163
Weyl, H., 129
Whitaker, E. T., 219, 220
Wilson, R. E., 214

Yih, C. S., 244
Yoshizawa, A., 145
Young, A. D., 45, 147, 148, 165, 168

Zarantonello, E. H., 245

# Subject Index

Acceleration, 3, 5
Accumulation of electric charge, 61, 215
Added inertia, 244
Aerothermochemistry, 209
Air, thermodynamic properties of, 212
Airfoil (in two dimensions), lift coefficient of, 242
Airfoil theory, three-dimensional, 245
 variable motion, 251
Alfvén speed, 292
Analysis of motion in a neighbourhood, 13, 223
Angular momentum, equation of, 38
Angular momentum of small portion of fluid, 17
Aspect ratio, small, plane airfoils of, 202
Asymptotic approximations at large Reynolds numbers, 132
Asymptotic solution for flow past a flat plate, 137
Axially symmetrical motion, 29, 30, 77, 121, 122

Bernoulli's equation, 65, 66
Blasius function, 123, 126, 138
Body forces, removal or neglect of, 44
Body of revolution, flow past, 29, 121, 122, 183, 198, 234
Boltzmann's equation, 93, 217
Book list, 271
Boundary conditions, 10, 99
 in slender-body theory, 190, 195, 197, 200, 201
Boundary layer along a flat plate, 123
 extended theory, 137
 solution in parabolic corrdinates, 125, 126

Boundary layer, displacement thickness of, 119
Boundary-layer equations, two-dimensional, 117, 121
 solutions with similar velocity distributions, two-dimensional, 127
 axisymmetrical, 130
Boundary-layer flow of a gas along a flat plate (zero pressure gradient), 162
Boundary-layer flow at a forward stagnation point, 109
Boundary-layer separation, 119
Boundary-layer theory for incompressible fluids, 115
Boundary layers, hypersonic, 148
Boundary layers on bodies of revolution, 121, 122
Boundary layers in gases, 147
 displacement thickness of, 161
 equations for, in axisymmetric flow, 152; Mangler's transformation to two-dimensional flow equations, 152
 in two-dimensional flow, 149; von Mises's transformation of, 154; transformation to equations for a fluid of constant properties, 158
 integrals of the energy equation in, with unit Prandtl number, 157
Bulk viscosity, coefficient of, 88, 96, 97, 98, 114

Cast-off vortex, 246
Cauchy's equations for the vorticity in an inviscid fluid, 74
Centered simple wave, 266
Characteristics, 86, 178, 181, 266
Charge, accumulation of electric, 61, 215

Charge density, electric, equation for, 60
Chemistry of air at high temperatures, 212
Circular cone, flow past, 181
Circulation, 18
Circulation, constancy of, for inviscid fluids, 70
Circulation in doubly (or multiply) connected spaces, 25
Circulation in viscous fluid, 91
Circulation, rate of change of, 23
Collision cross sections, 294, 296
Collision times, 297, 300 ff
Combustion processes, 209
Concentration gradients and currents in gas mixtures, 208
Condensation, 209
Conductivity, Electric, 47, 273, 276, 281, 293 ff
Conservation of mass, equation of, 5
Contact discontinuity, 78
Continuity, equation of, 5
Convective rate of change, in Eulerian specification, 5
Convex corner, supersonic flow round, 266
Corner, supersonic flow past, 82
Couette flow, 108
  of a viscous gas, 111
Couples on an immersed solid body (potential flow of an inviscid fluid of constant density), 238
Crocco's stream function, 70
Cylinder, two-dimensional steady flow of an inviscid gas past, 178
Deflection, expansive, 266
  small compressive, by a shock, 262
  small expansive (convex corner), 269
  small, in hypersonic flow in Prandtl-Meyer expansion, 269

  small, in hypersonic flow through a shock, 263
  through an oblique shock wave, 260
Diffusivity of heat, 96
Diffusion, 294
Dimensionless groups, 101
Displacement current, neglect of, 60, 273
Displacement thickness of a boundary layer, 118, 161
Dissipation of mechanical energy, rate of, 42, 89, 90, 91
Dissociation, 209, 211
Distortion of a fluid element, rate of, 13
Doublets in potential flow (incompressible fluid), 229
Doublets, method of, for cross-flow past a body of revolution (potential flow, incompressible fluid), 236
Drag, 102
  induced, 250
Drag coefficient, 102
Dynamical similarity, 102
Eckert number, 105
Electric charge accumulation, 61, 215
Electric charge density, equation for, 60
Electric conductivity, 47, 273, 276, 281, 293 ff
Electrically conducting fluids, 48, 58
Electrodynamics, 47, 274
  of a medium at rest, 50
  of moving media, 51
Electromagnetic energy, 50
Electromagnetic momentum, 50
Electrostatic waves in an ionized gas, 215
Ellipses, slender-body theory for flow past a body whose sections are, 195
Ellipsoid, flow past, 197
Energy, dissipation of mechanical, 43, 89

# SUBJECT INDEX

Energy, equation of, for general fluid, 39
  with electric and magnetic forces, 59
  integrals of, for gas boundary layers with unit Prandtl number, 157
Enthalpy-velocity relation across a boundary layer, 165
Entropy, gain in, through a shock wave, 114, 258
Equation of continuity, 5
Eulerian specification, 3
Exact solutions for the flow of a viscous liquid, 107
  of a viscous gas, 110
Expansion, coefficient of, 104
Expansive deflection, 266
Extended boundary-layer theory (for flow along a flat plate), 137
Filament lines, 4
Finite amplitude, sound waves of, 83
Flat plate, boundary layer in flow along, 123
  boundary-layer flow of a gas along, 162
  extended boundary-layer theory for flow along (asymptotic solution), 137
Force coefficients, 102
Forces on an immersed solid body (potential flow of an inviscid fluid of constant density), 238
Fourier transforms (slender-body theory for subsonic flows), 193
Frames of reference, 1, 52, 58
Free-streamline theory, 133, 177, 244
Froude number, 102
Frozen magnetic fields, 77
Gain in entropy through a shock wave, 114, 258
Grashof number, 104
Gravity, effect of, 103

Gravity forces, removal or neglect of, 44
Hall current, 49, 294 ff
Heat capacity lag, 95
Heat flux vector, 90, 93
Heat produced by compression and dissipation of energy, 147 151
Heat transfer, non-dimensional coefficients, 104
Heaviside operators, 199
  slender-body theory, supersonic flow, 200
Helmholtz's equations for the vorticity in an inviscid fluid, 73
Helmholtz's theorems on the vorticity in an inviscid fluid, 71
High temperature effects in gases, 207
Hodograph transformation, 176, 179
Hugoniot-Rankine relations for a shock, 78, 113, 285, 286
Hydrodynamic inertia, 244
Hydromagnetics, 47, 58, 273
Hydrostatic pressure, 44
Hypersonic approximation for small deflection, through a shock, 263
  by a Prandtl-Meyer expansion, 269
Hypersonic flow, 148, 182
Illingworth-Stewartson theorem for gas boundary layers, 158
Impulse of a motion of an incompressible fluid, 237
Impulse of a vortex pair, 249
Impulse of a vortex sheet, 248
Impulsive generation of motion, in incompressible fluid, 46
Incompressible fluid, 7
Induced drag, 250
Inertia, added or hydrodynamic, 244
Inviscid fluid, Bernoulli's equation for, 65, 66

conditions for steady motion of, 64, 66
constancy of circulation in, 70
equations of motion of, 63
  referred to moving axes, 67
persistence of irrotational motion in, 72, 75
vorticity in, 71, 73, 74
Inviscid gases, dynamics of, 171
Ionization, 209, 211
Ionized gas, 49, 293 ff
  longitudinal waves in, 215
Irrotational motion of an incompressible fluid, 229, 233
  forces and couples on an immersed solid body, 238
  impulse, 237
  kinetic energy, 236
Irrotational motion in an inviscid fluid, persistence of (Lagrange's theorem), 72, 75
Irrotational motions, 25
  in an inviscid gas with constant entropy, 68
  subsonic, uniqueness theorem for, 178
Jets, 177, 244, 245
Joukowski-Kutta condition, 241
Joukowski-Kutta formula for lift, 241
Joule heat, 49, 60
Karman-Tsien approximation, 177, 179
Kelvin's theorem on constancy of circulation, 70
Kinematics, 3
Kinematic viscosity, 89
Kinetic energy in potential flow of a fluid of constant density, 236
  connection with velocity potential at large distances, 243
Kinetic theory of gases, 92
Knudsen number, 99, 100, 207
Kutta-Joukowski condition, 241

Kutta-Joukowski formula for lift, 241

Lagrange's theorem, on persistence of irrotational motion in an inviscid fluid, 72, 75
Lagrangian specification, 3
Lags in adjustment of a gas molecule, 95, 96, 97, 98, 209
Laplace transforms, 199
  slender-body theory, supersonic flow, 200
Large Reynolds numbers, asymptotic approximations at, 132
Larmor frequency, 49, 301
Lift, 102
Lift coefficient, 102
Lift coefficient of airfoil (two dimensional), 242
Lift, Kutta-Joukowski formula for, 241
Lifting-surface theory, 202, 247, 248
Limits of solutions for steady flow at infinite Reynolds numbers, 133, 134
Line doublets in potential flow (incompressible fluid), 230
Line source of sound, 198
Line sources in potential flow (incompressible fluid), 230
Line vortices, 21
Line vortices in potential flow (incompressible fluid), 231
Linearized equation for the velocity potential in the irrotational motion of an inviscid gas with constant entropy, 69, 181, 183
Longitudinal waves in an ionized gas, 215
Lorentz transformation, 52
Low Reynolds numbers, flow at, references to, 110
Mach number, 103, 287
Magnetic induction, equation for, 60

# SUBJECT INDEX

Magnetic induction and vorticity, mathematical analogy between, 76, 77, 92, 273
Magnetogasdynamics, 47, 58, 273
Magnetohydrodynamics, 47, 58
Magnetosphere, 293
Mangler's transformation of an axisymmetric boundary-layer problem into a two-dimensional one, 122, 130, 152
Mass, virtual, 244
Maxwell stress tensor, 50,
Maxwell's equations, 47, 50, 52, 274
Mechanical energy, rate of dissipation of, 42
Minkowski's stress-energy tensor, 54
Mises's transformation, 154
Mixed subsonic and supersonic flows, 180, 181
Mixing-length theories of turbulent motion, 168
Mixtures of gases, 207
Molecular dissociation, 209, 211
Momentum, equation of, for general fluid, 38
 for Newtonian viscous fluid, 88
 with electric and magnetic forces, 58
Momentum thickness of a boundary layer (incompressible fluid), 161
Motion in a neighborhood, analysis of, 13, 223
Moving axes, 11
Moving axes, equation of motion of an inviscid fluid and integration thereof, 67
Navier-Stokes equations, 87
Neighborhood, analysis of motion in a, 13, 223
Newtonian fluid, 87
Notation, 1
Nusselt number, 105
Oblique shocks, deflection through and downstream Mach number for, 260

One-dimensional approximate theory of steady gas flow along a tube of varying cross section, 174

Parabolic coordinates, 125, 126
 boundary-layer theory of flow past a semi-infinite flat plate, 126
 extended boundary-layer theory, 137
Particle paths, 4
Péclet number, 105
Perfect gases, 42, 65, 66, 67, 253 265
Permanent translation, directions of, 240
Persistence of irrotational motion in an inviscid fluid (Lagrange's theorem), 72, 75
Pipe, flow of a viscous liquid in, 107, 108
Piston, vaves produced by accelerated motion of, 85
Plane airfoils of small aspect ratio, 202
Plane along a stream, boundary layer at, 124
Plane athwart a stream of viscous liquid, 109
Plasma frequency, 217
Poiseuille flow, 107, 108
Potential motions (see Irrotational motions)
Poynting vector, 50, 56
Prandtl-Meyer expansion, 265
Prandtl number, 95, 101

Radiation of heat, 166
Rankine-Hugoniot relations for a shock, 78, 113, 255
Rarefied gases, 99, 100, 207
Rate of change, local, convective, and total, in Eulerian specification, 4
Rate-of-strain components, 13, 223
 in general orthogonal coordinates, 225

Rate-of-strain tensor and stress tensor, relation between, 87, 93
Rayleigh numbers, 104
Reflection coefficient, 99, 281
Relativity, special, 52
Relaxation times in adjustment of a gas molecule, 95, 96, 97, 98, 209, 210
Reynolds number, 101
Rotating axes, 11
Rotating axes, equations of motion for an inviscid fluid and integration thereof, 67

Saha formula, 211
Second coefficient of viscosity, 88
Separation of a boundary layer, 119
Shear viscosity, coefficient of, 88, 93
   variation with temperature, 94, 101, 155, 156
Shearing flow, simple, 108
   of a viscous gas, 111
Shock, oblique, deflection through and downstream Mach number, 260
Shock-wave formation, 84, 85
Shock wave, gain in entropy through, 114, 258
   hypersonic approximation for small deflection through, 263
   structure of, in a viscous gas, 112
   relaxation-time effects on, 210
Shock waves, 78
   in air with dissociation, 212
   in perfect gases, 254
   strong, 81
   weak, 81, 262
   with magnetic fields, 275 ff
Similar velocity distributions, solutions of boundary-layer equations with, two-dimensional, 127
   axisymmetrical, 130
Simple shearing flow, 108
   of a viscous gas, 111
Simple wave, 86
   centered, 266

Skin friction, 36, 124
Slender-body theory, subsonic flow, 183
   supersonic flow, 197
Slip, 99
   in viscous flow of a liquid in a tube, 108
Small aspect ratio, plane airfoils of, 202
Solar wind, 293
Solidification of small portion of fluid, 17
Sound, line source of, 198
   speed of, 66, 289
Sound waves of finite amplitude, 83
Sources in potential flow (incompressible fluid), 229
Sources, method of, for flow past a body of revolution at zero yaw, 234
   in subsonic flow and in slender-body theory, 183
Special relativity, 52
Specific heats, 42, 94, 95
Spherically symmetrical expansion or contraction, 89

Stability of fluid flow, 169
Stagnation enthalpy, 64
Stagnation point, 64
   steady flow of a viscous liquid in a boundary layer at a forward, 109, 130
Stanton number, 105
Steady motion (defined), 4
Steady motion of an inviscid fluid, conditions for, 64, 66
Stewartson-Illingworth theorem for gas boundary layers, 158
Stokes's theorem, 19
Strain, rate of, 13
Stream functions, 27
   equations for, in steady irrotational motion of a gas, 70
   in motion of viscous, incompressible fluid, 107
   in a two-dimensional boundary layer, 120, 121

# SUBJECT INDEX 311

Streamlines (defined), 4
Stress, 35, 93
Stress tensor, Maxwell's, 50
Stress tensor and rate-of-strain tensor, relation between, 87 93
Stress-energy tensor, Minkowski's 54
Strong shocks, 81
Subsonic irrotational flow, uniqueness theorem for, 178
  approximate calculations of, 179
Summation convention, 1
Temperature, effects of high, in gases, 207, 211
Temperature boundary layers, 147, 150, 151
Temperature jump, 99
Thermal accomodation coefficient, 99
Thermodynamic properties of air, 212
Thermometric conductivity, 96
Total rate of change, in Eulerian specification, 5
Transonic flows, 179, 180, 197
Transverse curvature, effect of, in flow past bodies of revolution, 122
Tube, flow of a viscous liquid in, 107, 108
Tube of varying cross section, one-dimensional approximate theory of steady gas flow along, 174
Turbulence, 167
Two-dimensional motion, 26, 27, 28, 76, 77

Uniqueness theorem for subsonic irrotational flow, 178
Units, in electromagnetism, 47, 215

Vector potentials, 27, 274
Vectors, 1

Velocity potential, 25
  equation for in irrotational motion of an inviscid gas with constant entropy, 68
Velocity potential at large distances in incompressible fluid, connection with kinetic energy, 243
Virtual mass, 244
Viscosity coefficients, 88
Viscosity, effects of, 87
Viscosity, kinematic, 89
Viscous fluid, equations of momentum, 39, 88
  dissipation of energy, 43, 89, 90, 91
Viscous gas, exact solutions for the flow of, 110
  flow of through a shock wave, 112
Viscous liquid, exact solutions for the flow of, 107
Vortex, cast-off, 246
Vortex lines, 18
Vortex pair, impulse of, 249
Vortex sheet, impulse of, 248
Vortex sheets, 20, 72, 78, 244
Vortex tube, velocity field of an isolated, 32
Vortex tubes, 18
Vorticity, 13, 223
  diffusion of, 108, 115, 116
  equations for, in viscous fluid, 92
Vorticity in an inviscid fluid, Cauchy's equations for, 74
  Helmholtz's equations for, 73
  Helmholtz's theorems on, 71
Vorticity and magnetic induction, mathematical analogy between, 76, 77, 92, 273
Wave equation, two-dimensional, 198
Weak shocks, 81, 262
Wedge, supersonic flow past, 175

QA901 .G6
Goldstein / Lectures on fluid mechanics,